Recent Microwave Technologies

Edited by Ahmed Kishk and Kim Ho Yeap

Published in London, United Kingdom

Recent Microwave Technologies
http://dx.doi.org/10.5772/intechopen.97921
Edited by Ahmed Kishk and Kim Ho Yeap

Contributors
Bilel Hamdi, Taoufik Aguili, Faik G. Bogdanov, Roman Jobava, Lily Svanidze, Irina Chochia, Paul M. Meaney, Keith D. Paulsen, Somdotta Roy Choudhury, Ting-Tse Lin, Ting-Wei Wang, Di Lu, Nitin Muchhal, Shweta Srivastava, Mostafa Elkhouly, Dorra Bahloul, Ammar Kouki, Ines Amor, Tadashi Kawai, Ayumu Tsuchiya, Akira Enokihara, Ke Wu, Pankaj Jha, Giuseppe Addamo, Mauro Lumia, Oscar Antonio Peverini, Giuseppe Virone, Flaviana Calignano, Diego Manfredi, Piero Frediani, Marco Frediani, Gianni Innocenti, Irene Mellone, Roberto Simoni, Giampaolo Oteri, Jan Vrba Sr., David Vrba, Jan Vrba, Ondrej Fiser, Milan Babak, Jesus Cumana, Olga Gomonova, Serge Lefeuvre

First published in London, United Kingdom, 2022 by IntechOpen
IntechOpen is the global imprint of INTECHOPEN LIMITED, registered in England and Wales, registration number: 11086078, 5 Princes Gate Court, London, SW7 2QJ, United Kingdom

British Library Cataloguing-in-Publication Data
A catalogue record for this book is available from the British Library

Additional hard and PDF copies can be obtained from orders@intechopen.com

Recent Microwave Technologies
Edited by Ahmed Kishk and Kim Ho Yeap
p. cm.
Print ISBN 978-1-80355-927-8
Online ISBN 978-1-80355-928-5
eBook (PDF) ISBN 978-1-80355-929-2

For EU product safety concerns:
IN TECH d.o.o., Prolaz Marije Krucifikse Kozulić 3, 51000 Rijeka, Croatia,
info@intechopen.com or visit our website at intechopen.com.

Meet the editors

Ahmed A. Kishk is a professor and Tier 1 Canada Research Chair in Advanced Antenna Systems at Concordia University, Montréal, Québec, Canada. His research interests include millimeter-wave antenna arrays and passive devices. He is the recipient of several international awards. He was a technical program committee member for several international conferences. He was an Institute of Electrical and Electronics Engineers (IEEE) Antennas and Propagation Society (AP-S) Administrative Committee (AdCom) member from 2013 to 2015 and was the 2017 AP-S president. Dr. Kishk is a life fellow of the IEEE, a fellow of the Electromagnetic Academy, and a fellow of the Applied Computational Electromagnetics Society (ACES). He is a member of several IEEE Societies such as AP-S, MTT, EMC, SPS, ComS, EMBS, AESS, VTS, and Photonic Society. He is a senior member of the International Union of Radio Science, Commission B, and Applied Computational Electromagnetics Society.

Kim Ho Yeap is an associate professor at Universiti Tunku Abdul Rahman, Malaysia. He is an Institute of Electrical and Electronics Engineers (IEEE) senior member, a professional engineer registered with the Board of Engineers, Malaysia, and a chartered engineer registered with the UK Engineering Council. He is the external examiner and external course assessor of Wawasan Open University. From 2017 to 2022, he was editor-in-chief of the *Journal on Digital Signal Processing. He has also been a guest editor for the Journal of Applied Environmental and Biological Sciences* and *Journal of Fundamental and Applied Sciences.* He has also been a recipient of the university teaching excellence award and twenty-too research grants. He has published more than 100 research articles in electromagnetics, including refereed journal papers, conference proceedings, books, and book chapters.

Contents

Preface

Microwave bands range from 300 MHz to 300 GHz of the electromagnetic spectrum. Vast applications can be found in the signals that propagate in this band; they are employed in the fields of communication, networking, astronomy, and biomedical engineering, to name a few. This book provides a detailed elucidation on the physics of microwave signals, methods of modeling these signals, applications of these signals in various fields, and the underlying principles of some of the latest microwave devices.

Chapter 1 presents a comprehensive comparison between plane wave and orbital angular momentum (OAM) wave propagation using a patch antenna as a radiator at 2.45 GHz. This comparison allows the appreciation of the fundamental properties of the OAM wave when compared to its plane-wave counterpart. Two abbreviated terms are used for simplified comparison and discussion: PWPA for plane-wave patch antenna and OWPA for OAM wave patch antenna. PWPA refers to the planar patch antenna that produces plane waves in the far-field, whereas patch antennas that deliver OAM waves in the far-field are termed OWPA. All physical quantities for wave propagation such as electric field, magnetic field, wave impedance, wave vector, velocity, pitch, and propagation constant are theoretically studied for OAM waves and compared with plane waves. First, the OAM wave generation is studied through a widely used uniform circular antenna array (UCAA). Then, PWPAs and OWPAs are designed and verified through simulations and measurements. The OWPA is designed with characteristic mode analysis (CMA) based on a lossy substrate to excite a twisting wave at a determined patch location. With this in mind, a comparative investigation of PWPA and OWPA is conducted for different physical parameters. The cylindrical near-field scan clearly shows a helical wave motion for the OWPA, whereas a normal plane wave motion for the PWPA. Furthermore, the comparison of plane wave and OAM wave propagations is demonstrated using the combination of a pair of transmitting and receiving antennas. It is observed that the overall signal from the OWPA can be received with two PWPAs at an angle as OWPA has a dispersive beam. Moreover, the receiving antenna with a large aperture and plane wave horn antenna (PWHA) in the line of sight (LOS) range can also be used to receive the overall signal from OWPA. The received signals in the PWPA–PWPA, OWPA–OWPA, OWPA–PWPA–PWPA, and OWPA–PWHA transmitting and receiving pairs are thoroughly compared and studied. Measured and simulated results for transmission are -30 dB for 0 dB input signal in the OWPA–PWPA–PWPA and OWPA–PWHA cases, which are reasonably justified within the sensitivity/dynamic range of short-distance communication and radar sensing receivers.

Dielectric and conducting liquids with varying electromagnetic properties can offer novel alternatives for building tunable microwave passive components and antennas. Injecting these fluidics in or around microwave substrates alters their overall electrical characteristics, enabling circuit reconfigurability. Alternatively, changing the shapes and dimensions of conductors by using liquid metals can achieve similar reconfigurability. Chapter 2 provides an overview of different liquids and their electromagnetic properties as well as discusses the principles behind the reconfigurability of the electrical characteristics of typical guiding structures based on mode shape variation in the presence of fluids. It also describes an N-bit programmable impedance tuner in 3D LTCC technology based on these principles.

Chapter 3 describes multiscale auxiliary sources mainly used to solve complex electromagnetic problems, especially those that insert localized elements into circuits. Several equivalent relations (field-circuit) are established to simplify and make more accurate electromagnetic calculations by changing some characteristics of the localized elements known by their field representation as "voltage-current" representation and vice versa. Various examples are illustrated to show the effects of auxiliary sources in planar circuits containing localized elements (dipoles, diodes, transistors) in the millimeter and terahertz bands. An example of a graphene or gold dipole is demonstrated in this approach. Another typical example of a diode integrated with a radiating structure is also simulated.

Chapter 4 generalizes a recently proposed method of moment (MoM)-based approach to waveguide port excitation (WPE) problems on arbitrary conducting and composite geometries. This approach combines the canonical aperture coupling approach with the EFIE-PMCHWT formulation for composite structures. Each WPE problem in this approach is divided into equivalent sub-problems for internal and external regions solved using the MoM. Internal WPE problems are solved using waveguide modal expansion in the port plane, while external problems are solved using the equivalence principle to reduce these problems to the systems of algebraic equations for unknown electric and magnetic currents. The developed approach is validated on radiation and coupling problems for coaxial ports by comparing simulated results with those obtained by other approaches and measurements. An excellent agreement between the simulated and measured results is demonstrated. Finally, this approach is applied to practical EMC problems for microwave antennas fed by coaxial ports.

Chapter 5 deals with the modeling of microwave heating of a water drop. A drop model is reduced to its electric dipoles, masses, and charges. These components are then constructed using COMSOL Multiphysics and MATLAB software. The considered model proposes a microscopic point of view on microwave heating, which transforms electrical energy into heat.

Chapter 6 covers filter structures using defected ground structures (DGS). It discusses the limitations of electromagnetic band gap (EBG) structures and the development of DGS from the EBG structure.

DGS is an area of increasing interest in EBG technology. The chapter also discusses the features and physics of the well-known dumbbell DGS structures. It presents sew investigations on the choice of geometrical shapes for the DGS structure as an element for the proposed filters. All the proposed DGS structures used to implement different types of filters (lowpass, bandpass, and bandstop) are validated.

Tunable filters enable dynamic spectrum access for wireless systems, and tunable bandpass filters with constant bandwidth (BW) are most favorable for practical applications. Chapter 7 investigates the synthesis and realization techniques for tunable filters using the coupling matrix with variable entries. It presents the synthesis method and the matrix extraction procedures for the constant-BW bandpass filter and gives typical numerical examples. The chapter also discusses the relationship between the theoretical matrix and the physical circuits and then presents a planar tunable filter design to verify this relationship. Furthermore, the general approach to designing the constant-BW filters using the element variable coupling matrix is concluded. The planar circuit and 3D structure realizations are offered to demonstrate the synthesis design approach practically.

Chapter 8 presents novel compact metamaterial-based bandpass filters with improved stopbands. These filters have a compact size and a wider upper stopband resulting from the bandstop resonator characteristics. It evaluates several filters' design methodologies and performances using broadside-coupled complementary split-ring resonators (BC-CSRR) and edge-coupled complementary split-ring resonators (EC-CSRR) techniques. A comprehensive method to evaluate negative permittivity and permeability for designing the proposed metamaterial structure is also described.

With the remarkable progress in using the Internet of Things (IoT) and 5G, there is a demand for higher performance such as miniaturization, broadband/multi-band, low loss, and high integration for several microwave circuits. Chapter 9 treats microwave power dividers/combiners used in amplifiers, mixers, phase shifters, antenna feeding circuits, and so on. Here, the treated circuits are inductance-capacitance LC-ladder circuits and an absorption resistor. The chapter shows that multi-band (dual-band and tri-band) and broadband can be achieved by changing the number of stages of the LC-ladder circuit. In addition, the effectiveness of this design method is demonstrated by electromagnetic simulations and prototype experiments.

Chapter 10 describes recent research activities from the perspective of microwave technologies in medicine and biology. It brings new ideas about the possibilities of using microwaves in thermotherapy, particularly for hyperthermia in cancer treatment. It discusses the development of new hyperthermia applicators (e.g., on technologies like metamaterials, evanescent modes in waveguides and other types of transmission structures, etc.) and the use of microwaves in medical diagnostics. For example, microwave differential tomography, ultra-wideband UWB radar, and microwave radiometers will soon play an important role. Finally, the chapter highlights the experimental equipment necessary for research on the biological effects of electromagnetic fields.

Wireless power transmission (WPT) can provide an alternative for wireless power in implantable medical devices (IMDs), such as implantable pacemakers, optogenetic devices, and bio-impedance sensors. Chapter 11 comprehensively reviews recent WPT studies for emerging biomedical applications. It also outlines the specific key components for those applications and examines operating principles and system design.

Microwave imaging has long been proposed as an effective means for biomedical applications, such as breast cancer detection and therapy monitoring. Numerous numerical simulations have been presented demonstrating feasibility, but the translation to actual physical and clinical implementations is lacking. A team at Dartmouth has taken somewhat counterintuitive but fundamentals-based approaches to the problem, primarily addressing the confounding multipath signal corruption problem and exploiting core concepts from the parameter estimation community. The team has configured a unique system design that is a synergism of hardware and software worlds. Chapter 12 shares the team's approaches in the context of competing strategies and suggests reasons for why these techniques work, especially in 2D. Finally, the chapter presents data from actual neoadjuvant chemotherapy exams that confirm that the technique can image the tumor and visualize its progression during treatment.

The exponential growth of publications on additive manufacturing (AM) technologies in the microwave field proves the increasing interest of research institutions and industries in these techniques. Some advantages of AM with respect to conventional machining are weight reduction, design flexibility, and integration of different functionalities (electromagnetic, thermal, and structural) in a single part. Chapter 13 begins with an overview of the AM processes, focusing on material properties and post-processing. Then, it presents an extensive survey on microwave-guided components fabricated by AM processes published in the literature. It also discusses the most employed AM technologies for manufacturing RF waveguide components.

One of the greatest challenges for the European Union is the "Circular Economy Package." To achieve this goal, materials at the end of their life cycle must be recycled using a sustainable process. Chapter 14 examines the process of employing microwaves to recycle plastics to preserve their energy and chemical content. It illustrates microwave-assisted pyrolysis (MAP), which is an industrial novelty in plastic recycling. As a thermochemical treatment, pyrolysis represents a significant opportunity as long as it leads to the recovery of both the energy and chemical content of mixed, contaminated, or deteriorated plastics.

Ahmed Kishk
Department of Electrical and Computer Engineering,
Concordia University,
Montreal, Canada

Kim Ho Yeap
Department of Electronic Engineering,
Universiti Tunku Abdul Rahman,
Malaysia

Section 1

Physics of Electromagnetics

Chapter 1

Orbital Angular Momentum Wave and Propagation

Pankaj Jha and Ke Wu

Abstract

Orbital angular momentum (OAM) techniques are exploited for a wide range of potential radiofrequency (RF) and electromagnetic applications, including megahertz-through-terahertz wireless systems, fiber-based and free-space optical communications and sensing, just like acoustic and any other wave-based counterparts. In those RF and electromagnetic applications, OAM wave is set to enable the development of high-speed and high-capacity communications, radar imaging, and sensing systems, among many others. In this chapter, a comprehensive comparison between plane wave and OAM wave propagation using a patch antenna as a radiator at 2.45 GHz is presented and discussed. This comparison allows the appreciation of the fundamental properties of the OAM wave when compared against its plane wave counterpart. For simplified comparison and discussion, we will use two abbreviated terms: PWPA for plane-wave patch antenna and OWPA for OAM wave patch antenna. PWPA refers to as planar patch antenna that produces plane waves in far-field, whereas patch antenna that delivers OAM waves in far-field is termed as OWPA. In this context, all physical quantities for wave propagation such as electric field, magnetic field, wave impedance, wave vector, velocity, pitch, and propagation constant are theoretically studied for OAM waves and compared with plane waves. First, OAM wave generation is studied through widely used uniform circular antenna array (UCAA) in literature. Then, plane wave patch antenna (PWPA) and OAM wave patch antenna (OWPA) are designed and verified through simulation and measurement. OWPA is designed with characteristic mode analysis (CMA) based on a lossy substrate to excite a twisting wave at a determined patch location. With this in mind, a comparative investigation of PWPA and OWPA is conducted for different physical parameters. Cylindrical near-field scan clearly shows a helical wave motion for OWPA, whereas a normal plane wave motion for PWPA. Furthermore, the comparison of plane wave and OAM wave propagation is demonstrated using the combination of a Tx–Rx antenna pair. It is observed that the overall signal from OWPA can be received with two PWPAs at an angle as OWPA has a dispersive beam. Moreover, the receiving antenna with a large aperture and plane wave horn antenna (PWHA) in the line of sight (LOS) range can also be used to receive the overall signal from OWPA. The received signal in PWPA–PWPA, OWPA–OWPA, OWPA–PWPA–PWPA, OWPA–PWHA Tx–Rx pairs is thoroughly compared and studied. Measured and simulated results for transmission are −30 dB for 0 dB input signal in OWPA–PWPA–PWPA and OWPA–PWHA cases, which are reasonably justified within the sensitivity/dynamic range of short-distance communication and radar sensing receivers.

IntechOpen

Keywords: plane wave patch antenna (PWPA), OAM wave patch antenna (OWPA), plane wave propagation, OAM wave propagation, Gaussian and Bessel beam, OAM impedance, OAM pitch, vortex wave, propagation constant

1. Introduction

Electromagnetic (EM) waves have been known to human society since the sixteenth century when mariners observed that lightning strikes could deflect a compass needle. EM waves transport both energy and momentum, where the momentum is categorized into two types: linear momentum and angular momentum. Specifically, the angular momentum has one added component in connection with field polarization, namely spin angular momentum (SAM), and the other component associated with spatial field distribution, termed as orbital angular momentum (OAM). Allen et al. [1] first introduced the concept of OAM in the optical domain. The twisted radio waves are characterized by amplitude singularity along its beam axis, and helical phase front of $\exp(\pm jl\phi)$, where ϕ denotes azimuthal angle around the beam axis, $\pm$ indicates respectively clockwise and anticlockwise helical-phase front, and l corresponds to OAM index number, also called as topological charge, which suggests the number of twists in one wavelength [1]. OAM wave is widely investigated, and physical systems based on OAM wave have been implemented in the optical domain [2, 3]. Perhaps, the OAM research in microwave areas is focused on increasing the channel capacity of wireless communications by multiplexing different modes of OAM at a single frequency [4–6].

Current communications and satellite systems across the globe are based on linear and spin angular momentum (plane wave propagation). People have used all the dimensions of space, time, frequency, and polarization for multiplexing signals in the channel to achieve a maximum data rate. Foreseeing the nature of the emerging and connected world, scientists anticipate that we will need a data rate of terabit per second (Tbps) for transmission in a decade. Presently used means of wireless communication can support data rates only up to Gbps as we have utilized the maximum extent of resources to achieve this. So, there is the need for another degree of freedom to achieve the data rate of Tbps, which can be realized by multiplexing different OAM modes at a single frequency [7]. However, the practical adoption of OAM waves for long-distance wireless communication is unsuitable because of singularity (null) in the boresight direction of radiating fields [8, 9]. It has been shown through the aperture antenna theory that OAM capacity can be significantly increased for near-field applications by using an extended OAM receiving antenna [9]. In 2012, Tamburini et al. experimented through the transmission of a signal for 442 m using a helical reflector antenna to increase the channel capacity of a wireless communication link [10]. However, many controversies arose from whether OAM multiplexing is a subset of MIMO and OAM multiplexing cannot work in the far-field [11]. In 2018, NTT Japan demonstrated a 10 m OAM multiplexing experiment with uniform concentric circular array (UCCA) and OAM–MIMO multiplexing, achieving a milestone data rate of 100 Gbps [12]. Despite some controversies, the scientific community is searching for more mature research and development in OAM, OAM antenna, and its system-level implementation has gained momentum in the past decade.

Obviously, OAM has potential applications in optical fiber communication, free-space optical communication, radio communication, and acoustic communication [13]. Its applications mainly lie in high-speed and high-capacity communication, radar, imaging, and particle manipulation in the optical domain. Initially, cylindrical lenses were used to generate Hermite–Gaussian (HG) beam for the OAM wave by combining different states of the HG beam [1], then spiral phase plates were deployed to generate OAM waves in the optical domain [14]. In the radio domain, a similar trend was followed, and most research activities initially were focused on the generation of OAM waves using spiral phase plates [14–17]. Then, spiral reflectors were used to generate vortex beams [10, 18]. Also, planar antenna topology was deployed to generate OAM waves using uniform circular antenna arrays (UCCA) [19–25]. In [26], a circular time-switched antenna array was studied to generate an OAM wave. The time switched method uses high-speed RF switches to sequentially activate each array element, thus allowing the concurrent generation of OAM modes at harmonic frequencies of the switching period. Furthermore, a traveling wave circular slot was implemented to generate OAM waves [27, 28]. In addition, slot- and SIW-based OAM antennas were also considered [29, 30]. As in the optical domain, metamaterials have been used to generate OAM waves, similar to that has been implemented in the radio domain categorized as reflective metamaterials [31–40] and transmissive metamaterials [41]. OAM has also found its place in the guided-wave structure [42–47]. OAM has found applications in the acoustic domain, and analogous to antenna array in radio, uniform circular transducer antenna arrays are immensely used to generate acoustic vortex beam [48–50]. In nature, the vortex wave can well be observed as a twisting wave around a black hole [51], the most recent study of twisting wave around M87 black hole in radio astronomy [52], and the helical wave around the hurricane having a center point with null energy similar to the case of OAM [53]. **Figure 1** summarizes the state-of-the-art antenna types studied and researched for generating and detecting OAM waves from GHz to THz.

However, these previous works are dedicated to developing OAM waves using uniform circular antenna array (UCAA), spiral phase plates, and OAM multiplexing using UCCA. Minimal efforts have been invested in research to generate an OAM beam using a single patch antenna [60–62], which is set to relieve from complex feed network and space utilization. In this work, single patch-based plane wave patch antenna (PWPA) and OAM wave patch antenna (OWPA) are deployed as a demonstrator model to compare key performance parameters of PWPA and OWPA. Moreover, the main objective of this work is to show a comprehensive comparison of plane wave and OAM wave and appreciate the fundamental and distinguished properties of the OAM wave against those of the plane counterpart. Analytical, simulation and measurement comparisons between them are carried out thoroughly. In addition, it shows how the OAM wave can be useful for certain wireless applications even though point-to-point wireless connectivity with OAM is still a challenge because of its divergence pattern. Nevertheless, it is promising for short-range radar sensing applications. In this chapter, a detailed theoretical formulation of vector potentials, electric and magnetic fields for TE and TM modes, intrinsic impedances, wave vectors, propagation constants, wavelength, and pitch is devised and studied as a generic case for OAM waves. It shows clearly that the key performance parameters of plane waves are a special case of OAM waves. Furthermore, a narrative discussion of PWPA, OAM wave generation with single element patch using CMA, propagation, and characteristics comparison of PWPA and OWPA is carried out through simulations and validated through measurements.

Section 2 discusses the theoretical investigation of OAM wave and plane wave and their comparison in the rest of the chapter. Then, the OAM wave generation technique using UCCA is discussed in Section 3. This is followed by discussing simulated and measured results for PWPA and OWPA in Section 4. Subsequently, PWPA and

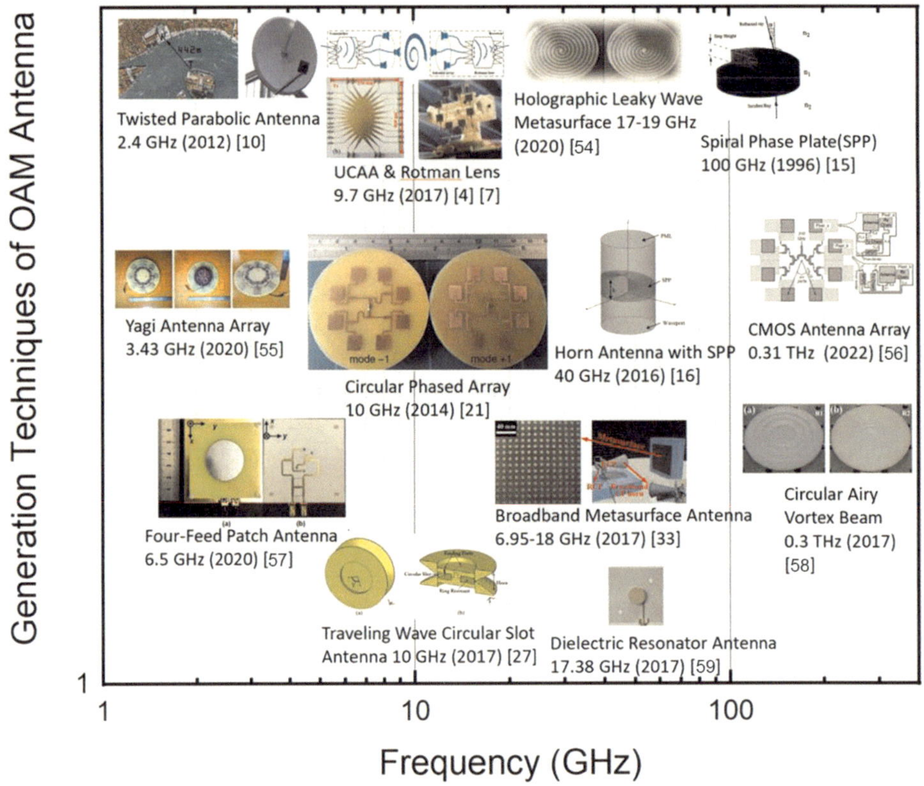

Figure 1.
State-of-the-art generation techniques of OAM wave using different antenna types from GHz to THz [54–59].

OWPA performances are compared critically in Section 5. Finally, plane wave propagation and OAM wave propagation are investigated and compared for different scenarios in Section 6. Then, concluding remarks are observed in Section 7.

2. Theoretical formulation of OAM and its comparison with plane wave

The set of nondiffractive Bessel beams solutions is based on the free-space differential Helmholtz equation, and their intensity profile has phase singularity with null amplitude at the center. The scalar form of Bessel beam skewing and propagating along the Z-axis in the cylindrical coordinates (ρ, ϕ, z) is given by [63]:

$$E\left(\vec{r}, t\right) = E_o J_l\left(k_\rho \rho\right) e^{j(k_z z - \omega t \pm l\phi)} \tag{1}$$

$$H\left(\vec{r}, t\right) = H_o J_l\left(k_\rho \rho\right) e^{j(k_z z - \omega t \pm l\phi)} \tag{2}$$

where E_o and H_o stand for amplitude; J_l is the lth order Bessel function of the first kind, k_ρ and k_z are the corresponding radial and longitudinal wave vectors satisfying the equation $k = 2\pi/\lambda = \sqrt{k_\rho^2 + k_z^2}$. The detailed derivation of the vector form of the Bessel beam is illustrated below.

2.1 Vector potential

The solution to wave equations in vector potential form is given by [64].

$$\nabla^2 \vec{A} - \frac{1}{c^2}\frac{\partial^2 \vec{A}}{\partial t^2} = 0 \tag{3}$$

Expanding it in the cylindrical coordinate (ρ, ϕ, z) yields

$$\begin{aligned}&\left[\frac{\partial^2 A_\rho}{\partial \rho^2} + \frac{1}{\rho}\frac{\partial A_\rho}{\partial \rho} + \frac{1}{\rho^2}\left(\frac{\partial^2 A_\rho}{\partial \phi^2} - A_\rho\right) + \frac{\partial^2 A_\rho}{\partial z^2} - \frac{2}{\rho^2}\frac{\partial A_\phi}{\partial \phi} - \frac{1}{c^2}\frac{\partial^2 A_\rho}{\partial t^2}\right]\hat{\rho} \\ &+ \left[\frac{\partial^2 A_\phi}{\partial \rho^2} + \frac{1}{\rho}\frac{\partial A_\phi}{\partial \rho} + \frac{1}{\rho^2}\left(\frac{\partial^2 A_\phi}{\partial \phi^2} - A_\phi\right) + \frac{\partial^2 A_\varphi}{\partial z^2} + \frac{2}{\rho^2}\frac{\partial A_\rho}{\partial \phi} - \frac{1}{c^2}\frac{\partial^2 A_\phi}{\partial t^2}\right]\hat{\phi} \\ &+ \left[\frac{\partial^2 A_z}{\partial \rho^2} + \frac{1}{\rho}\frac{\partial A_z}{\partial \rho} + \frac{1}{\rho^2}\frac{\partial^2 A_z}{\partial \phi^2} + \frac{\partial^2 A_z}{\partial z^2} - \frac{1}{c^2}\frac{\partial^2 A_z}{\partial t^2}\right]\hat{z} \\ &= 0\end{aligned} \tag{4}$$

where vector potential $\vec{A}$ is defined in the cylindrical coordinate system (ρ, ϕ, z) and (A_ρ, A_ϕ, A_z) are the scalar function. For the scenario when $A_\rho = A_\phi = 0$ and $A_z = \psi$, Eq. (4) reduces to

$$\frac{d^2\psi}{\partial \rho^2} + \frac{1}{\rho}\frac{d\psi}{d\rho} + \left(k_\psi^2 - k_z^2 - \frac{l^2}{\rho^2}\right)\psi = 0 \tag{5}$$

Hence, a set of Bessel beams is obtained by vector potential as

$$A_\rho = A_\varphi = 0; A_z\left(\vec{r}, t\right) = J_l\left(k_\rho \rho\right) e^{\,j(k_z z - \omega t \pm l\phi)} \tag{6}$$

where $k_z = k_\psi \cos(\alpha)$, $k_\rho = k_\psi \sin(\alpha)$, which can be deduced from **Figure 2**.

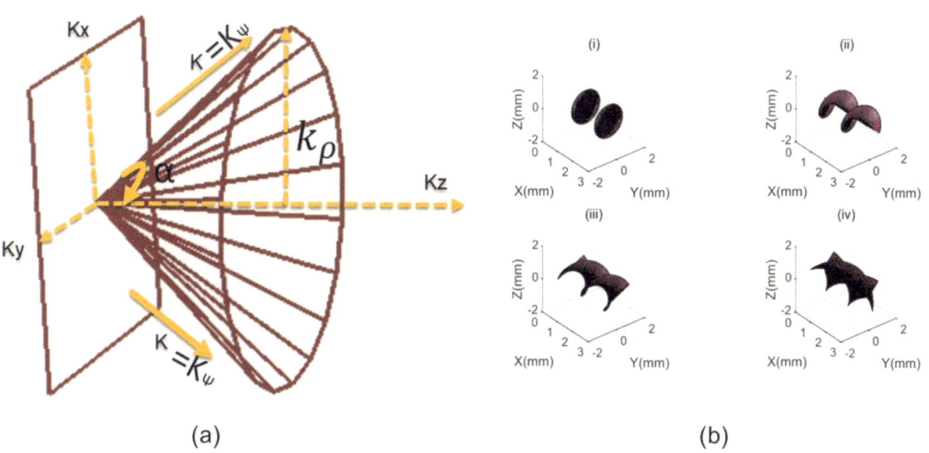

Figure 2.
(a) Plane-wave normal vectors making nondiffracting Bessel beam along the cone. A group of plane waves makes an angle α with the direction of propagation k_z to form an OAM wave. (b) 3-D phase front plot of OAM mode number (i) $l = 0$ (plane wave), (ii) $l = 1$, (iii) $l = 2$, (iv) $l = 3$.

2.2 Electric and magnetic field

The electric and magnetic fields can be obtained from vector potential $\vec{A}$ in the frequency domain form as

$$\vec{E} = -\nabla\Phi - \frac{1}{c}\frac{\partial A}{\partial t}; \vec{H} = \frac{1}{\mu}\nabla \times \vec{A} \tag{7}$$

After manipulating Eqs (6), (7), and using Lorentz condition, we obtain the electric and magnetic fields in a cylindrical coordinate system for TM and TE modes.

2.2.1 TM modes

$$\begin{bmatrix} E_\rho \\ E_\phi \\ E_z \end{bmatrix} = \begin{pmatrix} jk_\rho k_z \cos\alpha J_l'(k_\rho\rho) \\ \pm\frac{l}{\rho}k_z \cos\alpha J_l(k_\rho\rho) \\ k_\rho^2 \sin\alpha J_l(k_\rho\rho) \end{pmatrix} e^{j(k_z z - \omega t \pm l\phi)} \tag{8}$$

$$\begin{bmatrix} H_\rho \\ H_\phi \\ H_z \end{bmatrix} = \begin{pmatrix} \pm\frac{l}{\rho}\omega\varepsilon J_l(k_\rho\rho) \\ jk_\rho\omega\varepsilon J_l'(k_\rho\rho) \\ 0 \end{pmatrix} e^{j(k_z z - \omega t \pm l\phi)} \tag{9}$$

2.2.2 TE modes

Using the principle of duality $(E_{TE} = H_{TM};\ H_{TE} = -E_{TM})$

$$\begin{bmatrix} E_\rho \\ E_\phi \\ E_z \end{bmatrix} = \begin{pmatrix} \pm\frac{l}{\rho}\omega\varepsilon J_l(k_\rho\rho) \\ jk_\rho\omega\varepsilon J_l'(k_\rho\rho) \\ 0 \end{pmatrix} e^{j(k_z z - \omega t \pm l\phi)} \tag{10}$$

$$\begin{bmatrix} H_\rho \\ H_\phi \\ H_z \end{bmatrix} = \begin{pmatrix} jk_\rho k_\rho \cos\alpha J_l'(k_\rho\rho) \\ \pm\frac{l}{\rho}k_z \cos\alpha J_l(k_\rho\rho) \\ k_\rho^2 \sin\alpha J_l(k_\rho\rho) \end{pmatrix} e^{j(k_z z - \omega t \pm l\phi)} \tag{11}$$

Clearly, all the electric and magnetic field components in TE and TM modes show azimuthal phase dependence with the OAM index number l and addition of extra elements to the exponential term. This is one of the possible solutions when $A_z = \psi$ and $A_\rho = A_\phi = 0$, and other possible solution can be considered as $A_\rho = \psi, A_\phi = \pm jA_\rho$ and $A_z = 0$. In this work, our focus is on the first case as it deals with linearly polarized OAM waves and the latter case corresponds to circularly polarized OAM waves, which is not the scope of this chapter as the antenna designed in further sections is also focused on a linearly polarized OWPA.

2.3 Key performance parameters for plane wave and OAM wave

In OAM, the wave travels helically around its longitudinal propagation direction, and this leads to define the pitch of OAM waves similar to the screw thread pitch

$$\text{Pitch of the OAM wave } (T_p) = \frac{2\pi l}{k_\psi \cos(\alpha)} = \frac{l\lambda_\psi}{\cos(\alpha)} \tag{12}$$

where $\lambda_\psi = \frac{2\pi}{k_\psi} = \frac{2\pi c}{\omega} = \lambda_{plane}$

As $l = 0$ the pitch of vortex waves vanishes, which can be seen from **Figure 2(b)**, indicating the formation of a plane wave phase front.

The wave vector is the position gradient of the phase of the Bessel beam

$$\vec{k} = \nabla(k_z z \pm l\phi) = k_z \hat{z} \pm \frac{l}{\rho}\hat{\phi} \tag{13}$$

where

$$k_{OAM}^2 = k_z^2 + \frac{l^2}{\rho^2} \rightarrow k = \sqrt{k_z^2 + \frac{l^2}{\rho^2}} \tag{14}$$

When Bessel beams with zero orbital angular momentum ($l = 0$; plane wave), the angular velocity is given by

$$\omega = k_\psi c = \frac{k_z c}{\cos\alpha} \tag{15}$$

So, the group velocity of OAM waves is obtained by the phase gradient of plane wave angular velocity, using Eq. (15)

$$\vec{V}_g = \nabla_k \omega = \frac{c}{\cos\alpha}\hat{z} \tag{16}$$

The intrinsic impedance of OAM waves in both TM and TE cases is obtained using Eq. (8) to (11)

$$\eta_{OAM\_TM} = \frac{E_\rho}{H_\phi} = \frac{-E_\phi}{H_\rho} = \frac{k_z}{\omega\varepsilon}\cos\alpha = \eta_{plane}\cos\alpha \tag{17}$$

$$\eta_{OAM\_TE} = \frac{\eta_{plane}}{\cos\alpha} \tag{18}$$

For plane-wave case $\alpha = 0$, which gives the classical value of the impedance of 377 Ω. This suggests that the generic form of the wave is actually an OAM wave, whereas the plane wave is a special case of OAM wave when $l = 0$ and $\alpha = 0$.

2.4 Propagation constant of plane wave and OAM wave in free space

The wavenumber in free space for OAM as in Eq. (14) and $k_z = k_\psi \cos\alpha$; This leads to

$$k_{OAM} = \sqrt{k_\psi^2 \cos^2(\alpha) + \frac{l^2}{\rho^2}} = \sqrt{\omega^2 \mu \varepsilon \cos^2(\alpha) + \frac{l^2}{\rho^2}} \tag{19}$$

In free space, the attenuation constant and phase constant of OAM waves are obtained as

$$\alpha_{OAM} = 0; \quad \beta_{OAM} = \sqrt{\omega^2 \mu_o \varepsilon_o \cos^2(\alpha) + \frac{l^2}{\rho^2}} \tag{20}$$

Special case: $\alpha = 0; \;\; l = 0$ (plane wave)

$$\alpha_{plane} = 0; \;\; \beta_{plane} = \omega\sqrt{\mu_o \varepsilon_o} = \frac{\omega}{c} \tag{21}$$

This gives us the attenuation and phase constant of the plane waves as a special case of the OAM waves.

We have derived all the physical quantities associated with wave propagation such as electric field, magnetic field, intrinsic impedance, wave vector, propagation constant, velocity, wavelength, and pitch of the OAM wave and compared it with the plane wave. By virtue of the derivation presented above, it brings us to the theoretical

Physical quantities	Plane wave propagation	OAM wave propagation
Electric field	$l = 0, k_p = 0, J_o(0) = 1$ $E\left(\vec{r}, t\right) = E_o e^{\,j(k_z z - \omega t)}$	$E\left(\vec{r}, t\right) = E_o J_l(k_\rho \rho) e^{\,j(k_z z - \omega t \pm l\phi)}$
Intrinsic impedance	$\eta_{plane} = \sqrt{\frac{\mu}{\varepsilon}}$ $\eta_{plane\_FS} = \sqrt{\frac{\mu_o}{\varepsilon_o}} = 377$	$\eta_{OAM\_TM} = \eta_{plane} \cos\alpha$ $\eta_{OAM\_TE} = \frac{\eta_{plane}}{\cos\alpha}$
Wave vector and propagation constant	$k_\psi = k_z = \omega\sqrt{\mu\varepsilon} = \beta = \frac{2\pi}{\lambda}$ $k_\psi^2 = k_x^2 + k_y^2 + k_z^2 = \left(\frac{2\pi}{\lambda}\right)^2$ For free space, $k_\psi = \omega\sqrt{\mu_o\varepsilon_o} = \frac{\omega}{c}$ $\alpha_{plane} = 0$	$K_z = \sqrt{\omega^2\mu\varepsilon - (l/\rho)^2} = K_\psi \cos(\alpha)$ $k_\rho = k_\psi \sin(\alpha)$ $k_{OAM}^2 = k_z^2 + k_\rho^2 = k_z^2 + \frac{l^2}{\rho^2}$ $k_{OAM} = \sqrt{\omega^2\mu\varepsilon\cos^2(a) + \frac{l^2}{\rho^2}}$ For free space, $\alpha_{OAM} = 0$ $\beta_{OAM} = \sqrt{\omega^2\mu\varepsilon\cos^2(\alpha) + \frac{l^2}{\rho^2}}$
Velocity	Phase velocity, $V_p = \frac{\omega}{k}\hat{k}$ Angular velocity, $\omega = \frac{k_z c}{\cos(\alpha)}\hat{z}$	Group velocity, $V_g = \frac{c}{\cos(\alpha)}\hat{z}$
Wavelength	$\lambda_z = \frac{2\pi}{k_z}$	$\lambda_z = \frac{2\pi}{k_z \cos(\alpha)} = \frac{\lambda_{FS}}{\cos(\alpha)}$
Wave vector	$\vec{k} = k_z\hat{z}$	$\vec{k} = k_z\hat{z} \pm \frac{l}{\rho}\hat{\phi}$
Pitch	0	$T_p = \frac{2\pi l}{k_\psi \cos(\alpha)} = \frac{l\lambda_\psi}{\cos(\alpha)}$

Table 1.
Comparison of plane wave and OAM wave propagation.

comparison of plane wave propagation (Gaussian beam propagation; zero-order Bessel beam) and OAM wave propagation (Bessel beam propagation; higher-order Bessel beam) in **Table 1**. After careful investigation and comparison, it is understandable that the plane wave is the special case of the OAM wave when $\alpha = 0,\ l = 0$. **Figure 2(a)** shows that the conical beam pattern formed by the Bessel beam propagating along the z-axis and group of plane waves at an angle combines to create an OAM wave. Moreover, once we have all the derivations of physical quantities for the OAM wave, respective physical quantities of the plane wave can be deduced by substituting $\alpha = 0,\ l = 0$ as presented in **Table 1**. The phase front plot of the OAM wave for different index numbers is shown in **Figure 2(b)**. This plot clearly indicates that the OAM wave of zero-order has no pitch, but as the order of OAM increases, the number of pitches increases same as the order of OAM per wavelength. The number of helixes formed for the OAM wave depends on the OAM order.

3. Generation of OAM vortex beam using uniform circular antenna array

First, we have analyzed a circular antenna array with eight elements for generating the OAM wave propagation at a frequency (f = 2.45 GHz), which has been widely used in literature, as shown in **Figure 3**. The number of elements in UCCA determines the largest mode (l) generated by the array. The mode number is governed as per theory $-N/2 < l < N/2$, where N is the number of elements in the array. Therefore, for the eight elements array, the mode number that can be produced is $-4 < l < 4$. For an N-element UCCA, each patch element is fed by the same amplitude but with an incremental phase shift, as shown in **Figure 4**. The generation of a particular mode in UCAA is achieved by an interelement phase shift as $2\pi l/N$, where l is the OAM winding number.

The simulated reflection coefficient is well below -15 dB at $f = 2.45$ GHz, as depicted in **Figure 5**. A simultaneous excitation of each element with constant amplitude and differential phase shift, as shown in **Figure 4**, is incorporated in simulation to obtain the reflection coefficient. The simulated 3D radiation pattern in both dBi and linear scale is shown in **Figure 6**. It can clearly be visualized the amplitude null in the boresight direction along the z-axis.

Figure 7(a) depicts the orthographic view of the power flow of an antenna, which shows annular intensity cross sections with null energy at the center, and maximum

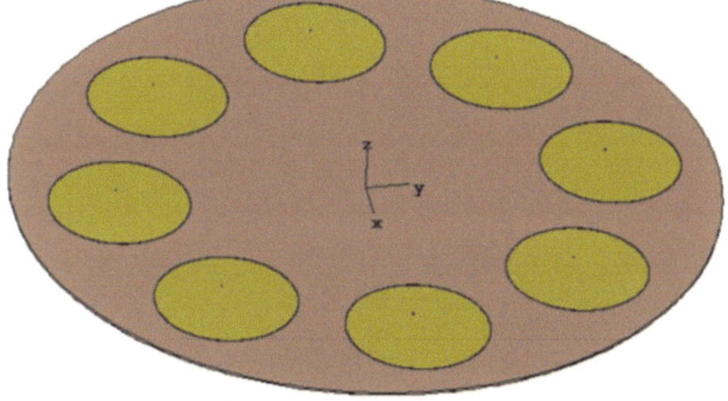

Figure 3.
Eight-element uniform circular antenna array (UCAA) for OAM antenna design.

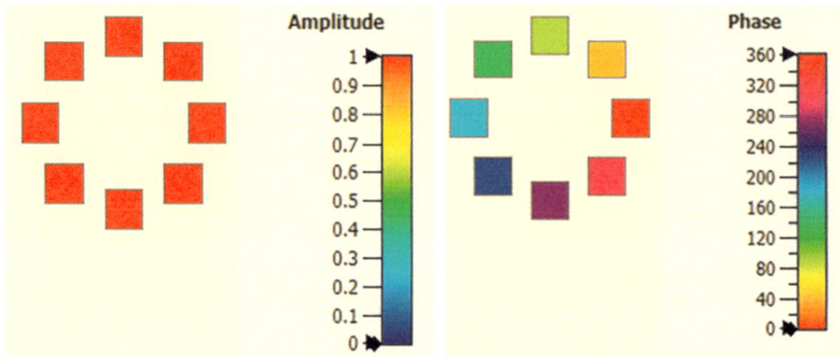

Figure 4.
Amplitude and phase excitation pattern of an eight-element array.

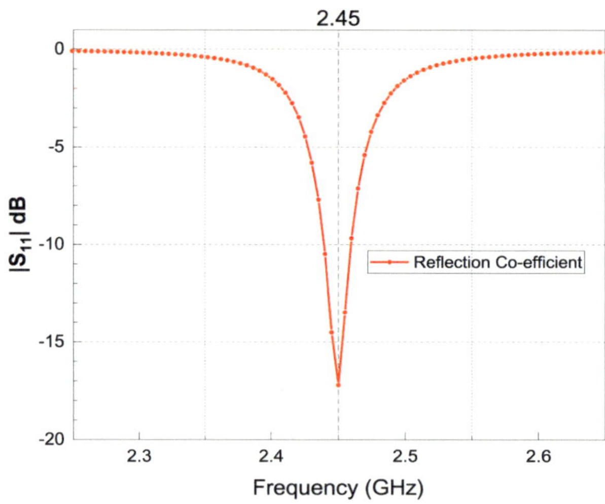

Figure 5.
Reflection coefficient of eight-element UCAA.

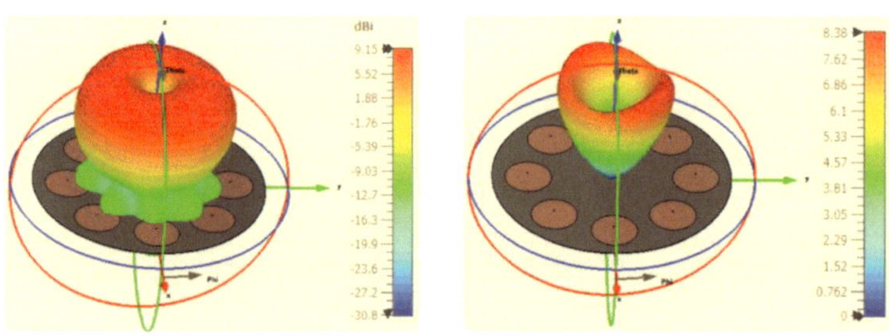

(a) (b)

Figure 6.
3D radiation pattern. (a) Realized gain (dBi) and (b) realized gain (linear).

energy is exhibited on the ring, satisfying the key characteristics of an OAM wave. This is shown for OAM mode $l = 1$, whereas for mode $l = 0$, beam power is at the center, forming the plane wave case. The twisted radio beams can be observed from the phase fronts of an OAM, as shown in **Figure 7(b)**.

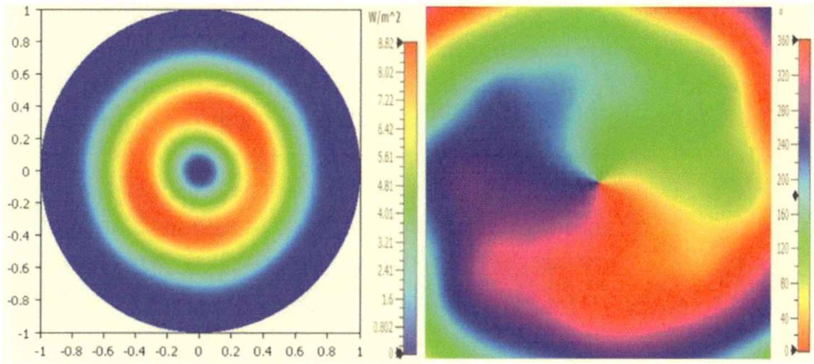

Figure 7.
(a) 2D plot of power flow of an antenna and (b) 2D plot of phase fronts of OAM mode 1 at cutplane z = 32 mm.

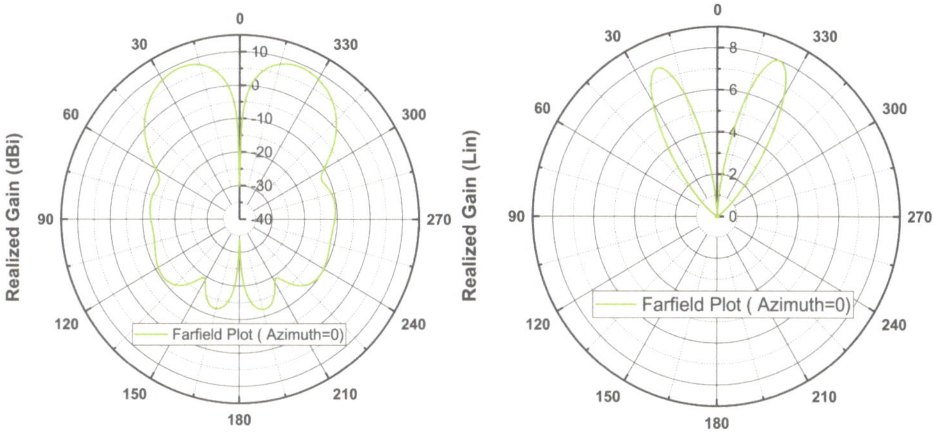

Figure 8.
1D polar plot of gain of a UCCA antenna. (a) Realized gain (dBi) and (b) realized gain (linear).

It is clearly observed in **Figure 8** that half-power null size from −33° to +33°, and maximum gain of 9.15 dBi or 8 (linear scale) at 23°. The design of a uniform circular antenna array (UCCA) and uniform concentric circular antenna array (UCCAA) for generating particular OAM modes require a complex feed network. Its design becomes further complicated if multimodes OAM is generated simultaneously. Henceforth, in another section of this chapter, our discussion is on generating OAM waves using a patch antenna, which reduces the complexity of feed design.

4. PWPA and OWPA results and discussion

4.1 PWPA (plane wave patch antenna)

PWPA has been widely used in communications because it is one of the efficient radiators, with easy and low-cost fabrication and smooth integration with microwave circuits, and it supports linear and circular polarizations. In this work, the patch antenna is designed at frequency $f = 2.45$ GHz for demonstration and comparison purposes on Rogers RT5880 substrate having dielectric constant ($\varepsilon_r = 2.2$), thickness ($t = 0.787\ mm = 30\ mils$), and loss tangent ($\delta = 0.0009$). The antenna has a coaxial probe feed from the ground plane, and it is located 6.4 mm in X-direction from the center point (0, 0, 0), as shown in **Figure 9**.

The designed procedure outline is based on the well-established cavity model formulation with a simulation model of circular microstrip patch antenna for the dominant mode TM^z_{lp0}. The radius (r) of the circular patch is determined for the desired resonant frequency at a given value of dielectric constant ε_r [65].

$$r = \frac{c}{2\pi\sqrt{\varepsilon_r}}\left(\frac{\chi'_{lp}}{f_{lp0}}\right) \tag{22}$$

where χ'_{lp} represents the zeros of the derivative of the Bessel function $J_l(x)$. A circular patch antenna has one degree of freedom, i.e., increasing the patch radius will decrease the resonant frequency and vice versa.

The CST Studio Suite® is utilized for the analysis of the patch antenna. 3D CAD is analyzed using a time-domain solver (FIT) and frequency-domain solver (FEM). The convergence of the S-parameter result is ensured from both the numerical techniques so that the simulated results match with measured results. The FIT solver technique on hexahedral mesh with Perfect Boundary Approximation® (PBA), to achieve a conformal meshing of the patch without having staircase approximation of mesh in 3D, is used for fast and accurate computation of fields in the patch, and the FEM solution is based on the tetrahedral mesh. The FIT solver is typically suited for electrically large and broadband problems as the requirement of hardware memory scale linearly with the number of mesh cells. In contrast, the FEM solver is suited for electrically small, narrowband, and resonant problems as the memory scale-up quadratically with the number of mesh cells [66].

The simulated and measured reflection coefficients of PWPA are plotted in **Figure 10**. PWPA is simulated with both FIT and FEM, which suggest a nearly perfect agreement with measurements. All the values are well below −30 dB, as depicted in **Figure 10**. The 1D polar plot and 3D radiation pattern plot of realized gain (dBi and linear) are depicted in **Figure 11**. All the plots show 7 dBi of realized gain in simulation, whereas 5 dBi in measurement, offset of merely 2 dB in measurement versus

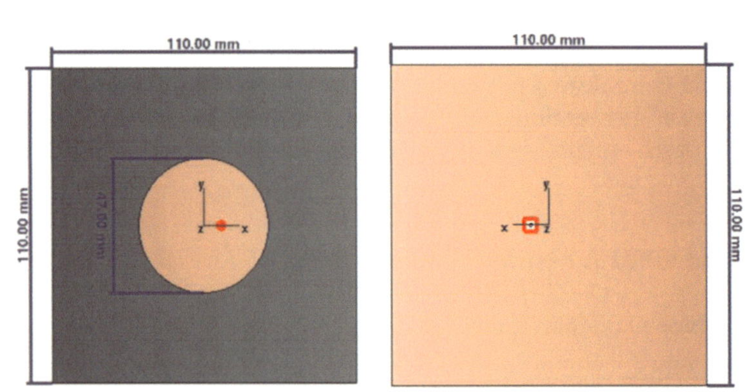

Figure 9.
Front and back view of simulated plane wave circular patch antenna in CST studio suite®.

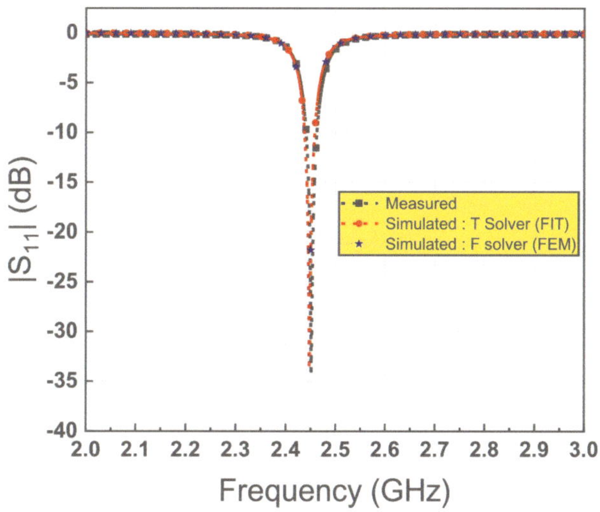

Figure 10.
Simulated and measured reflection coefficient of PWPA.

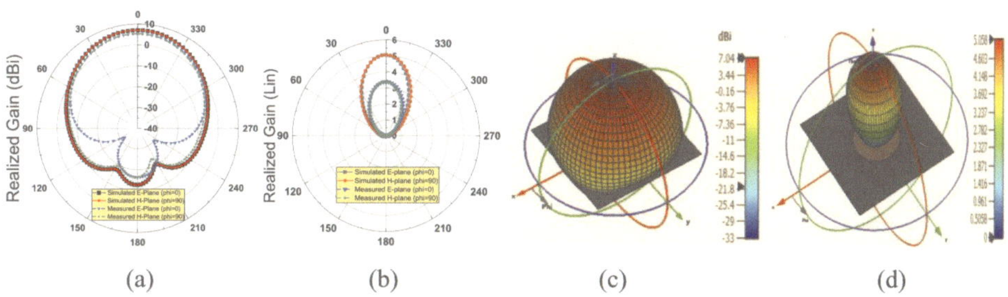

Figure 11.
(a) 1D polar plot of simulated and measured realized gain for E-plane and H-plane realized gain in dBi, and (b) realized gain in linear. (c) Plot of simulated 3D radiation pattern realized gain in dBi, and (d) realized gain in linear.

simulation. This offset is attributed to surface wave loss, dielectric loss, and exclusion of the connector model in simulation. On a linear scale, the realized gain in simulation is 5, while the measurement is 3.5.

4.2 OWPA (OAM wave patch antenna)

In literature, the OAM wave antenna has generally been designed with a UCCA and complex feed structure. In this work, we designed an unsymmetrical patch antenna [67, 68] by identifying a particular feed location where the patch will produce the OAM wave. The OWPA parameters are the same as those of PWPA. The red dot indicates the position of coaxial feed at X=8.75 mm, Y=21 mm from the center point of the antenna, as shown in **Figure 12**. Before a full-wave analysis of this antenna, characteristic mode analysis (CMA) is performed to examine the distribution of electric field and surface current on the antenna and excite the particular mode such that it is set to generate the desired OAM wave. The modal approach has been widely popular in the research community to design novel antennas for various applications, whereas CMA is gaining its popularity because it provides insight for improved

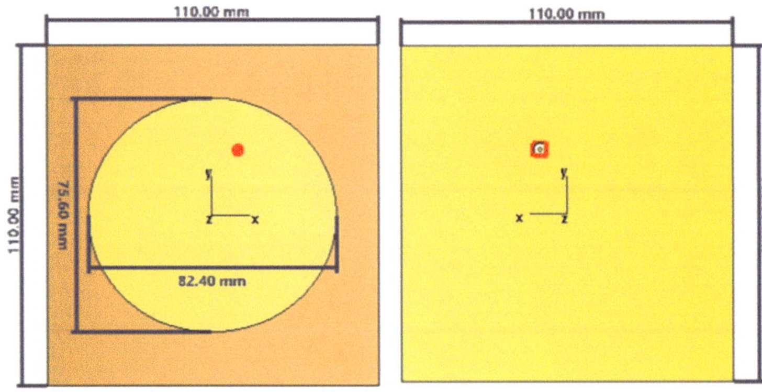

Figure 12.
Front and back view of simulated OWPA in CST studio suite®.

antenna design. CMA is a modal analysis (without excitation) for radiating structures or infinite open structures, whereas eigenmode analysis is for cavity structure or cavity filter design. The main advantage of CMA is to provide a better physical understanding of the surface current and field behavior of an antenna, which helps improve antenna design or postulate new antenna ideas. CMA provides valuable information such as resonant frequencies of inherent modes, far-field modal radiation patterns, modal surface currents on the analyzed structure, and modal significance at given frequencies. CMA can provide insight into the physical phenomena of an antenna of arbitrary shape and thus facilitate the analysis, synthesis, and optimization of antennas.

In the CMA theory, a generalized eigenvalue problem is solved. The generalized eigenvalue problem is given as follows [69]:

$$XJ_n = \lambda_n RJ_n \tag{23}$$

where R & X are real and imaginary Hermitian parts. J_n and λ_n are real eigenvectors and eigenmodes of the nth order mode. There are two main parameters that should be investigated while performing the CMA, which is given in the following:

Model significance (MS): MS is an intrinsic property that signifies the coupling capability of each characteristic mode with external sources. Individual mode contribution for the EM response of a given source is pointed by modal significance. It is easier to use MS rather than eigenvalue to examine the resonance of a structure. For perfectly radiating mode, MS = 1 for λ_n = 0.

$$MS = \left|\frac{1}{1 + j\lambda_n}\right| \tag{24}$$

Characteristic angle (α_n): It sets forth the comprehension of the mode behavior near resonance. Mathematically, $\alpha_n = 180^\circ - \arctan(\lambda_n)$.

- For $\alpha_n = 180^\circ \Rightarrow$ Resonant Mode.
- For $\alpha_n = 90^\circ - 180^\circ \Rightarrow$ Mode associated will be inductive.
- For $\alpha_n = 180^\circ - 270^\circ \Rightarrow$ Mode associated will be capacitive.

In general, for a resonant mode, the following three criteria should be satisfied as

$$\lambda_n = 0, MS = 1, \alpha_n = 180^{\circ}.$$

In this work, CMA is performed for five modes tracking in CST Studio Suite® between 2 and 3 GHz, and it is found that mode 1 (f = 2.3345 GHz) and mode 3

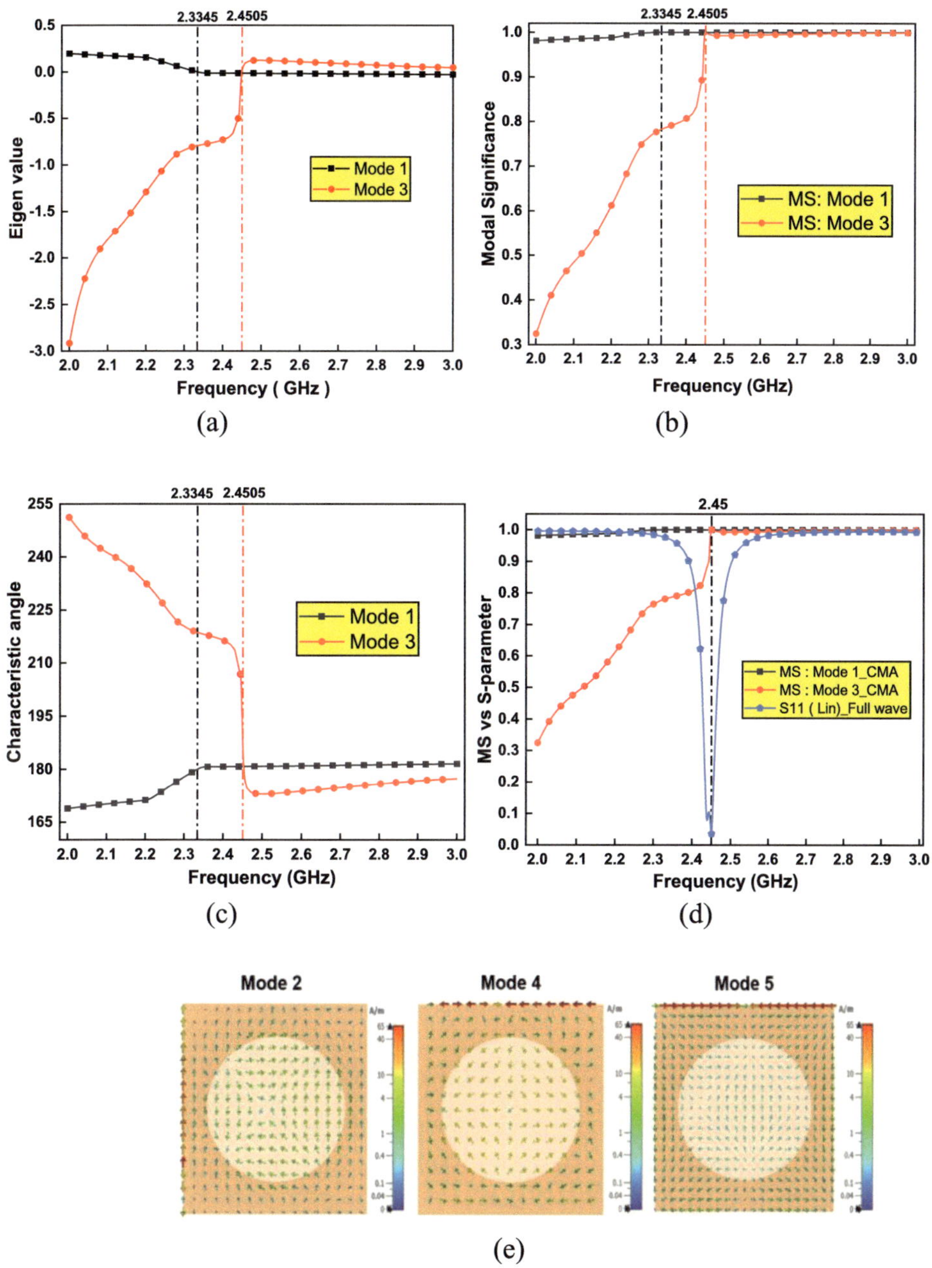

Figure 13.
Simulated (a) eigenvalue, (b) modal significance, (c) characteristic angle, and (d) MS vs. S-parameter using CMA in CST–MWS, (e) surface current of mode 2, mode 4, and mode 5.

(f = 2.4505 GHz) are the primary contributing modes for radiation. Mode 2, mode 4, and mode 5 have MS approaching 1 but do not satisfy all three resonant mode criteria. Moreover, the surface current of mode 2, mode 4, and mode 5 does not have null energy at the center, as shown in **Figure 13(e)**; the desired characteristic for twisting wave generation, henceforth, has been removed from further analysis. In addition, among mode 1 to mode 5, which are the most contributing modes, are determined by studying the nature of surface currents, electric fields, and far-field radiation behaviors of different CMA modes. Two orthogonal CMA modes are combined to form a full-wave analysis mode. The modal radiation quality factor is calculated as in Eq. (25). **Table 2** shows all five modes sorted at 2.45 GHz, resonance frequencies, Q-factors, and MS. Upon performing CMA, the plot of the significance of the modes sorted over frequency, in other words, which mode is contributing as the strongest to the radiation from the antenna at a given frequency (f = 2.45 GHz) is obtained in **Figure 13(b)** [70].

$$Q_n = \frac{\omega}{2} \left| \frac{d\lambda_n}{d\omega} \right| \Bigg|_{\omega=\omega_{res}} \tag{25}$$

We can also directly obtain the surface current distribution, electric and magnetic field, and far-field plot for all the modes at a given frequency of interest at 2.45 GHz using CMA.

In addition, overlaying the full-wave S-parameter curve on the modal significance plot as depicted in **Figure 13(d)** immediately shows which CMA mode is contributing to the resonances observed in the driven S parameter curve. In fact, for a given feed point, we can see a strong correlation between the driven mode currents and the CM currents. It is worth noting that when we drive the antenna, there is a contribution from all the modes at this frequency. Therefore, we cannot expect a perfect correlation. CMA is performed with metallic and dielectric together, giving better information than the classical approach of calculating CMA without lossy dielectric substrate. **Figure 14** clearly signifies the full-wave mode at f = 2.45 GHz is a combination of two CMA modes, mode 1 (f = 2.3345 GHz) and mode 3 (f = 2.4505 GHz). Overall, this gives rise to zero surface current, electric field, and far-field radiation at the central axis of the antenna. Besides, it generates a spiral motion of the electric field, which is the desired characteristic for twisting waves. With this understanding through CMA, we determine the location of the feed of an antenna, which will contribute to mode 1 and mode 3. We choose the feed as a common location for the surface current of both mode 1 and mode 3 of an elliptical patch antenna.

In this way, we will tame a simple plane wave patch antenna to an OAM wave patch antenna. Simulated and measured reflection coefficients of OWPA are shown in **Figure 15**. OWPA are simulated with FIT and FEM, showing a nearly perfect agreement with measurements. All the reflection coefficient values are well below −30 dB,

Mode	Resonance frequency	Q-factor	MS
1	2.3345	1.20	0.999628
2	2.3505	1.28	0.999370
3	2.4505	33.52	0.998896
4	2.7915	1.00	0.995350
5	2.8519	0.98	0.994957

Table 2.
Parameters of CMA in CST studio suite®.

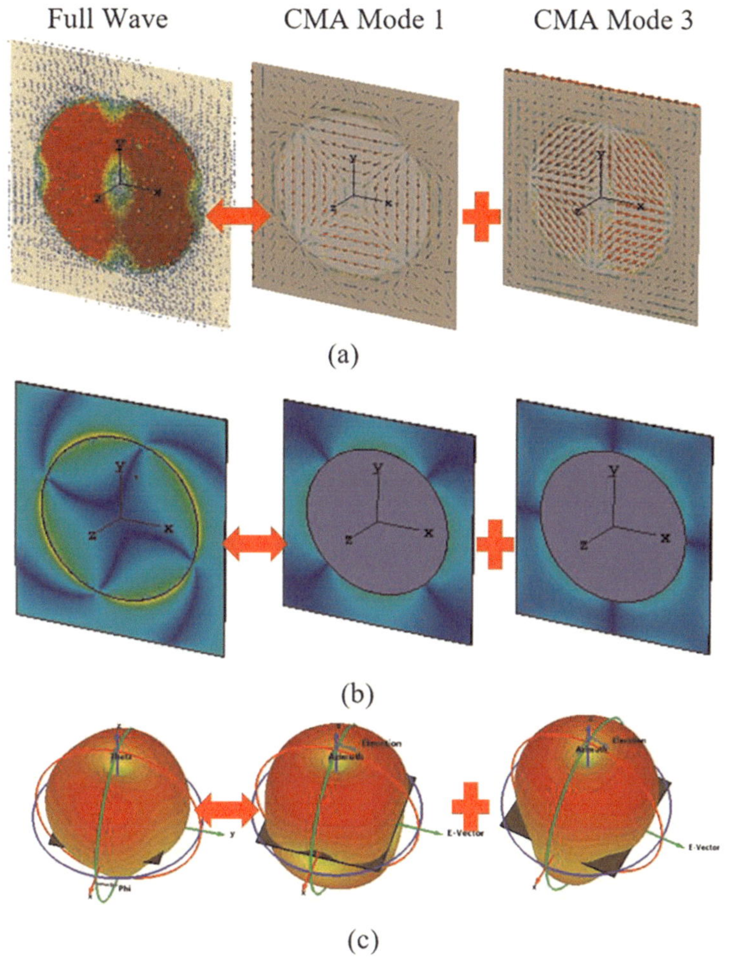

Figure 14.
Simulated (a) surface current, (b) electric field, (c) 3D radiation pattern for full-wave analysis, CMA mode 1 and mode 3 analysis.

as shown in **Figure 15**. The 1D polar plot of realized gain (dBi and linear) is depicted in **Figure 16**. The plots show 3.8 dBi of realized gain in simulation, whereas 1.8 dBi in measurement offset of merely 2 dB in simulation and measurement. On a linear scale, the realized gain in simulation is 2.5, whereas in measurement is 1.5.

5. Comparison of PWPA and OWPA

Until this point, the behavior of PWPA and OWPA is well understood, and it will be worth comparing the performance of these antennas. Let us begin with comparing the near-field electric field as depicted in **Figure 17(a)**, which shows a spiral motion of the electric field with null energy at the center for OWPA. At the same time, PWPA has energy at the central axis through an isoline pattern. **Figure 17(b)** also predicts similar behavior for PWPA and OWPA while showing their near-field magnetic field. It is worthwhile to visualize the E-near-field scan in **Figure 17(c)** on a cylindrical surface, which clearly indicates a helical motion of the electric field along the Z-axis

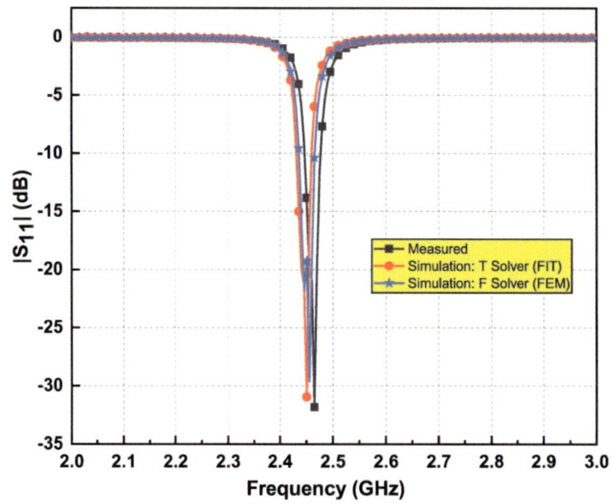

Figure 15.
Simulated and measured reflection coefficients of OWPA.

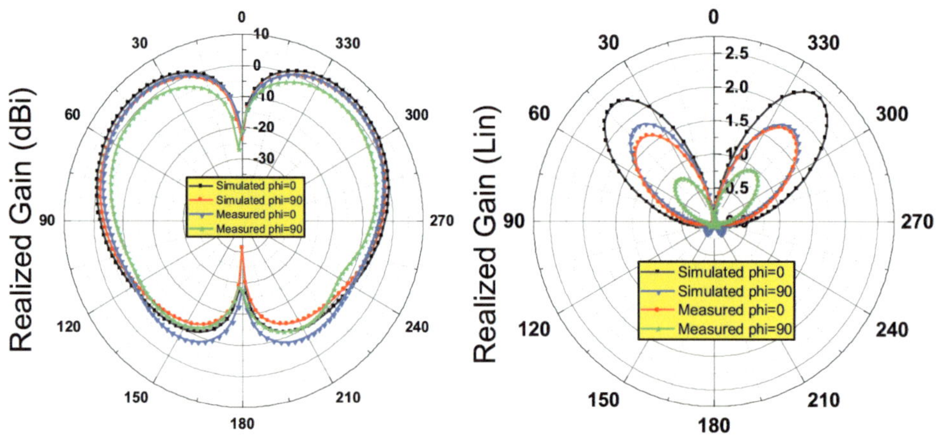

Figure 16.
1D polar plot of simulated and measured realized gain for E-plane and H-plane: (a) realized gain in dBi and (b) realized gain in linear.

(propagation axis) while the wave is advancing forward from OWPA. In contrast, the cylindrical scan for PWPA shows a typical plane wave pattern progressing normally in the direction of propagation. If we visualize the scenario in the cylindrical coordinate system (ρ, ϕ, z), E-field is swirling along the $\vec{\rho}$ and $\vec{\phi}$ orientation while advancing along the z-direction, precisely verified from Eq. (9) in theory.

This phase profile of the OAM beam is orthogonal to the propagation axis. The cylindrical scan is created for $h = 1$ m, $r = 5$ cm, and step size 1° in CST Studio Suite®. This shows us that PWPA has a uniform phase front, whereas the phase front of OWPA is changing with respect to $\exp(\pm jl\phi)$. This phase change pattern is also noted in **Figure 17(e)**, which shows twice the change 2π over a wavelength for OWPA, suggesting that OAM winding number or mode (l) 2 be produced. Comparing PWPA and OWPA through **Figure 17(d), (f)** and **(g)**, E-pattern in far-field (head-on), power flow (head-on), and 3D radiation pattern, respectively, simulated in CST-MWS, all shows that Gaussian amplitude pattern of PWPA, whereas ring-like amplitude pattern for OWPA. Measured and simulated radiation patterns for E-plane and H-plane are demonstrated in **Figure 17(h)**, intriguingly showing a Gaussian pattern for PWPA, on the other hand, second-order Bessel beam pattern for OWPA. The radiation pattern is measured in our SATIMO starlab setup, as shown in **Figure 22**. It can be seen that merely 2 (lin) difference of the simulated and measured realized gain caused by the surface waves loss, dielectric loss in patch antenna, and 0.5 dB error in measurement in the SATIMO setup. The realized gain of PWPA is twice as much as OWPA, as the beam in OWPA is ring-like, and the gain is divided. Although PWPA has higher gain and power flow, OWPA power flow covered area is larger.

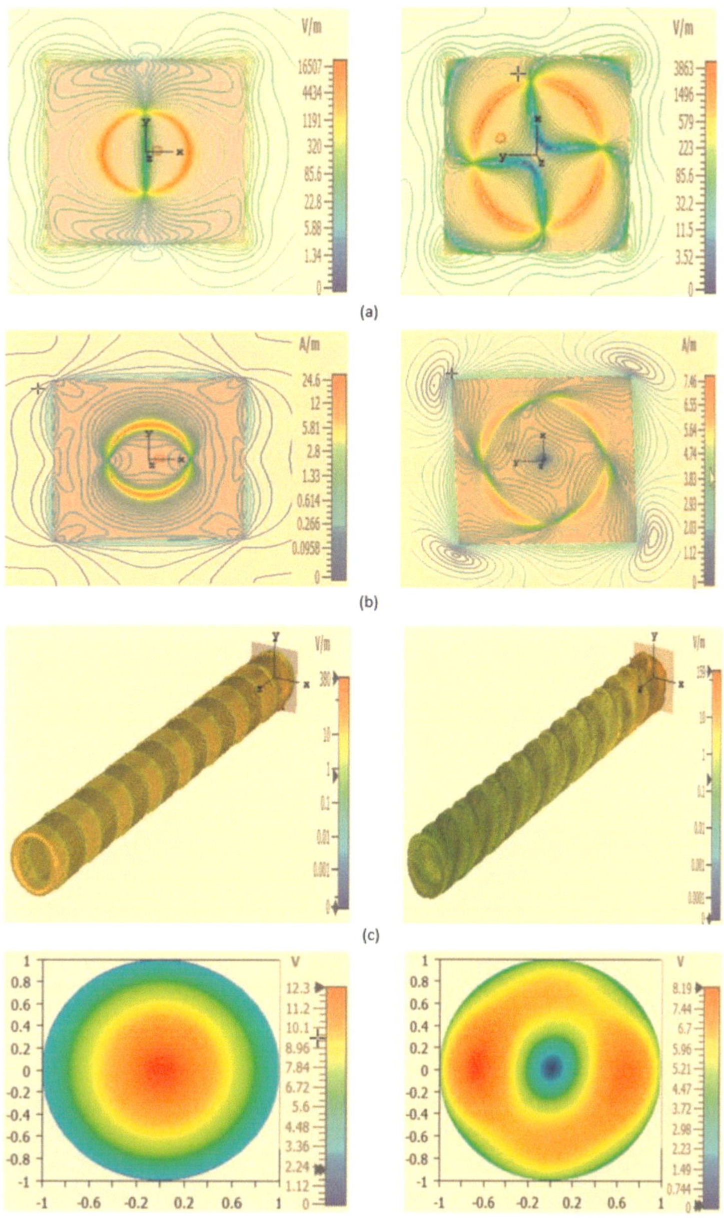

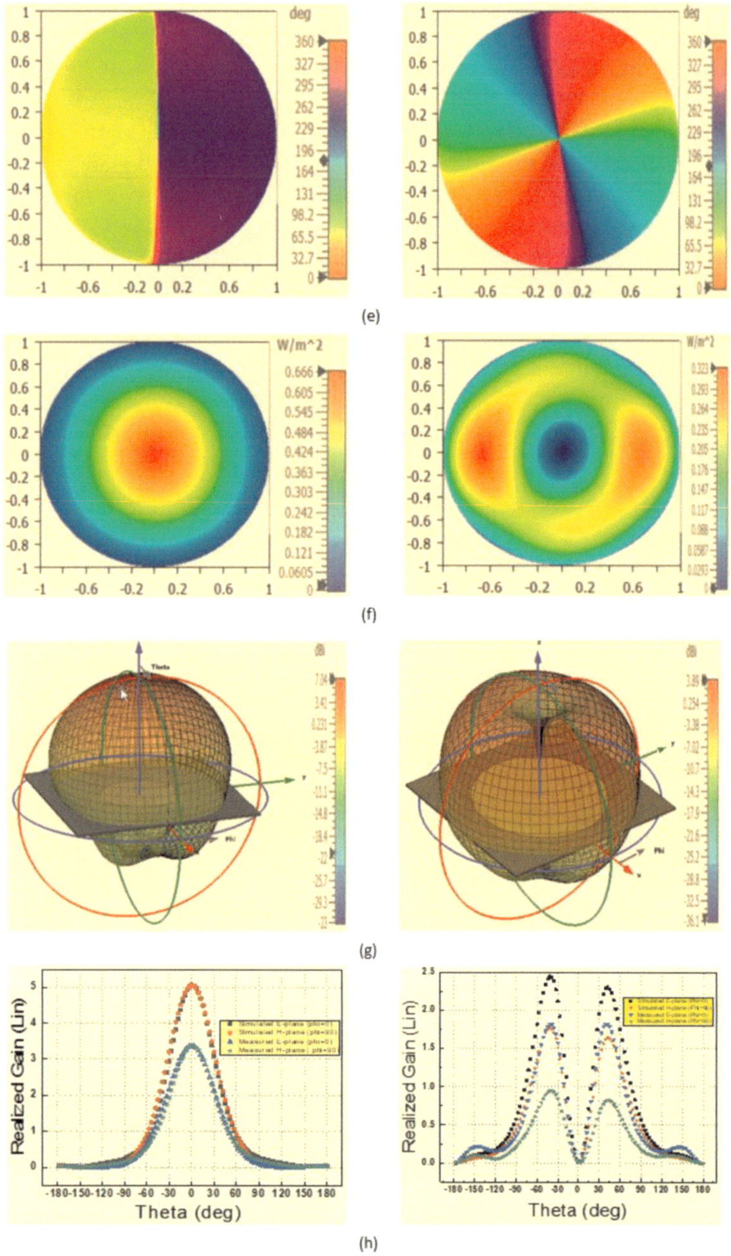

Figure 17.
Simulated PWPA (left) and OWPA (right). (a) E-field, (b) H-field, (c) cylindrical E-near-field scan, (d) E-pattern far-field, (e) phase pattern far-field, (f) power flow far-field, (g) 3D realized gain. Also, the 1D-polar plot of measured and simulated (h) realized gain (Lin) of PWPA and OWPA.

6. Comparison of plane wave and OAM wave propagation

This section compares plane wave and OAM wave propagation using PWPA and OWPA designed at frequency 2.45 GHz. The electrical size of PWPA or OWPA at the center frequency is 0.9λ. The PWPA or OWPA is enclosed in radiating bounding box of size $1\ \text{m} \times 1\ \text{m} \times 1\ \text{m}\ (8\lambda \times 8\lambda \times 8\lambda; 512\lambda^3)$, as shown in **Figure 18(a)**. The extra open and added space radiating boundary ensures the far-field behavior in free space for both Tx–Rx antennas. The nature of the electric field, magnetic field, and power flow of individual PWPA, OWPA, and combined Tx–Rx pairs is monitored. Three cases simulated in the work are: (i) Tx–Rx: PWPA–PWPA, (ii) Tx–Rx: OWPA–OWPA, (iii) Tx–Rx–Rx: OWPA–PWPA–PWPA. In **Figure 18(a)**, two PWPA antennas, such as Tx and Rx, are separated by 1 m in their far-field distance.

Similarly, in the second case, **Figure 18(a)**, OWPA–OWPA Tx–Rx pair analysis is performed within the same boundary box as another simulation setup. **Figure 18(b)** illustrates the third case of OWPA–PWPA–PWPA configuration. The electric field carpet plots for individual PWPA and OWPA are depicted in **Figure 18(c)** and **(d)**, respectively. It clearly shows no field components along the z-axis for OWPA, whereas a maximum of the electric field for PWPA takes place along the z-axis. In fact, the OWPA electric field is skewed and is at an angle of 41° at x–z, y–z, and y–x plane. It also suggests that the Poynting vector for OWPA is not along the z-axis, but it is skewed along the z-axis.

It is evident that OWPA does not have a beam in direct LOS, so in **Figure 18(b)**, two PWPA at an angle of 41° are placed as a receiver antenna to pick up the signal from OWPA. This ensures a maximum reception of the signal from the OWPA antenna. The angle of 41° alignment is chosen as the main lobe direction of OWPA is at ±41°. In this presentation, O1 is the OAM wave patch antenna, and P2 and P3 are the plane wave patch antennas. Moreover, the RF propagation path modeling for OWPA(O1)–PWPA(P2)–PWPA(P3) clearly shows the direction of power flow and the establishment of a communication path, as shown in **Figure 19(c)**. The OWPA–OWPA configuration depicted in **Figure 19(b)** shows that maximum energy reception does not happen if we consider choosing both Tx and Rx antennas as OWPA in direct line of sight (LOS). In contrast, this is not the case for plane wave propagation, as portrayed in **Figure 19(a)**.

It allows us to monitor the simulated and measured transmission and reflection in both plane wave and OAM wave propagation scenarios. Firstly, PWPA–PWPA simulated and measured reflection coefficients are shown in **Figure 20(a)**. Reflection coefficients are well below −20 dB for both Tx–Rx pairs, while the transmission coefficient is just above −30 dB. In **Figure 20(b)**, the results are presented for OWPA (O1)–PWPA (P2)–PWPA (P3) configurations, which are one of the desired arrangements for the OAM receiving scenario. Both simulated and measured transmission

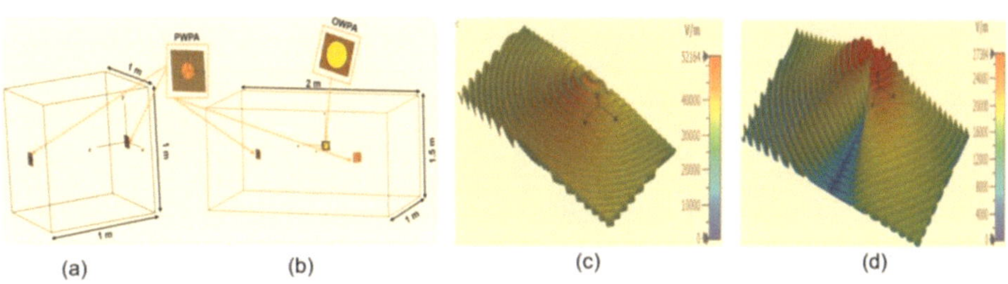

Figure 18.
(a) Far-field radiation condition in simulation to capture fields for PWPA–PWPA; (b) far-field radiation condition in simulation to capture fields for OWPA–PWPA–PWPA; (c) carpet plot of the electric field for PWPA and (d) OWPA on xz-plane gives visualization that PWPA has a maximum of propagation along the z-axis, whereas OWPA has no propagation along the z-axis. Propagation of OWPA is at an angle from the center of the z-axis.

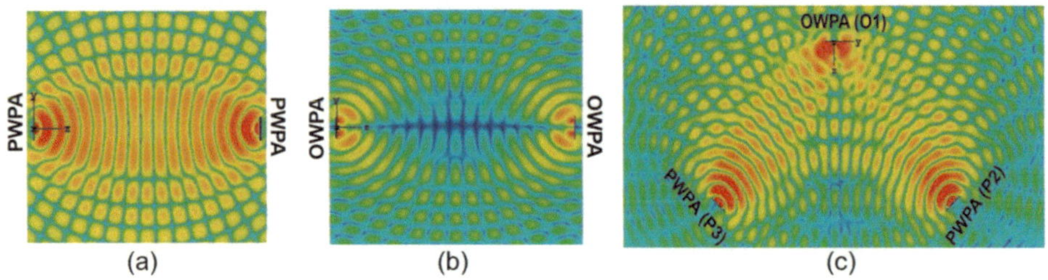

Figure 19.
RF propagation path modeling for (a) PWPA–PWPA, (b) OWPA–OWPA, (c) OWPA (O1)–PWPA (P2)–PWPA (P3). All the antennas are at the far-field distances, E-field is considered in depiction where red indicates maximum value, and blue represents minimum value.

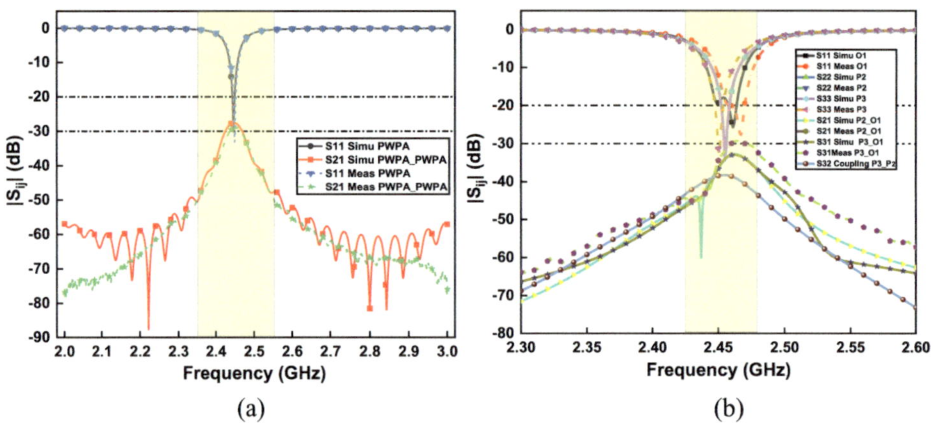

Figure 20.
Simulated and measured transmission and reflection coefficient of Tx–Rx pair (a) PWPA–PWPA and (b) OWPA (O1)–PWPA (P2)–PWPA (P3).

parameters for O1–P2 and O1–P3 are just above −30 dB, whereas all the reflection parameters are below −20 dB for all O1, P2, and P3. The frequency shift between the measured and simulated results is approximately 5 MHz, whereas the magnitude difference is negligible. All the measured results follow as designed and expected through simulation. The power received from two PWPA is combined, and the phase-detection method has to be used further in the system to know the order of OAM [70].

Furthermore, a comprehensive measured comparison of different configurations OWPA–OWPA, OWPA–PWPA, and OWPA–PWHA is shown in **Figure 21**. Among all these configurations, one of the least received signals is observed in OWPA–OWPA and the RF propagation path model, whereas the most robust signal is received in the case of OWPA–PWHA at −30 dB. OWPA–PWPA configuration has received a signal level just above OWPA–OWPA at −50 dB, as depicted in **Figure 21**.

It is worth noting that the acquired transmission behavior in **Figure 21** clearly shows the pattern of two peaks with a valley for all configurations, forming the M-shape as the desired characteristics for vortex reception, which cannot be observed in the O1–P2–P3 reception configuration in **Figure 20(b)**. The signal received at individual two PWPA should go through further a signal power combining technique to get the desired OAM pattern. The measurement setup for two configurations, PWPA–PWPA and OWPA–PWHA, is presented in **Figure 22**. The rest configuration is adopted similarly in measurement.

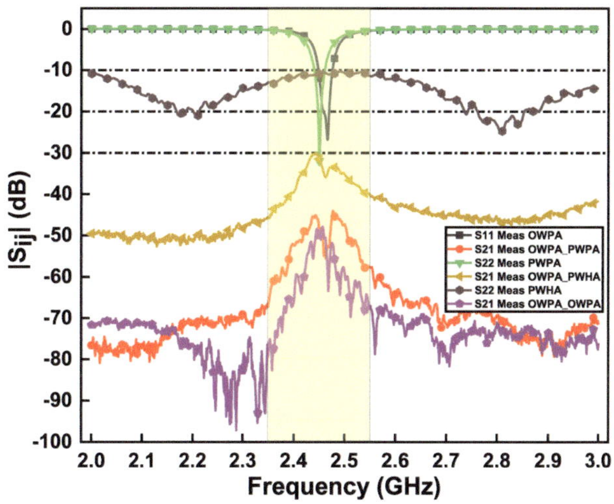

Figure 21.
Measured transmission and reflection coefficient of Tx–Rx pair OWPA–PWPA, OWPA–PWHA, OWPA–OWPA.

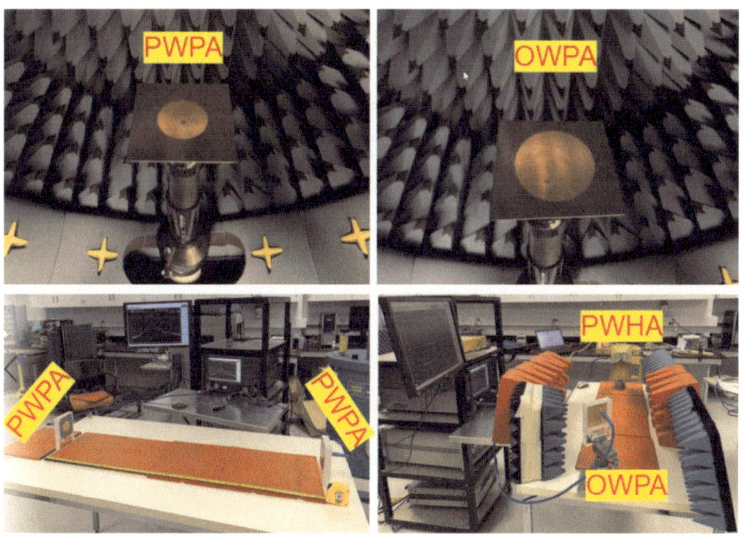

Figure 22.
Measurement setup in SATIMO lab and for Tx–Rx configuration.

7. Conclusions

This chapter provides a thorough analysis and comparison of the fundamental properties of plane wave and OAM wave in different Tx–Rx combinations using simulation, RF path modeling, and measurement to understand OAM reception techniques. The analytical equations for the electric field, magnetic field, and other physical parameters with respect to wave propagation have been derived and studied. Moreover, a comprehensive comparison of plane wave and OAM wave is examined in theory, designed PWPA and OWPA, and their propagation and reception mechanisms. First of all, analytical equations of OAM wave have been developed and compared with those of plane wave. It is found that all the plane wave parameters can effectively be deduced from the modeling of OAM waves, which presents an excellent insight for understanding the differences between OAM wave and plane wave. Cylindrical near-field scan shows that the pitch of plane wave becomes zero while the OAM wave exhibits nonzero pitch through a helical motion along the direction of propagation. Subsequently, the designed PWPA and OWPA characteristics comparison is discussed in detail to show and appreciate the behavior of the patch antenna for the generation of plane wave and OAM wave. Moreover, a comparison of plane wave and OAM wave propagation is discussed and demonstrated that OWPA–PWHA and OWPA (O1)–PWPA (P2)–PWPA (P3) are favorable mechanisms for the reception of OAM signals.

Nevertheless, these comparisons provide a detailed understanding of propagation behavior, theory, and modeling for plane wave and OAM wave using a simple patch antenna concept. Furthermore, OAM wave can be used for short-range radar sensing and opportune the different paradigm in vortex wave.

Acknowledgements

The authors would like to thank Traian Antonescu, Maxime Thibault, Jules Gauthier, and Louis Philippe of the Poly-Grames Research Center, Polytechnique Montreal, Montreal, QC, Canada, for their support in fabrication and measurements.

Author details

Pankaj Jha and Ke Wu*
Department of Electrical Engineering, Poly-Grames Research Center, Polytechnique Montréal, Montréal, QC, Canada

*Address all correspondence to: ke.wu@ieee.org

References

[1] Allen L, Beijersbergen MW, Spreeuw RJC, Woerdman JP. Orbital angular momentum of light and the transformation of Laguerre–Gaussian laser modes. Physical Review A. 1992; **45**(11):8185-8189. DOI: 10.1103/PhysRevA.45.8185

[2] Allen L, Barnett SM, Padgett MJ. Optical Angular Momentum. New York, NY, USA: CRC Press, Taylor & Francis Group; 2003

[3] Zettili N. Quantum Mechanics: Concepts and Applications. 2nd ed. Chichester, UK: Wiley; 2009

[4] Lee D, Sasaki H, Fukumoto H, Hiraga K, Nakagawa T. Orbital angular momentum (OAM) multiplexing: An enabler of a new era of wireless communications. IEICE Transactions on Communications. 2017;**E100.B**(7): 1044-1063. DOI: 10.1587/transcom.2016SCI0001

[5] Xu C et al. Free-space radio communication employing OAM multiplexing based on Rotman lens. IEEE Microwave and Wireless Components Letters. 2016;**26**(9): 738-740. DOI: 10.1109/LMWC.2016.2597262

[6] Oldoni M et al. Space-division demultiplexing in orbital-angular-momentum-based MIMO radio systems. IEEE Transactions on Antennas and Propagation. 2015;**63**(10): 4582-4587. DOI: 10.1109/TAP.2015.2456953

[7] Lee D, Sasaki H, Fukumoto H, Shiba H. Orbital angular momentum multiplexing technology towards the realization of Tbit/s wireless transmission. NTT Technical Review. 2017;**15**(9):9

[8] Zhao Y, Zhang C. Distributed antennas scheme for orbital angular momentum long-distance transmission. IEEE Antennas and Wireless Propagation Letters. 2020;**19**(2):332-336. DOI: 10.1109/LAWP.2019.2962199

[9] Morabito AF, Di Donato L, Isernia T. Orbital angular momentum antennas: Understanding actual possibilities through the aperture antennas theory. IEEE Antennas and Propagation Magazine. 2018;**60**(2):59-67. DOI: 10.1109/MAP.2018.2796445

[10] Tamburini F, Mari E, Sponselli A, Thidé B, Bianchini A, Romanato F. Encoding many channels on the same frequency through radio vorticity: First experimental test. New Journal of Physics. 2012;**14**(3):033001. DOI: 10.1088/1367-2630/14/3/033001

[11] Tamagnone M, Craeye C, Perruisseau-Carrier J. Comment on 'encoding many channels on the same frequency through radio vorticity: First experimental test'. New Journal of Physics. 2012;**14**(11):118001. DOI: 10.1088/1367-2630/14/11/118001

[12] Lee D et al. An experimental demonstration of 28 GHz band wireless OAM-MIMO (orbital angular momentum multi-input and multi-output) multiplexing. In: 2018 IEEE 87th Vehicular Technology Conference (VTC Spring), June. 2018. pp. 1-5. DOI: 10.1109/VTCSpring.2018.8417790

[13] Chen R, Zhou H, Moretti M, Wang X, Li J. Orbital angular momentum waves: Generation, detection and emerging applications. IEEE Communication Surveys and Tutorials. 2020;**22**(2):840-868. DOI: 10.1109/COMST.2019.2952453

[14] Beijersbergen MW, Coerwinkel RPC, Kristensen M, Woerdman JP. Helical-wavefront laser beams produced with a spiral phaseplate. Optics Communications. 1994;**112**: 321-327. DOI: 10.1016/0030-4018(94) 90638-6

[15] Turnbull GA, Robertson DA, Smith GM, Allen L, Padgett MJ. The generation of free-space Laguerre–Gaussian modes at millimetre-wave frequencies by use of a spiral phase plate. Optics Communications. 1996;**127**: 183-188. DOI: 10.1016/0030-4018(96) 00070-3

[16] Wei W, Mahdjoubi K, Brousseau C, Emile O. Horn antennas for generating radio waves bearing orbital angular momentum by using spiral phase plate. IET Microwaves, Antennas and Propagation. 2016;**10**(13):1420-1427. DOI: 10.1049/iet-map.2016.0064

[17] Shi CB, Li YB, Wu W, Wu RY, Cui TJ. An ultrathin spiral phase plate for generation of OAM radio waves. In: 2016 Progress in Electromagnetics Research Symposium, PIERS 2016; August 8, 2016-August 11, 2016; Shanghai, China. 2016. pp. 3740-3743. DOI: 10.1109/ PIERS.2016.7735415

[18] Tamburini F, Mari E, Thidé B, Barbieri C, Romanato F. Experimental verification of photon angular momentum and vorticity with radio techniques. Applied Physics Letters. 2011;**99**:1-4. DOI: 10.1063/1.3659466

[19] Gong Y et al. Generation and transmission of OAM-carrying vortex beams using circular antenna array. IEEE Transactions on Antennas and Propagation. 2017;**65**(6):2940-2949. DOI: 10.1109/TAP.2017.2695526

[20] Spinello F, et al. Experimental near field OAM-based communication with circular patch array. July 2015. [Online]. arXiv:1507.06889 [physics]. Available from: http://arxiv.org/abs/1507.06889 [Accessed: 01 November 2020]

[21] Tennant A, Bai Q, Allen B. Experimental circular phased array for generating OAM radio beams. Electronics Letters. 2014;**50**(20): 1414-1415. DOI: 10.1049/el.2014. 2860

[22] Deng C, Chen W, Zhang Z, Li Y, Feng Z. Generation of OAM radio waves using circular vivaldi antenna array. International Journal of Antennas and Propagation. 2013 Available from: h ttps://www.hindawi.com/journals/ijap/ 2013/847859/ [Accessed: 02 November 2020]

[23] Brousseau C, Mahdjoubi K, Emile O, Wei W. Generation of OAM waves with circular phase shifter and array of patch antennas. Electronics Letters. 2015; **51**(6):442-443. DOI: 10.1049/ el.2014.4425

[24] Chen J. Performance evaluation of electromagnetic OAM waves with circular plate antenna array. In: 2019 International Conference on Communications, Information System and Computer Engineering (CISCE), July 2019, Haikou, China. 2019. pp. 226-229. DOI: 10.1109/ CISCE.2019.00059

[25] Guo Z-G, Yang G-M. Radial uniform circular antenna array for dual-mode OAM communication. IEEE Antennas and Wireless Propagation Letters. 2017; **16**:404-407. DOI: 10.1109/ LAWP.2016.2581204

[26] Tennant A, Allen B. Generation of OAM radio waves using circular time-switched array antenna. Electronics Letters. 2012;**48**(21):1365. DOI: 10.1049/ el.2012.2664

[27] Zhang Z, Zheng S, Jin X, Chi H, Zhang X. Generation of plane spiral OAM waves using traveling-wave circular slot antenna. IEEE Antennas and Wireless Propagation Letters. 2017;**16**: 8-11. DOI: 10.1109/LAWP.2016.2552227

[28] Zhang W et al. Four-OAM-mode antenna with traveling-wave ring-slot structure. IEEE Antennas and Wireless Propagation Letters. 2017;**16**:194-197. DOI: 10.1109/LAWP.2016.2569540

[29] Gui L, Akram MR, Liu D, Zhou C, Zhang Z, Li Q. Circular slot antenna systems for OAM waves generation. IEEE Antennas and Wireless Propagation Letters. 2017;**16**:1443-1446. DOI: 10.1109/LAWP.2016.2641458

[30] Ke B, Li L, Wu Z, Bai J. Generation of dual-modes OAM beams based on SIW circular slot antenna. In: 2017 Sixth Asia-Pacific Conference on Antennas and Propagation (APCAP), October 2017, Xi'an. 2017. pp. 1-3. DOI: 10.1109/APCAP.2017.8420560

[31] Wu J, Zhang Z, Ren X, Huang Z, Wu X. A broadband electronically mode-reconfigurable orbital angular momentum metasurface antenna. IEEE Antennas and Wireless Propagation Letters. 2019;**18**(7):1482-1486. DOI: 10.1109/LAWP.2019.2920695

[32] Chen MLN, Jiang LJ, Sha WEI. Artificial perfect electric conductor-perfect magnetic conductor anisotropic metasurface for generating orbital angular momentum of microwave with nearly perfect conversion efficiency. Journal of Applied Physics. 2016;**119**(6): 064506. DOI: 10.1063/1.4941696

[33] Xu H, Liu H, Ling X, Sun Y, Yuan F. Broadband vortex beam generation using multimode Pancharatnam–berry metasurface. IEEE Transactions on Antennas and Propagation. 2017;**65**(12): 7378-7382. DOI: 10.1109/TAP.2017.2761548

[34] Yu S, Li L, Shi G, Zhu C, Zhou X, Shi Y. Design, fabrication, and measurement of reflective metasurface for orbital angular momentum vortex wave in radio frequency domain. Applied Physics Letters. 2016;**108**: 121903. DOI: 10.1063/1.4944789

[35] Yu S, Li L, Shi G. Dual-polarization and dual-mode orbital angular momentum radio vortex beam generated by using reflective metasurface. Applied Physics Express. 2016;**9**(8):082202. DOI: 10.7567/APEX.9.082202

[36] Yu S, Li L, Shi G, Zhu C, Shi Y. Generating multiple orbital angular momentum vortex beams using a metasurface in radio frequency domain. Applied Physics Letters. 2016;**108**(24): 241901. DOI: 10.1063/1.4953786

[37] Shi Y, Zhang Y. Generation of wideband tunable orbital angular momentum vortex waves using graphene metamaterial reflectarray. IEEE Access. 2018;**6**:5341-5347. DOI: 10.1109/ACCESS.2017.2740323

[38] Chen X, Zhou H, Liu M, Dong J. Measurement of orbital angular momentum by self-interference using a plasmonic metasurface. IEEE Photonics Journal. 2016;**8**(1):1-8. DOI: 10.1109/JPHOT.2015.2509859

[39] Zhang Y, Lyu Y, Wang H, Zhang X, Jin X. Transforming surface wave to propagating OAM vortex wave via flat dispersive metasurface in radio frequency. IEEE Antennas and Wireless Propagation Letters. 2018; **17**(1):172-175. DOI: 10.1109/LAWP.2017.2779269

[40] Yang L, Sun S, Sha WEI. Ultrawideband reflection-type

metasurface for generating integer and fractional orbital angular momentum. IEEE Transactions on Antennas and Propagation. 2020;**68**(3): 2166-2175. DOI: 10.1109/TAP.2019. 2948711

[41] Qin F, Wan L, Li L, Zhang H, Wei G, Gao S. A transmission metasurface for generating OAM beams. IEEE Antennas and Wireless Propagation Letters. 2018; **17**(10):1793-1796. DOI: 10.1109/ LAWP.2018.2867045

[42] Mao F, Li T, Shao Y, Huang M, Dong N. Generation of radio beams carrying OAM basing on coaxial waveguide. In: 2016 Progress in Electromagnetic Research Symposium (PIERS), August 2016, Shanghai, China. 2016. pp. 1267-1270. DOI: 10.1109/ PIERS.2016.7734635

[43] Li Y, Xiao G, Li H. Generating radio OAM waves in circular waveguide with coaxial line feed structure. In: 2019 IEEE International Conference on Computational Electromagnetics (ICCEM), March 2019, Shanghai, China. 2019. pp. 1-3. DOI: 10.1109/ COMPEM.2019.8779191

[44] Wei WL, Mahdjoubi K, Brousseau C, Sharaiha A, Emile O. Horn antenna for generating orbital angular momentum (OAM) waves. In: 2015 Loughborough Antennas & Propagation Conference (LAPC); November 2015; Loughborough, Leicestershire, United Kingdom. pp. 1-3. DOI: 10.1109/ LAPC.2015.7365994

[45] Yang Y et al. Study of a water-immersed orbital angular momentum horn antenna. In: 2018 International Conference on Microwave and Millimeter Wave Technology (ICMMT); May 2018; Chengdu. 2018. pp. 1-3. DOI: 10.1109/ICMMT.2018. 8563931

[46] Wei WL, Mahdjoubi K, Brousseau C, Emile O, Sharaiha A. Two monopole antennas for generating radio OAM waves in circular waveguide. In: 2016 10th European Conference on Antennas and Propagation (EuCAP); April 2016; Davos, Switzerland. 2016. pp. 1-4. DOI: 10.1109/EuCAP.2016. 7481623

[47] Huang M, Zong X, Nie Z. Method to generate electromagnetic field with orbital angular momentum in circular waveguide. In: 2016 IEEE International Symposium on Antennas and Propagation (APSURSI); June 2016; Fajardo, PR, USA. 2016. pp. 1897-1898. DOI: 10.1109/APS.2016.7696655

[48] Deep-level stereoscopic multiple traps of acoustic vortices. Journal of Applied Physics;**121**(16) Available from: https://aip.scitation.org/doi/full/ 10.1063/1.4981122 [Accessed: 03 November 2020]

[49] Volke-Sepúlveda K, Santillán AO, Boullosa RR. Transfer of angular momentum to matter from acoustical vortices in free space. Physical Review Letters. 2008;**100**(2): 024302. DOI: 10.1103/PhysRevLett. 100.024302

[50] Yang L, Ma Q, Tu J, Zhang D. Phase-coded approach for controllable generation of acoustical vortices. Journal of Applied Physics. 2013;**113**(15):154904. DOI: 10.1063/1.4801894

[51] Tamburini F, Thidé B, Molina-Terriza G, Anzolin G. Twisting of light around rotating black holes. Nature Physics. 2011;**7**(3) Art. no. 3. DOI: 10.1038/nphys1907

[52] Tamburini F, Thidé B, Valle MD. OAM radio astronomy—A new way to study the universe and its black holes. Royal Astronomical Society. 2020. p. 11

[53] Pan S, Wang L, Wang W, Zhao S. An effective way for simulating oceanic turbulence channel on the beam carrying orbital angular momentum. Scientific Reports. 2019;**9**(1) Art. no. 1. DOI: 10.1038/s41598-019-50465-w

[54] Oraizi H, Emamian H. Generation of orbital angular momentum modes via holographic leaky-wave metasurfaces. Scientific Reports. 2020;**10**(1):7358. DOI: 10.1038/s41598-020-64278-9

[55] Liu XY, Zhu Y, Xie W, Peng GH, Wang W. Generation of plane spiral orbital angular momentum waves by microstrip Yagi antenna array. IEEE Access. 2020;**8**:175688-175696. DOI: 10.1109/ACCESS.2020.3026241

[56] Khan MIW et al. A 0.31-THz orbital-angular-momentum (OAM) wave transceiver in CMOS with bits-to-OAM mode mapping. IEEE Journal of Solid-State Circuits. 2022:1-1. DOI: 10.1109/JSSC.2022.3141366

[57] Huang Y et al. Generation of broadband high-purity dual-mode OAM beams using a four-feed patch antenna: Theory and implementation. Scientific Reports. 2019;**9**(1):12977. DOI: 10.1038/s41598-019-49377-6

[58] Liu C, Liu J, Niu L, Wei X, Wang K, Yang Z. Terahertz circular airy vortex beams. Scientific Reports. 2017;7(1):Art. no. 1. DOI: 10.1038/s41598-017-04373-6

[59] Pan Y et al. Generation of orbital angular momentum radio waves based on dielectric resonator antenna. IEEE Antennas and Wireless Propagation Letters. 2017;**16**:385-388. DOI: 10.1109/LAWP.2016.2578958

[60] Barbuto M, Toscano A, Bilotti F. Single patch antenna generating electromagnetic field with orbital angular momentum. In: 2013 IEEE Antennas and Propagation Society International Symposium (APSURSI); July 2013. pp. 1866-1867. DOI: 10.1109/APS.2013.6711591

[61] Liu Y, Li W, Li Q, Zhu J. Rectangular patch antenna generating second-order vortex wave. In: 2019 Computing, Communications and IoT Applications (ComComAp), October 2019. 2019. pp. 358-360. DOI: 10.1109/ComComAp46287.2019.9018819

[62] Singh S, Upadhayay MD, Pal S. OAM wave generation using square-shaped patch antenna as slot array equivalence. IEEE Antennas and Wireless Propagation Letters. 2020;**19**(4):680-684. DOI: 10.1109/LAWP.2020.2976611

[63] McGloin D, Dholakia K. Bessel beams: Diffraction in a new light. Contemporary Physics. 2005; **46**(1):15-28. DOI: 10.1080/001075104 2000275259

[64] Hayt WH, Buck JA. Engineering Electromagnetics. 6th ed. Boston, MA, USA: Mc-Graw Hill; 2001

[65] Balanis CA. Antenna Theory Analysis and Design. 4th ed. Hoboken, New Jersey, USA: John Wiley & Sons, Inc.; 2016

[66] Studio CST. Training Manual. Darmstadt, Germany: Dassault Systemes Deutschland GmbH; 2021

[67] Shen L. The elliptical microstrip antenna with circular polarization. IEEE Transactions on Antennas and Propagation. 1981;**29**(1):90-94. DOI: 10.1109/TAP.1981.1142545

[68] Barbuto M, Trotta F, Bilotti F, Toscano A. Circular polarized patch antenna generating orbital angular momentum. PIER. 2014;**148**:23-30. DOI: 10.2528/PIER14050204

[69] Harrington R, Mautz J. Theory of characteristic modes for conducting bodies. IEEE Transactions on Antennas and Propagation. 1971;**19**(5):622-628. DOI: 10.1109/TAP.1971.1139999

[70] Mohammadi SM et al. Orbital angular momentum in radio: Measurement methods: OAM in radio-measurement methods. Radio Science. 2010;**45**(4):n/a-n/a. DOI: 10.1029/2009RS004299

Chapter 2

Fluidics for Reconfigurable Microwave Components

Dorra Bahloul, Ines Amor and Ammar Kouki

Abstract

Dielectric and conducting liquids with varying electromagnetic properties can offer novel alternatives for building tunable microwave passive components as well as antennas. Injecting these fluidics in or around microwave substrates alters their overall electrical characteristics, enabling circuit reconfigurability. Alternatively, changing the shapes and dimensions of conductors by using liquid metals can achieve similar reconfigurability. An overview of different liquids and their electromagnetic properties is first given. The principles behind the reconfigurability of the electrical characteristics of typical guiding structures based on mode shape variation in the presence of fluids are discussed. The realization of an N-bit programmable impedance tuner in 3D LTCC technology based on these principles is presented.

Keywords: liquid metal, liquid dielectric, cavities, impedance tuners, RF reconfigurability, 3D substrate, LTCC, transmission lines

1. Introduction

The increasing importance of microwave components and antenna reconfigurability stems from the fact that such functionality can significantly improve the performance of various radio systems while promoting sustainability. Indeed, it allows maximizing hardware re-use, thereby reducing the size and component count while avoiding crosstalk and noise issues that arise in multi-band/wideband systems. While a considerable effort has been dedicated to MEMS- and semiconductor-based reconfigurability, fluidics have emerged as a viable alternative for the same purpose owing to their relative abundance, suitability for flexible electronics, and power handling capabilities. However, they may not be as performing in terms of speed and high-frequency coverage.

2. Properties of dielectric and conducting liquids

2.1 Dielectric liquids

Thousands of liquid dielectrics, either organic or synthetic, are available and can be considered for microwave component reconfigurability. Their relative abundance and characteristics, such as toxicity, viscosity, electrical, and thermal properties, depending on their chemical composition. They have been widely used in different fields, including electrical, digital, and microwave circuits [1]. For microwave component reconfigurability, the key parameter of interest is the relative permittivity, which is generally complex as well as frequency and temperature-dependent. However, these liquids are typically identified by their static permittivity ($\varepsilon_s = \varepsilon_o\varepsilon_{rs}$). **Table 1** summarizes the static relative permittivity of the most commonly used dielectric liquids under ambient temperature, i.e., 20°C [2, 3].

For polar liquids with single Debye relaxation behavior, the frequency-dependent complex permittivity can be calculated using the Debye relaxation formula [3]:

$$\varepsilon = \varepsilon_\infty + \frac{(\varepsilon_s - \varepsilon_\infty)}{1 + j\omega\tau} = \varepsilon' + j\varepsilon'' \tag{1}$$

where ε_∞ is the high-frequency permittivity limit, τ is the relaxation time, and the loss tangent δ is given by $\varepsilon''\varepsilon'$. For instance, for DI (DeIonized) water at 20°C, the frequency dependency of the permittivity can be given in terms of its dielectric constant and loss tangent explicitly by [4]:

$$\begin{cases} \varepsilon_r = 6.28 + \dfrac{73.91}{1 + 0.26 \times 10^{-20} f^2} \\ \delta = \dfrac{4.48 \times 10^{-9} f}{80.19 + 2.26 \times 10^{-20} f^2} \end{cases} \tag{2}$$

Figure 1 shows the significant variation of these parameters versus frequency up to 20 GHz and illustrates how this liquid becomes increasingly lossy, which may limit its usability at very high frequencies. It should also be noted that ε_∞, ε_s, and τ are temperature dependent [4].

2.2 Liquid metals

Metals are abundant materials on earth. Most of them are present in a solid state at room temperature. Only a few metals are liquids under ambient temperature.

Liquid	Static relative permittivity (ε_{rs})
DI water	80.29
Acetone	21.13
Methanol	33.64
Butanol	18.19
Ethanol	25.16
Propanol	21.15
Ethanediol	41.89
Benzene	2.3
Chloroform	4.8
Silicone fluid	2.3

Table 1.
Static permittivity of most common liquid dielectrics at 20°C.

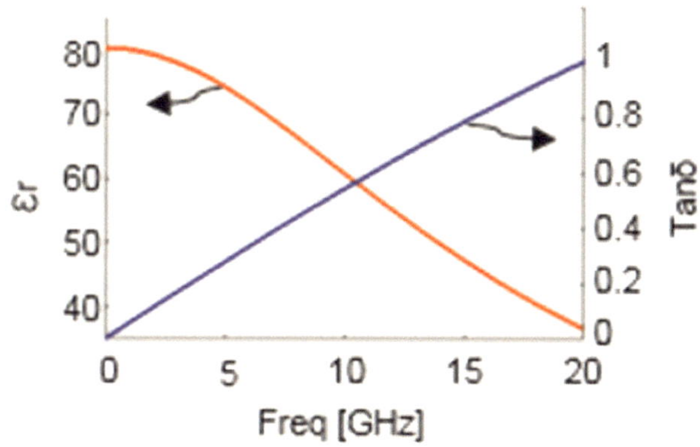

Figure 1.
DI water dielectric constant and loss tangent versus frequency at 20°C.

Liquid metal	Melting point (°C)	Electrical conductivity (S/m)	Thermal conductivity (W/m/K)	Viscosity (kg/m/s)
Gallium	29.76	6.73×10^6	29.3	1.37×10^{-3}
EGaIn	15.5	3.4×10^6	26.6	1.99×10^{-3}
Galinstan	−19	3.46×10^6	16.5	2.4×10^{-3}
Mercury	−38.8	1.04×10^6	8.5	1.526×10^{-3}

Table 2.
Key physical properties of most commonly used liquid metals.

Francium (Fr), cesium (Cs), rubidium (Rb), mercury (Hg), and gallium (Ga)-based metals are the known liquid metals [5]. The radioactivity of the Francium and the reactivity of cesium and rubidium are the main reasons these liquid metals are avoided in microwave component reconfigurability applications. On the other hand, mercury is considered a toxic fluid that cannot be manipulated safely and should be avoided or used under extremely well-controlled conditions. Gallium is a risk-free metal that melts at 29.7°C, which is slightly higher than room temperature. However, the melting points of Gallium-based alloys such as Eutectic Gallium Indium (EGaIn), Eutectic Gallium Tin (EGaSn), and Eutectic Gallium Indium Tin (known as Galinstan) are lower, and they are around 15, 21, and −19°C, respectively [5]. **Table 2** summarizes the key physical properties of the most suitable liquid metals for use in microwaves.

3. Reconfigurablity with dielectric liquids

Many different guiding structures can be used for microwave component designs. In most cases, these structures are used and modeled as transmission lines whose electrical characteristics are given by their characteristic impedance, Z_o, and effective dielectric constant, ε_{reff}. These parameters result from the mode shape propagated by the line and determine the propagation constant β and wave speed ν by:

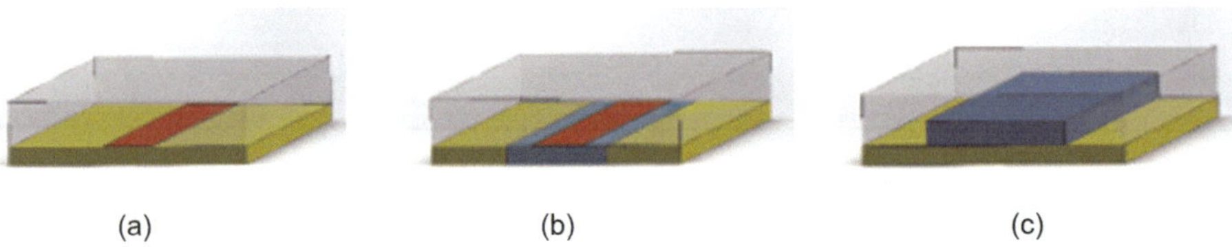

Figure 2.
Microstrip transmission line structures. (a) Conventional; (b) fluidic channel below; (c) fluidic channel above.

$$\begin{cases} \beta = \sqrt{\varepsilon_{\text{reff}}}k_o \\ v = \dfrac{c}{\sqrt{\varepsilon_{\text{reff}}}} \end{cases} \tag{3}$$

Therefore, altering the propagated mode shape using liquid dielectrics will change the electrical characteristics, thereby enabling reconfigurability. To illustrate this, consider one of the most commonly used guiding structures in microwave circuits: the microstrip line. **Figure 2a** shows the conventional microstrip line. **Figure 2b** and **c** show modified microstrip line structures with a fluidic channel included in the substrate (b) and a fluidic channel on top of the microstrip substrate (c).

Assuming a Duroid dielectric substrate having ε_r = 2.33 and a height of 4.8 mm with a strip width of 3 mm, the propagation characteristics for these three structures can be computed for different fluidic fillings in **Figure 2b** and **c**. **Table 3** summarizes the obtained results at 1 GHz for selected dielectric liquids and illustrates how fluidics allows for reconfiguring the transmission line's characteristics. For a given physical length, l, of the reconfigurable structures, the electrical length, βl, and, consequently, the propagation delay will vary along with Z_o. This principle has been employed in many applications, such as [6, 7].

Structure (2a, 2b or 2c)	**Zo (Ω)**	$\varepsilon_{\text{reff}}$	**Mode shape**
2a—Regular microstrip	113.2	1.56	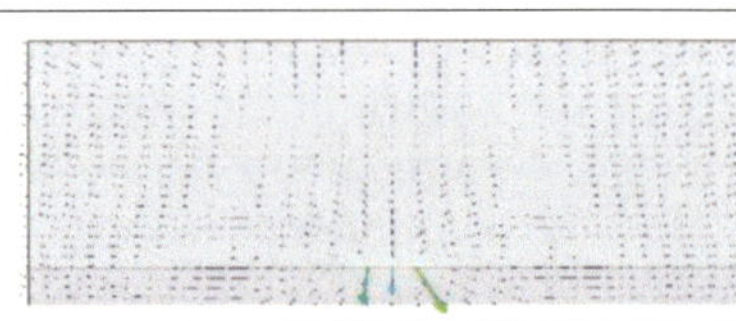
2b—Air-filled	132.5	1.15	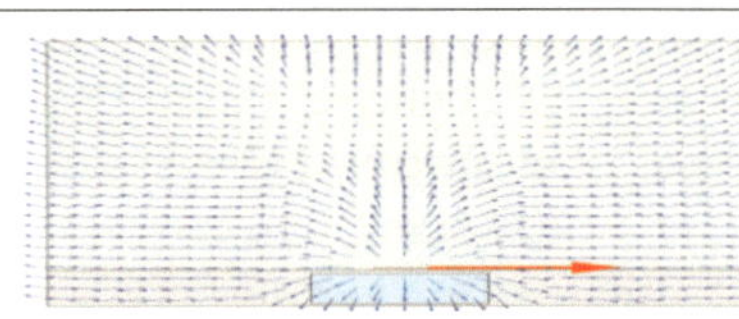
2b—Ethanol $\varepsilon_{\text{rEthanol}}$ = 12.93	84.0	2.88	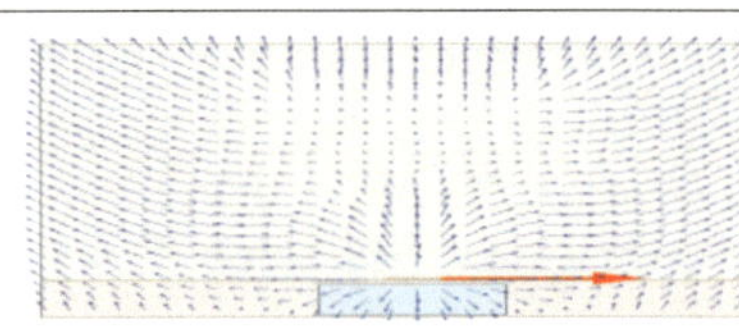
2b—Acetone $\varepsilon_{\text{rAcetone}}$ = 21.13	83.0	2.95	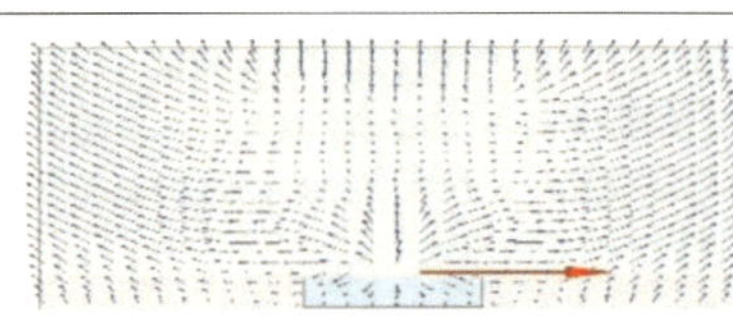

Structure (2a, 2b or 2c)	Zo (Ω)	ε_{reff}	Mode shape
2b—Methanol $\varepsilon_{Methanol}$ = 30.17	80.3	3.13	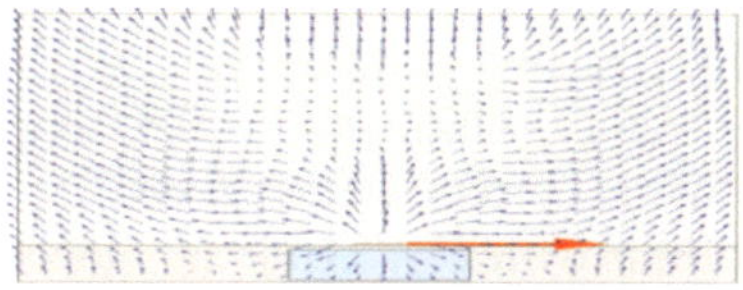
2b—DI water $\varepsilon_{rDI\ water}$ = 80.29	74.9	3.42	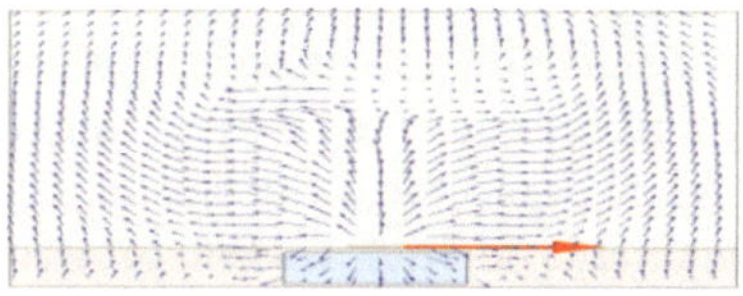
2c—Air-filled	113.5	1.57	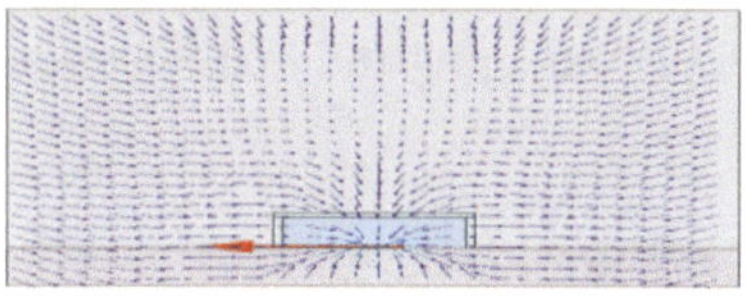
2c—Ethanol $\varepsilon_{rEthanol}$ = 12.93	69.7	4.07	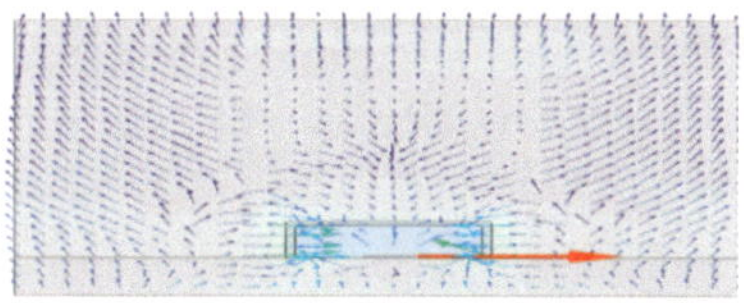
2c—Acetone $\varepsilon_{rAcetone}$ = 21.13	67.6	4.32	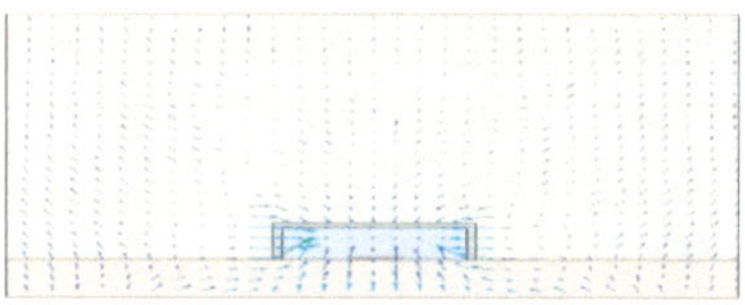
2c—Methanol $\varepsilon_{Methanol}$ = 30.17	63.8	4.73	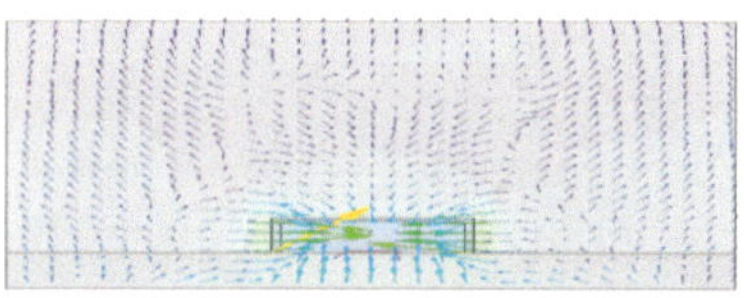
2c—DI water $\varepsilon_{rDI\ water}$ = 80.29	51.8	6.00	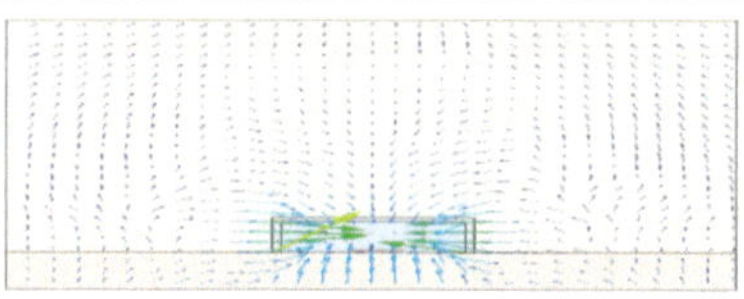

Table 3.
Reconfigurable propagation characteristics of a microstrip line.

4. Reconfigurability with liquid metals

There are two ways liquid metals can be used for microwave circuit reconfigurability: contacting and contactless [8]. These refer to whether or not the liquid metal enters in contact with conductors in the microwave circuit, the contacting approach, or is brought in close proximity to these conductors without touching them, the contactless method. This latter method is similar in principle to using dielectric fluids, whereby the liquid metal can behave as a capacitive load on top of or below the transmission line's conductors, thereby altering the propagation characteristics. This requires thin dielectric layers between the liquid metal's hosting cavities and the rest of the circuit's conductors to avoid contact resulting in undesirable reactions, e.g., corrosion. This approach can be highly effective, particularly when the liquid metal is placed near the maximum electrical field value resulting in a good tuning range [9, 10]. In the contacting approach, liquid metals are used to short connections between conductors, i.e., like contact switches [11–13] in a microwave circuit or extend a single conductor with a custom shape determined by the encapsulating cavity [14, 15]. This allows altering signal paths in a circuit, thereby enabling the reconfiguration of their electrical length and propagation delays, which reconfigures the overall component.

While liquid metals can offer significant advantages for reconfigurable microwave components and deformable printed electronics, they are relatively complex to handle and suffer from fast-oxidation when exposed to oxygen. Therefore, they require specific packaging and manipulation guidelines to overcome these limitations [16, 17].

5. Example: LTCC-based fluidic microwave impedance tuners

The implementation of fluidic reconfigurable components is inherently three-dimensional in nature, while the circuits are mostly planar. This has led to the use of various 3D printed PDMS structures that are assembled onto microwave circuits [18]. Low-Temperature Cofired Ceramics (LTCC) technology is three-dimensional circuit fabrication technology that is well suited for fluidic applications. This section uses this technology to illustrate the microwave reconfigurability of impedance tuners using dielectric fluids and liquid metal.

5.1 Design approach

5.1.1 Two-state cell

As discussed in Sections 3 and 4, if we alter the materials surrounding a signal conductor in a microwave circuit by injecting a dielectric fluid or a liquid metal in a nearby cavity, we can generate a two-state cell (**Figure 3a**): one state for the empty cavity and the other for when the cavity is filled. The characteristic impedance and the electric length are ($Z_c = Z_1$, $E_L = E_{L1}$) when the cell is in state 1, whereas when it is under state 2, it is characterized by ($Z_c = Z_2$, $E_L = E_{L2}$). Therefore, considering a 50 Ω load, we generate two different input impedances, as illustrated by the blue and the red dots in **Figure 3b** for the two states.

5.1.2 N-state tuner

Reconfigurable impedance tuners present a promising alternative to fixed ones in the design of reconfigurable circuits as they can offer multiple impedances at different frequency bands. Here, we consider distributed impedance tuners where N identical uniformly distributed two-state RF cells, C1-CN, cascaded, as shown in **Figure 4a** [18]. The *i*th cell, C_i, has a characteristic impedance Z_{ci} and an electric length E_{Li}. When the cell is in state 1, its characteristics are ($Z_{ci} = Z_1$, $E_{Li} = E_{L1}$) whereas when it is under state 2 it is characterized by ($Z_{ci} = Z_2$, $E_{Li} = E_{L2}$). Here, we are interested in the scenario where the tuner is loaded with a 50 Ω impedance, and Z_1 is set to 50 Ω in a 50 Ω system such that when all cells are in state 1, we have a 50 Ω line. Consequently,

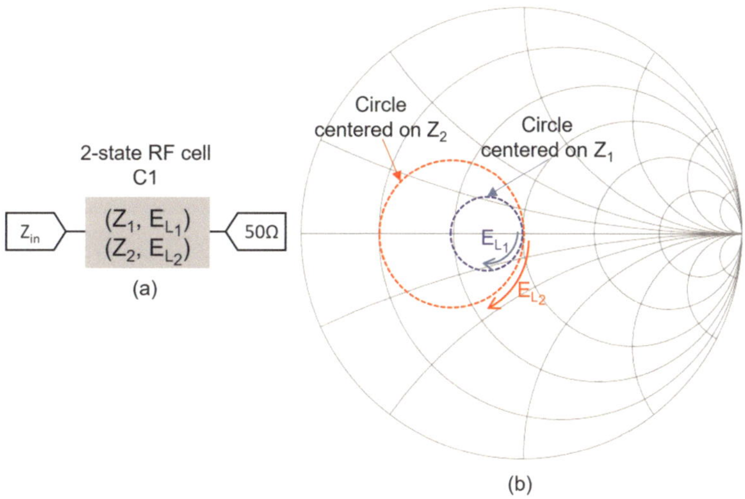

Figure 3.
Two-state cell: (a) illustrative view, (b) impedance distribution.

for a given combination of cell states, the resulting Z_{in}'s impedance consists of successive movements on the circles centered on Z_1 and Z_2 by an electrical length E_{L1} and E_{L2}, respectively, starting from the Smith chart center, as shown in **Figure 4b**. A wide coverage at a particular frequency corresponds to one where the 2^N synthesized impedances are distributed uniformly throughout the Smith chart. We define, therefore, the impedance ratio R and the tuner's total electric length E_{Lt} by:

$$R = \frac{Z_1}{Z_2} \tag{4}$$

$$EL_t = N.E_{l_1} \tag{5}$$

To synthesize impedances with high reflection coefficients, i.e., close to the Smith chart edge, Z_2 should be minimized, and losses should be reduced. Therefore, for a Z_1 set to 50 Ω, we seek to increase R. We also note that the generated impedances Z_{in} are located within a region of the Smith chart limited by E_{Lt}, as shown in **Figure 4b**. For an E_{Lt} close to a multiple of half-wavelength ($\lambda/2$), a large coverage with well-distributed impedances is expected. However, this condition may be satisfied only for some frequencies. Apart from its impact on E_{Lt}, the cell number N also controls the coverage resolution, i.e., a low value leads to scattered impedances points while a high one leads to a crowded coverage with a larger size and an increased complexity. Therefore, the tuner's coverage depends on the cells' electrical parameters and their number.

In the next sub-sections, we propose two types of RF fluidic impedance tuners based on the distributed architecture of **Figure 4a**, where reconfigurability is enabled through liquid dielectrics or metals. Here, the proposed tuners are built in LTCC substrate. Therefore, they leverage the LTCC's inherent 3D nature and its ability to realize buried transmission lines and cavities inside a multilayer substrate to form fluidic channels of varying shapes, sizes, and positions in a few standard fabrication steps.

5.2 LTCC dielectric fluidic impedance tuner

This section proposes the design of a new RF fluidic impedance tuner in a 3D LTCC substrate for RF frequency applications [19]. To do so, the design of a single

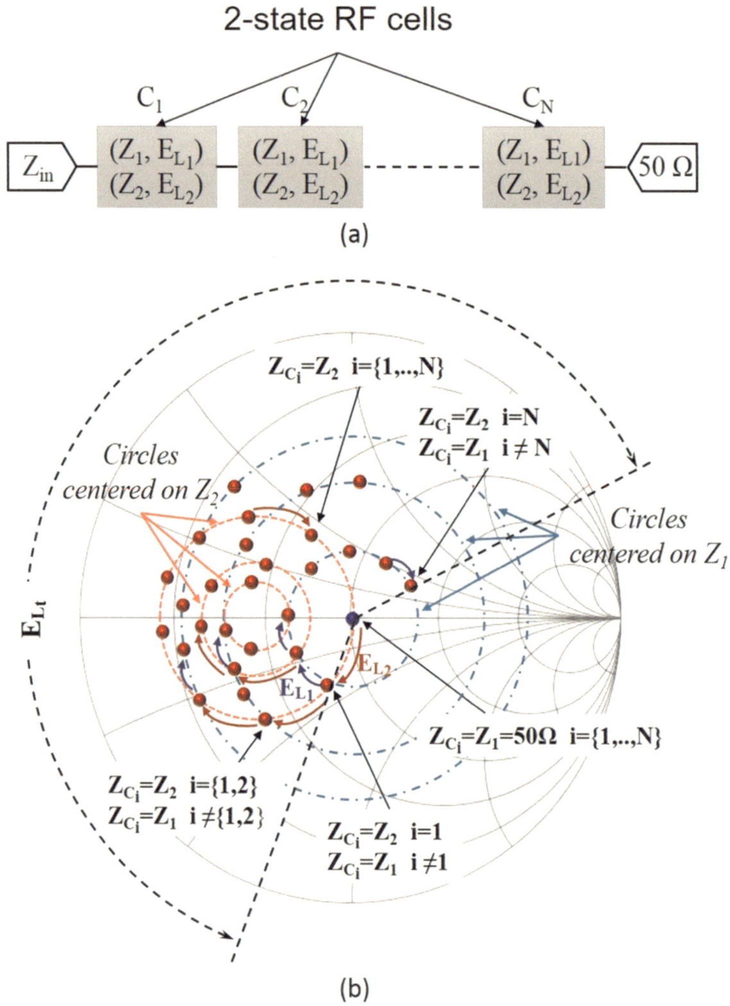

Figure 4.
Impedance tuner illustrative. (a) Scheme, (b) smith chart coverage.

fluid cell is detailed. Then, we introduce an eight-cell LTCC fluidic impedance tuner covering the 0.9–2.4 GHz frequency range. Its coverage is characterized based on EM simulations.

5.2.1 Two-state cell

A 3D view of the proposed two-state RF fluidic cell is shown in **Figure 5**. It is implemented in a multilayer LTCC substrate. It consists of a grounded coplanar waveguide (CPWG) transmission line buried in an LTCC multilayer substrate with an empty cavity, i.e., a channel, above a part of the line. An inlet and an outlet to inject/extract the desired gas or liquid, along with two tapered transitions, for on-wafer probe measurement, are also integrated into the cell design.

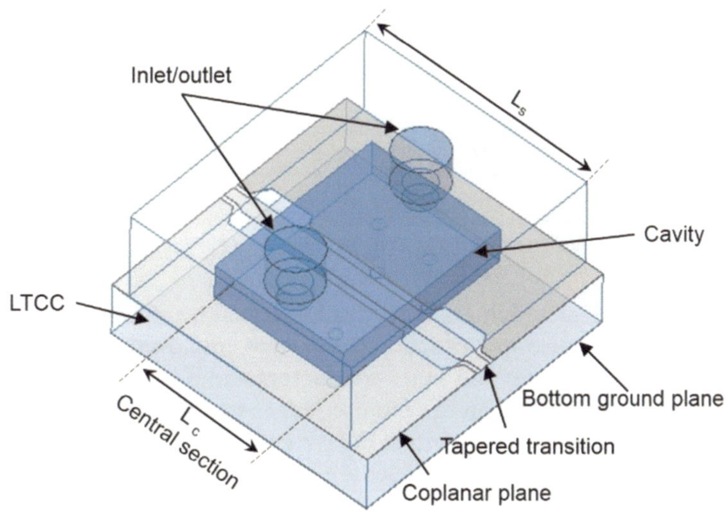

Figure 5.
Two-state RF dielectric liquid cell 3D view.

Reconfigurability of the proposed RF fluidic cell is accomplished by changing the fluidic content of the cavity, thereby varying the dielectric constant of the central section, i.e., the transmission line section where the cavity is added. This results in changing the propagation constant. Therefore, based on the fluid filling the cavity and its length L_c (**Figure 5**), the equivalent impedance and the electrical length of the complete fluidic cell can be toggled between the two states (Z_1, E_{L1}) and (Z_2, E_{L2}).

For wide reconfigurability, we should increase R, as demonstrated in Section 5.1.2. Considering the fixed physical parameters of the different CPWG sections, this can be achieved through a wide change in the cavity dielectric constant, generating the highest change in the characteristic impedance Z_c and the propagation constant β3 of the central section. To quantify this change, we introduce the impedance ratio r_z and the propagation ratio r_β as follow:

$$r_z = \frac{Z_c(\varepsilon_c = 1)}{Z_c(\varepsilon_c = \varepsilon_{r\text{liquid}})} \tag{6}$$

$$r_\beta = \frac{\beta_c(\varepsilon_c = \varepsilon_{r\text{liquid}})}{\beta_c(\varepsilon_c = 1)} \tag{7}$$

where we require that Z_C (ε_c = 1) = 50 Ω. In this manner, the fluidic cell presents a 50 Ω impedance if the cavity is empty. Here, different sections of the cells are therefore dimensioned using multiple layers of DuPont 951 to provide 50 Ω under the empty state. For maximum ratios, high values of $\varepsilon_{\text{rliquid}}$ must be chosen. Here, DI-water is chosen as the filling fluid given its high relative permittivity, as stated in Section 1.1. Consequently, based on the central section physical parameters and DI-water permittivity, we obtain a change in the electrical characteristics, as shown in **Table 4**.

To complete the design of the fluidic cell, several (cavity length: L_c; cell length L_s) combinations were considered and simulated. We aim to minimize DI-water-related loss while maintaining considerable change in the reflection coefficient; we selected an L_c of 1.5 mm and an L_s of 3 mm.

f (GHz)	Empty		DI Water		r_z	r_β
	Z_{03} (Ω)	β_3 (rad/m)	Z_{03} (Ω)	β_3 (rad/m)		
0.9	50	39.5	16.5	122	3.03	3.08
2.4	50	105	16.6	322	3.01	3.06

Table 4.
Characteristics for the central section when empty and filled with DI water.

The designed cell depicted in **Figure 5** was fabricated using Lacime in-house LTCC process flow [20]. Multiple tapes of DuPont 951 were employed. Particularly, silver paste is used to print the outer and inner conductors and fill the vias. Rectangular and circular shapes are laser drilled on the required tapes to form the necessary fluidic channels, inlets, and outlets. To maintain the structural integrity during the stacking and laminating steps, fugitive carbon tapes are used to fill the inner and outer cavities, respectively. They sublimate at 600°C during the sintering step leaving behind empty channels. **Figure 6** shows the fabricated fluidic circuit.

The designed fluidic cell was simulated and measured between 0.9 and 2.4 GHz for filled and empty cavity cases. As can be seen, the measured and simulated insertion loss and input impedances track very well (**Figure 7**). For instance, an input impedance of 50 Ω is obtained for the empty cell, whereas when DI-water is injected into the cavity, the cell impedance and its electrical length change and show variation with frequency. Consequently, a part of the signal is reflected, resulting in an insertion loss increase (**Figure 7a**) and a change in the reflection magnitude and phase (**Figure 7b**).

5.2.2 Eight-cell tuner

Eight fluidic cells were cascaded to form the tuner, as shown in **Figure 8**. Two 50 Ω vertical CPW transitions to upper CPWG lines were added at the input and output to measure the fabricated device easily. The overall tuner size, including the vertical transitions, is 26 mm × 10 mm × 1.5 mm.

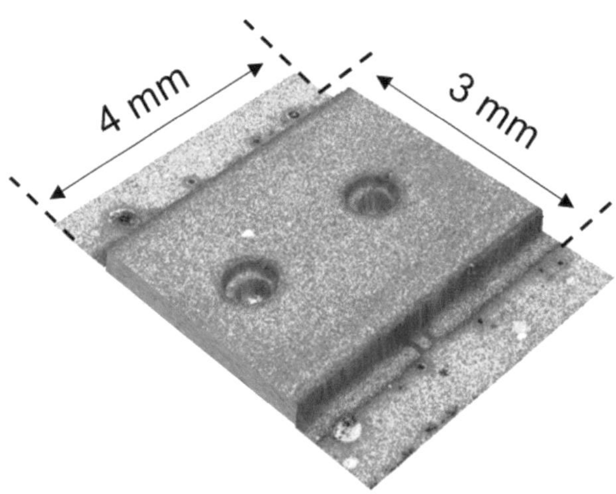

Figure 6.
Fabricated RF fluidic cell.

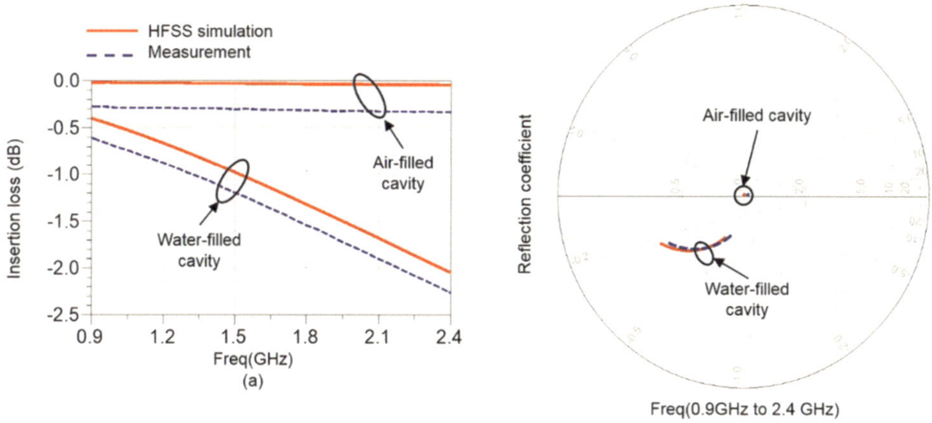

Figure 7.
Liquid dielectric cell measured and simulated: (a) insertion loss; (b) reflection coefficient.

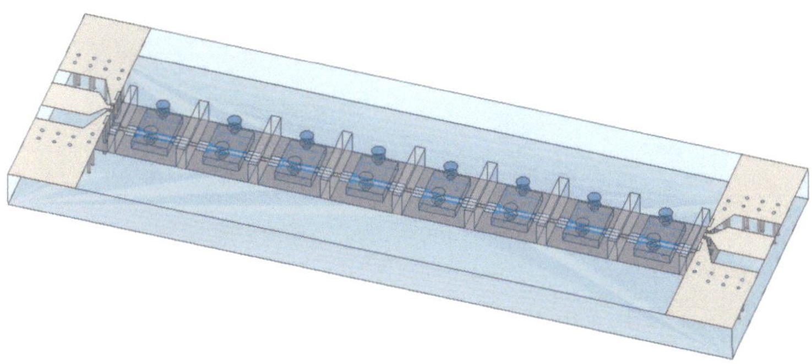

Figure 8.
Fabricated fluidic tuner.

The operation of this tuner was characterized through ADS/HFSS simulations. The obtained input impedances when a 50 Ω load terminates its output port as shown with the red points in **Figure 9**. As can be seen, the tuner offers reasonable coverage of the Smith chart, particularly at higher frequencies. However, the coverage tends to be concentrated away from the edge of the Smith chart. This is attributed to the value of the fluidic cell's impedance Z_2, which depends basically on the fluid parameters. Still, despite the retracted coverage, the impedance matching capability of the proposed tuner is quite good, even at such low frequencies. In fact, based on (8) as introduced in [21], each point in the Smith chart, representing ideal matching with a given tuner state, would become a circle if lower matching levels are permitted. For example, an impedance coverage area larger than the 256 individual points is reached at a 10 dB matching level at the considered frequencies, as shown in **Figure 9**.

$$\Gamma_{in} = \frac{\Gamma_s - S_{11}}{\Gamma_s S_{11} - 1} \tag{8}$$

where Γ_{in} is the reflection coefficient at the required matching level, Γ_s is the reflection coefficient at a particular point in the Smith chart, and S_{11} is the reflection coefficient at a point from the tuner constellation.

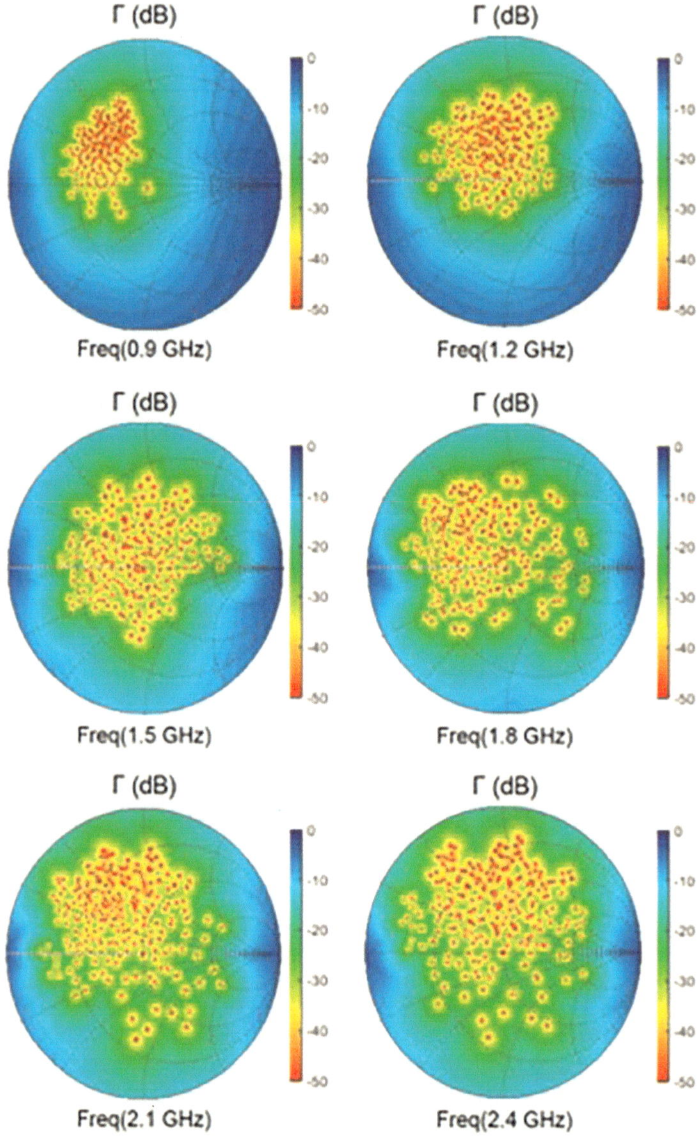

Figure 9.
Liquid dielectric tuner simulated reflection coefficient at different frequencies.

The fluidic tuner was also fabricated using the Lacime LTCC process (**Figure 10**). To measure the entire eight-cell impedance tuner, we measure a single cell in both states and cascade 8 measurement results in various combinations in ADS. The tuner's impedance coverages based on cascaded measurements and HFSS simulation are depicted in **Figure 11** for six frequency points in the range of [0.9–2.4 GHz]. The measurement results show a contracted coverage compared to HFSS simulations, particularly at low frequencies. This is attributed to the non-conformity of the fabricated and designed circuits due to the LTCC fabrication errors, as layer misalignment. Still, the tuner offers good impedance coverage and can provide reconfigurability at low RF frequencies.

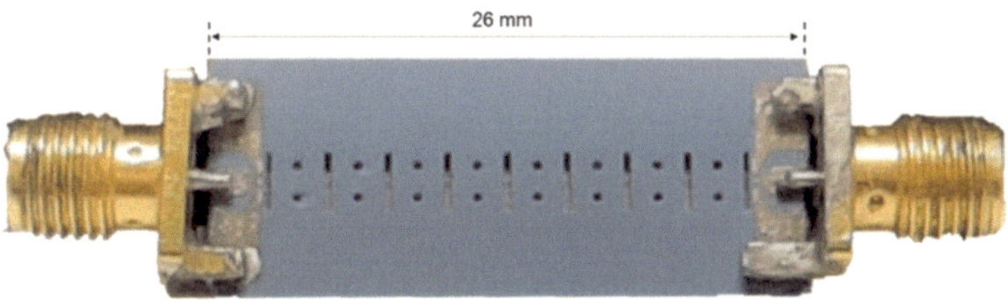

Figure 10.
Fabricated fluidic tuner.

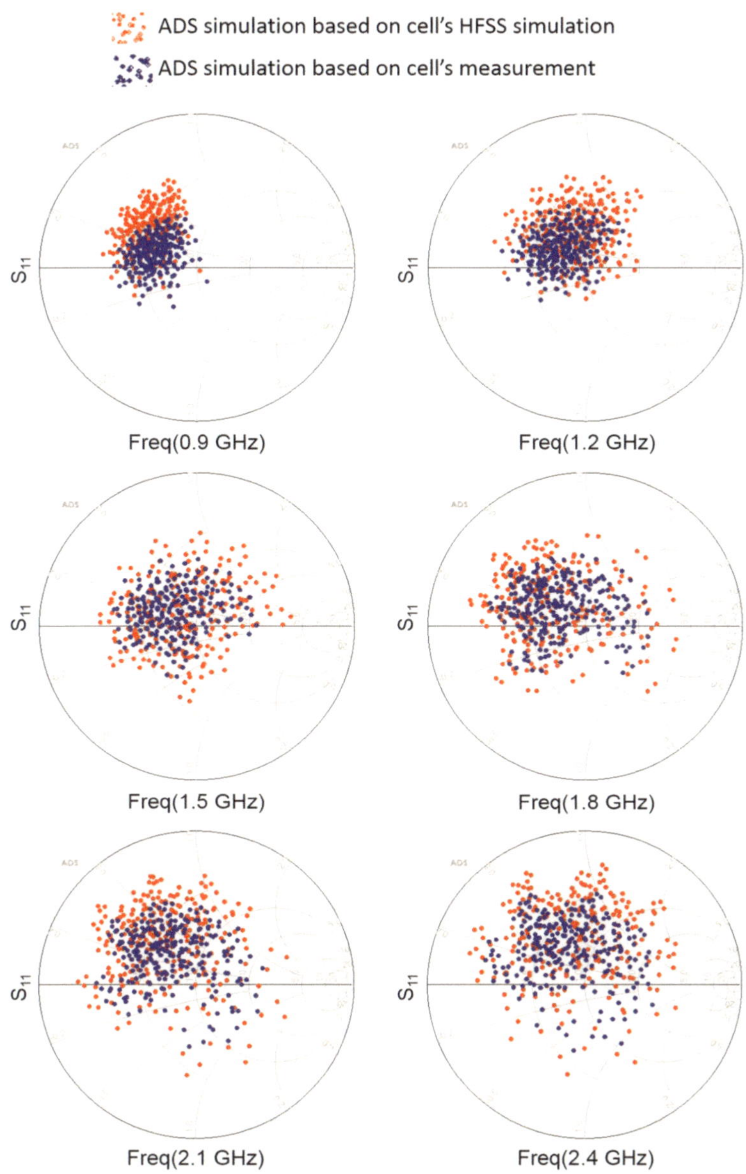

Figure 11.
Liquid dielectric tuner simulated and measured impedance coverage at different frequencies.

5.3 LTCC liquid metal imoedance tuner

A dielectric fluidic-based RF tuner with good Smith chart coverage up to 2.4 GHz has been introduced in Section 5.1. However, as explained in Section 2, DI-water is a lossy material at higher frequencies. Therefore, this tuner is not recommended to be used at upper microwave frequencies. Most dielectric liquids have either low dielectric constants and/or high losses [2]. Therefore, they are not suitable to be used in impedance tuning. Liquid metals with good electrical conductivity and deformable capability, as shown in Section 2, have gained attention in the RF field and seem to be good candidates as a filling liquid for our LTCC tuner. In this section, we will prove how liquid metals may enable reconfigurability based on the dielectric fluid two-state cell tuner [22].

5.3.1 Two-state cell

In the two-state liquid dielectric design of **Figure 12**, the cavity content is in direct contact with the signal line as no barrier between the two is used. If the same cavity were to be used with liquid metal, a short circuit would occur between the signal line in silver, and the liquid metal, Galinstan. In this case, there will be a risk of chemical reactions between the two that may lead to corrosion of the signal lines. Therefore, the cavity design has been altered to include a thin dielectric insulating layer, as shown in green in **Figure 12**, to prevent any contact.

The cell's reconfigurability is achieved by changing its corresponding capacitance: empty and liquid metal-filled states corresponding to low and high capacitance, respectively. A big ratio between these two capacitances results in a higher R ratio and, thereby, a wider Smith chart coverage of an n-state impedance tuner. Here we kept the same physical parameters of the liquid dielectric cell (**Figure 5**), and we added a 2-mil thin LTCC layer. **Figure 13** shows the insertion loss and the complex reflection coefficient of the designed and fabricated cell and demonstrates a good agreement between HFSS simulation and measurements for both states in the frequency band [1–10 GHz]. For instance, as expected, the empty cell is perfectly matched to 50 Ω, and Ga filled cavity cell shows an insertion loss and a variation in the reflection coefficient.

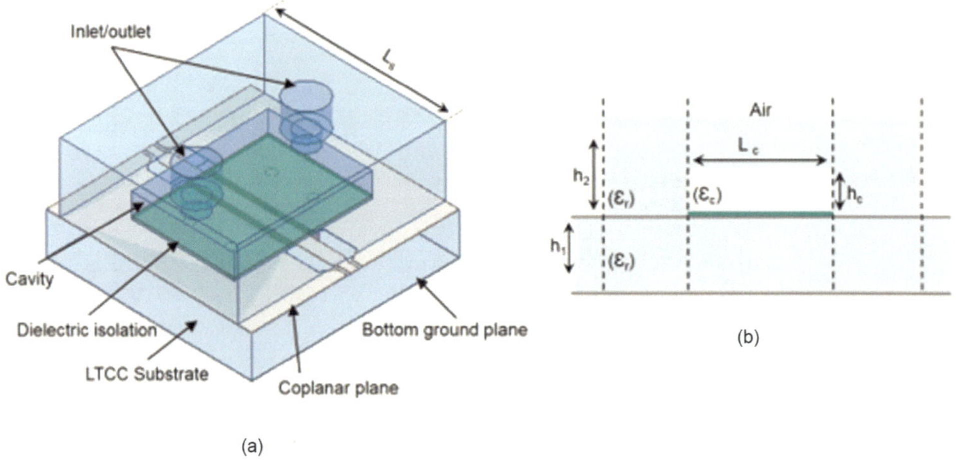

Figure 12.
Two-state RF liquid metal cell: a) 3D view; b) side view.

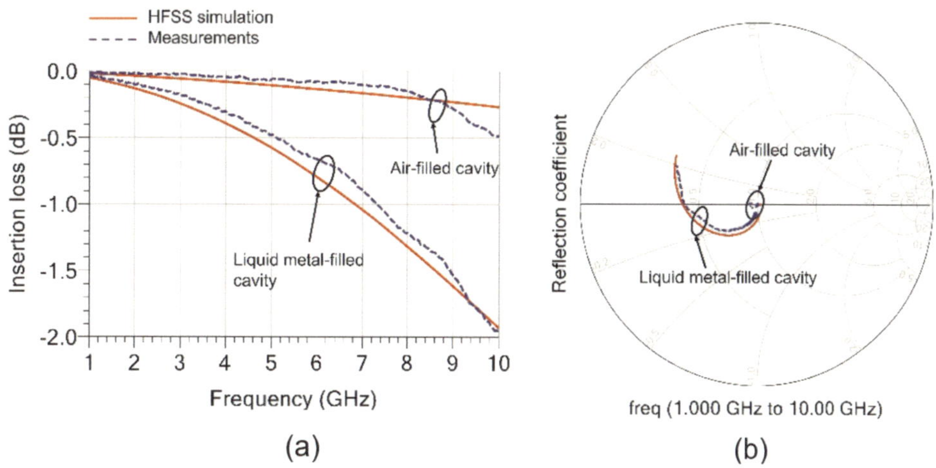

Figure 13.
Liquid metal cell measured and simulated: (a) insertion loss; (b) reflection coefficient.

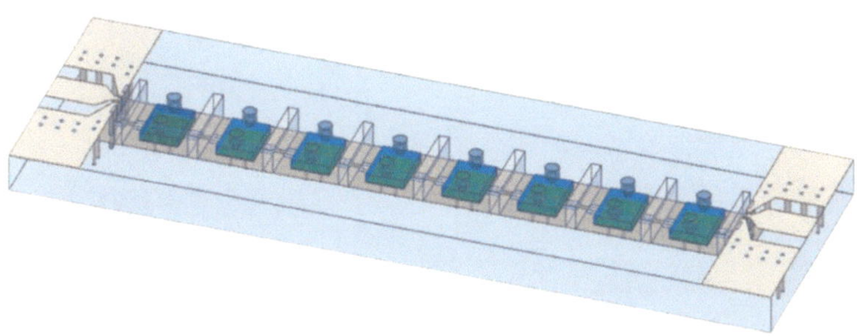

Figure 14.
Liquid metal impedance tuner: 3D view.

5.3.2 Eight-cell tuner

Similar to the dielectric-fluidic tuner, eight liquid-metal cells were cascaded, as shown in **Figure 14**. Two 50 Ω vertical CPW transitions to upper CPWG lines were added at the input and output for interconnection and measurement purposes. The final tuner has the same dimension as the liquid dielectric one, i.e., 26 mm × 10 mm × 1.5 mm. **Figure 14** shows the designed and fabricated tuner.

Figure 15 shows the simulated and measured Smith chart coverage when a 50 Ω load terminates the output port. Good impedance distributions are obtained at different frequency points in the frequency range [1 GHz, 10 GHz], particularly at higher frequencies. Like the fluidic tuner discussed previously, a well-distributed and uniform coverage cannot be obtained for all the frequencies because of the frequency-dependent impedance and electrical length of the unit cell.

6. Conclusion

Conductive and dielectric liquids offer viable options for achieving microwave circuits' reconfigurability. Both types of fluids can alter the wave propagation

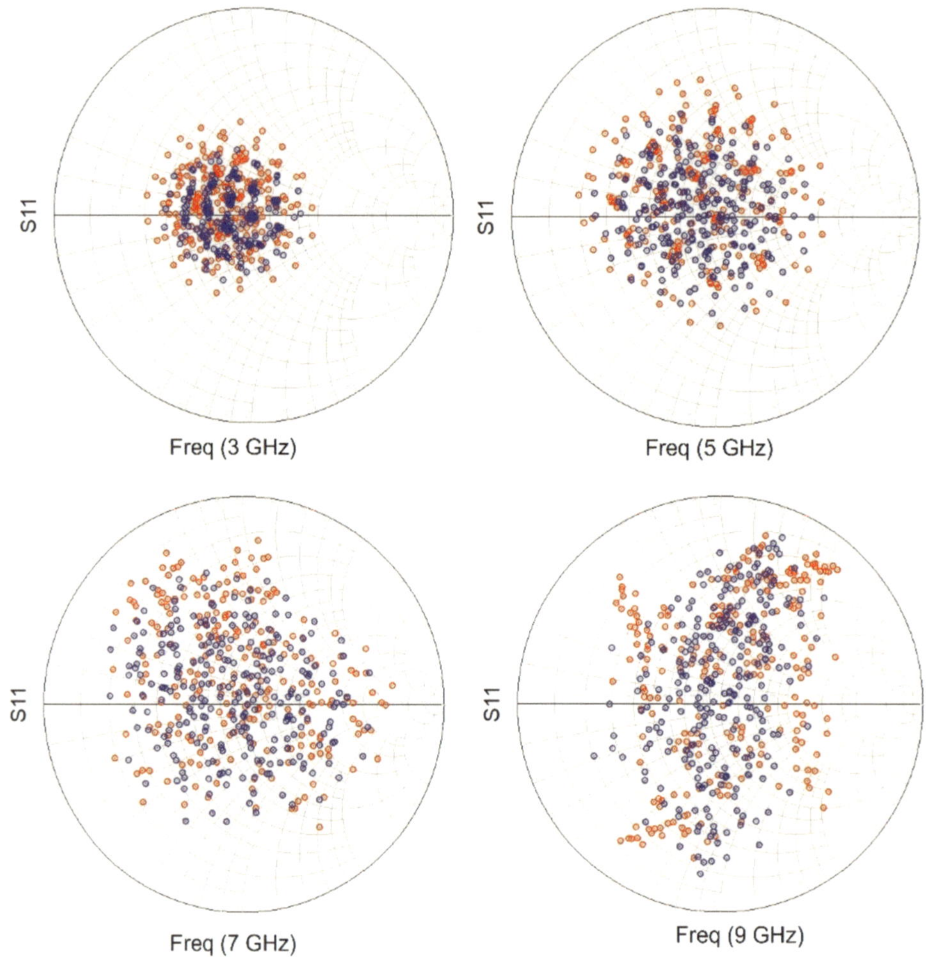

Figure 15.
Liquid metal-based tuner simulated and measured impedance coverage at different frequencies.

characteristics in guiding structures by injecting and removing the fluids from cavities in the close vicinity of conventional conductors. Conducting fluids offer the additional capability of reconfiguring the path of a guided wave by changing the shape of the conductors. The principles behind these reconfiguration techniques were presented and discussed. The design and realization of reconfigurable microwave impedance tuners using dielectric and conductive fluidics in LTCC technology were used to detail possible applications of the introduced reconfiguration. The results presented in this chapter, and those reported by others in the open literature, are promising and will support further research into this area.

Author details

Dorra Bahloul*, Ines Amor and Ammar Kouki
École de Technologie Supérieure, Montreal, Canada

*Address all correspondence to: dorra.bahloul.1@ens.etsmtl.ca

References

[1] Entesari K, Saghati AP. Fluidics in microwave components. IEEE Microwave Magazine. 2016;**17**(6):50-57

[2] Gabriel C, Gabriel S, Grant EH, Halstead BSJ, Mingo DMP. Dielectric parameters relevant to microwave dielectric heating. Chemical Society Reviews. 1998;**27**:214

[3] Gregory, Andrew, Clarke R. Tables of the complex permittivity of dielectric reference liquids at frequencies up to 5 GHz. National Physical Laboratory Report. 2012

[4] Andryieuski A, Kuznetsova SM, Zhukovsky SV, Kivshar YS, Lavrinenko AV. Water: Promising opportunities for tunable all-dielectric electromagnetic metamaterials. Scientific Reports. 2015;**5**:7-8

[5] Khoshmanesh K, Tang SY, Zhu JY, Schaefer S, Mitchell A, Kalantar-Zadeh K, et al. Liquid metal enabled microfluidics. Lab on a Chip. 2017;**17**(6): 974-993

[6] Chen R-S, Wong S-W, Lin J-Y, Yang Y, Li Y, Zhang L, et al. Reconfigurable cavity bandpass filters using fluid dielectric. IEEE Transactions on Industrial Electronics. 2021;**68**(9): 8603-8614

[7] Zhou W-J, Tang H, Chen J-X. Novel microfluidically tunable differential dual-mode patch filter. IEEE Microwave and Wireless Components Letters. 2017; **27**(5):461-463

[8] Entesari K, Saghati AP. Fluidics in microwave components. IEEE Microwave Magazine. 2016;**May**:50-75

[9] Mumcu G, Dey A, Palomo T. Frequency-agile bandpass filters using liquid metal tunable broadside coupled split ring resonators. IEEE Microwave and Wireless Components Letters. 2013; **23**:187-189

[10] Li M, Yu B, Behdad N. Liquid-tunable frequency selective surfaces. IEEE Microwave and Wireless Components Letters. 2010;**20**: 423-425

[11] Chen CH, Peroulis D. Electrostatic liquid-metal capacitive shunt MEMS switch. 2006 IEEE MTT-S International Microwave Symposium Digest. 2006: 263-266

[12] Chen C-H, Whalen J, Peroulis D. Non-toxic liquid-metal 2-100 GHz MEMS Switch. 2007 IEEE/MTT-S International Microwave Symposium. 2007:363-366

[13] King AJ, Patrick JF, Sottos NR, White SR, Huff GH, Bernhard JT. Microfluidically switched frequency-reconfigurable slot antennas. IEEE Antennas and Wireless Propagation Letters. 2013;**12**:828-831

[14] Khan MR, Hayes GJ, Zhang S, Dickey MD, Lazzi G. A pressure responsive fluidic microstrip open stub resonator using a liquid metal alloy. IEEE Microwave and Wireless Components Letters. 2012;**22**:577-579

[15] Lei BJ, Hu W, Ohta AT, Shiroma WA. A liquid-metal reconfigurable double-stub tuner. 2012 IEEE/MTT-S International Microwave Symposium Digest. 2012

[16] Koo C, LeBlanc B, Kelley M, Fitzgerald H, Huff G, Han A. Manipulating liquid metal droplets in microfluidic channels with minimized skin residues toward tunable RF

applications. Journal of Microelectromechanical Systems. 2015; **24**(4):1069-1076

[17] Diebold AV, Watson AM, Holcomb S, Tabor C, Mast D, Dickey MD, et al. Electrowetting-actuated liquid metal for RF applications. Journal of Micromechanics and Microengineering. 2017;**27**(2):1-9

[18] Hwang Y, Paydar OH, Candler RN. 3D printed molds for non-planar PDMS microfluidic channels. Sensors and Actuators A: Physical. 2015;**226**:137-142

[19] Bahloul D, Kouki AB. LTCC-based fluidic tuners for low microwave frequency reconfigurable circuits. IEEE Transaction and Microwave Theory and Techniques. 2020;**60**(8):3308-3317

[20] LTCC@ÉTS. (2018). LTCC@ÉTS Process & Capabalities Available from: https://www.etsmtl.ca/Unites-de-recherche/LTCC/Services-offerts/processV2.pdf

[21] Domingue F, Fouladi S, Kouki AB, Mansour RR. Design methodology and optimization of distributed MEMS matching networks for low-microwave-frequency applications. IEEE Transactions on Microwave Theory and Techniques. 2009;**57**(12):3030-3041

[22] Amor, Ines (2021). LTCC-Based Impedance Tuner Using Liquid Metal for Reconfigurable Microwave Circuit. Master report, École de technologie supérieure; 2021

Section 2

Numerical Modeling

Chapter 3

Multiscale Auxiliary Sources for Modeling Microwave Components

Bilel Hamdi and Taoufik Aguili

Abstract

This chapter presents multiscale auxiliary sources mainly used to solve complex electromagnetic problems, especially those that insert localized elements into circuits. Several equivalence relations (field-circuit) are established to simplify and make more accurate electromagnetic calculations by changing some characteristics of the localized elements known by their field representation as "voltage-current" representation and vice versa. Various examples are illustrated to show the effects of auxiliary sources in planar circuits containing localized elements (dipoles, diodes, transistors) in the millimeter and terahertz bands. An example of a graphene or Gold dipole is demonstrated in this approach. Another typical example of a diode integrated in a radiating structure is also simulated.

Keywords: auxiliary sources, multiscale circuits, surface impedances, located element, microwave diode, microwave transistor, equivalent circuits, active microwave components, MM and terahertz waves

1. Introduction

Microwave circuits are now most often made in planar technology on various substrates: Silicon for medium frequencies of mobile telephony, Alumina for hybrid circuits, gallium arsenide for millimeter waves. The active parts have dimensions in the micron range for the lines; we pass to the ten of microns as far as transverse dimensions are concerned; as, for the other dimensions (lines or selves), they can be of the order of a few millimeters. For many years, it appeared that it was inappropriate to simulate such an ensemble with a single software that, by a very tightly meshed, would describe the details of the multiple heterogeneities of a localized element, then, with a looser mesh, would take into account the excitation lines and the box (waveguide).

The first idea to avoid the unnecessary approach of simulating small elements that can be isolated from the surrounding circuit to measure them has resided the introduction of localized elements ("localized elements") in electromagnetic calculation software and electromagnetic calculations: Finite Elements, finite differences, method of moments [1, 2].

The second idea is more rigorous. It simulates the different parts of the circuit by software adapted to each function (method of moments for the homogeneous parts and finite differences for the strongly heterogeneous elements, for example). In a second step, the values of the fields are equalized to the limits of the domains [3, 4].

In the many studies published on the subject, one can note that the coupling problem is always expressed in a matrix form that Maxwell's equations take after a spatial or spectral truncation operation. The "compression method" [5] is, in this sense, clear and efficient. It applied to planar circuits modeled by a method of moments. The "basis functions" or test functions are triangular functions (or "roof-top" functions). The problem is then discretized, and the relations between voltage and current in a localized element are then introduced as relations between neighboring roof-tops. We use the edges as impedance ports to introduce localized elements in the finite element method [6, 7]. The approach is similar for the finite difference time domain methods and transmission line method (FDTD and TLM).

Coupling two sub-domains of a circuit, one being the "external" circuit and the other, a "localized" or "interior" part, is an objective that must be achieved independently of the mesh or the chosen numerical method. This objective is achieved in the context of integral methods by applying the "equivalence principle": An electric field on a given surface is replaced by two opposing magnetic current sources [8]. We can also reason by duality with sources of electric currents. Therefore, the model exists but must be redefined in the sense of a change of scale until it can integrate localized elements. The difficulties appear then because the passage to the limit of the zero dimensions for an element or a source is not possible in general electromagnetism under the penalty of divergences in the series. However, it is easy to admit that the meticulous description of the internal functioning of a transistor is not indispensable to predict its behavior in a circuit. The only thing that counts is its extrinsic electrical characteristics, i.e., those accessible to measurement. Then, we can affirm that the electromagnetic effects in the external domain depend only on certain parameters. No matter how the active element is manufactured, only a few average values of the electromagnetic fields count. Rather than approaching the active element by going to the zero-dimensional limit, which is impossible, we should approach the problem by defining an equivalence between elements by identifying their electrical characteristics.

To clarify these notions and evaluate the limits of validity of the hypotheses, it was judged simpler to detail the calculation on an analytical example after recalling the definition of the different types of sources used in the electromagnetic simulation. In the context of integral methods for studying planar circuits, we place them in the spectral domain (or simply in "transverse resonance").

2. Different forms of sources in planar circuits

We consider a plane P on which a metal circuit is drawn with surface impedances and active elements. This plane is excited by sources that can be described by giving an electric field or a surface current on a specific surface. We can consider an incident wave on the circuit. These sources will be called "main sources." The "auxiliary sources" are intended to be used as intermediary in the calculation to pass from the external domain to the internal domain (active element), it is the translation of the equivalence principle: The diffraction of a plane wave by a slot requires, for example, the use of an auxiliary source of magnetic current in the slot to obtain the radiation pattern. If this source is given, it allows for calculating an impedance or a quadripole. Finally, the "virtual sources" are contrary to the auxiliary sources chosen arbitrarily with electromagnetic quantities defined on a conductor or an aperture but conditioned by continuity. It is, for example, the current density on a microstrip line for which we want to calculate the propagation constant of the fundamental mode. This current density that the electric field is zero in its domain of definition. Taking the example of the radiating slot again, from the moment we want to calculate the amplitude of the electric field in the slot, given the amplitude of the incident plane wave, by writing the continuity of the magnetic field; then this electric field is a virtual source. These distributions can be interpreted well from an equivalent scheme [9–11].

2.1 Principal sources

Three types can be distinguished.

2.1.1 Excitation of a circuit by an incident wave

This source is used for free-space diffraction problems or obstacles in the guide. The latter case is defined as follows: Let S be the straight section of an excitation guide, enclose TE and TM modes whose transverse electric fields are orthonormal functions with two components f_n. The fundamental mode f_0, the transverse magnetic field, is given by:

$$\boldsymbol{J}_0 = \boldsymbol{H}_0 \times \hat{n}_0 = \boldsymbol{I}_0 f_0 \tag{1}$$

$\hat{n}_0$ being the normal unit vector to S directed towards the load. In the opposite direction, the higher-order modes on S behave as if the guide is infinite, **Figure 1**.

For these modes, we can write (they are by hypothesis TE or TM modes):

$$\boldsymbol{J}_n = \boldsymbol{H}_n \times \hat{n} = -Y_{M_n} \boldsymbol{E}_n \tag{2}$$

Y_{M_n} is the mode n admittance that is equal, respectively, to $\gamma_n / j\omega\mu$ and $j\omega\varepsilon/\gamma_n$ for TE and TM modes [11]. The minus sign comes from the orientation $\hat{n}_0$. In Eq. (2), $\boldsymbol{J}_n$ and $\boldsymbol{E}_n$ are the transverse current and fields with a rotation of $\pi/2$ for the magnetic field of mode order n. Eqs. (1) and (2) can be written compactly by using the operators of projection. For that, we pose by definition, f being a function with two arbitrary components:

$$\hat{P}_n f = f_n \langle f | f_n \rangle \tag{3}$$

$\hat{P}_n$ is also written $\hat{P}_n = |f_n\rangle\langle f_n|$: The scalar product is commonly used. It is written as

$$\langle f_n | f \rangle \overset{\text{Definition}}{=} \int_S f_n^{*t} f dS \tag{4}$$

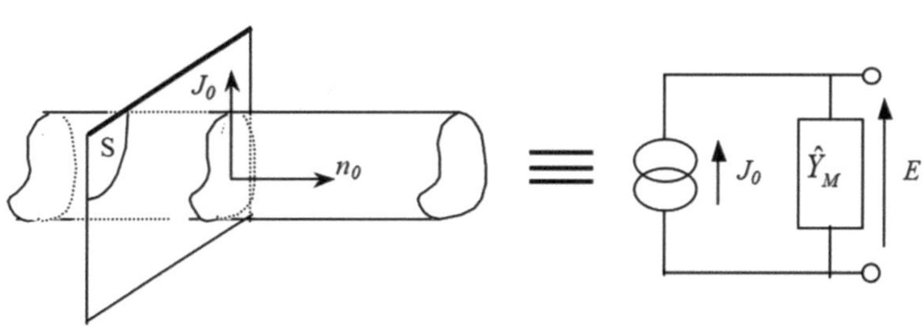

Figure 1.
Source model in a homogenous guide.

Thus, we can write Eq. (2) and Eq. (1) as follows:

$$J = -\sum_{n>0} Y_{M_n} \hat{P}_n E + J_0 \quad (5)$$

J and E are the transverse current and fields, respectively, at S, Eq. (5) can also be written as

$$J = -\hat{Y}_{M_n} E + J_0 \quad (6)$$

This formula can be represented by a scheme analogous to a dipole source in the circuit theory. The two borders represent the surface S, E is a function of two variables with two components, and J behaves as a line current oriented in the direction $\hat{n}_0$.

2.1.2 Excitation of a circuit by a very fine cable

Figure 2 shows the excitation of a metal line by a coaxial crossing the circuit box.

The dimension AB is tiny in wavelength, so we can assume that its surroundings satisfy the quasi-electrostatic hypothesis. In principle, a rigorous study requires calculating the circuit with a current source in the aperture C. However, since C is assumed to be small, we can directly evaluate the boundary conditions between the planes P_1 and P_2 enclose the excitation. This approximation has the advantage of allowing planar integral modeling without having to resort to a three-dimensional approach in the proximity of AB. A second hypothesis concerns the field fluctuations around the aperture and the excitation line influence the diffraction of the waves from the coaxial cable. This influence will be neglected as long as the average value of the corresponding fields on the source extent is zero. This assumption must be justified and clarified. If it is not valid, then parasitic elements will have to be introduced in the equivalent scheme.

Any field fluctuation is eliminated by arbitrarily limiting the source to the microstrip line's limits. The result between A and B is a constant electric field and a constant surface current density J_s, it corresponds to the conduction current in the wire distributed over the whole width of the source. In this operation, as the dimensions of sources a and b (**Figure 3**) are arbitrary, it is important to ensure that these

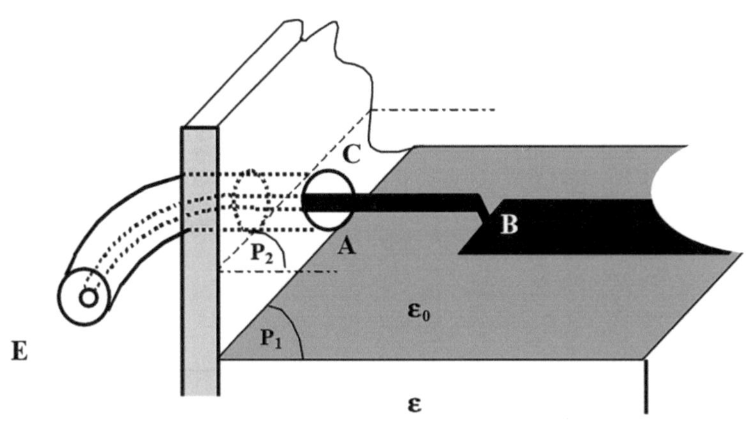

Figure 2.
Excitation of a line by a coaxial.

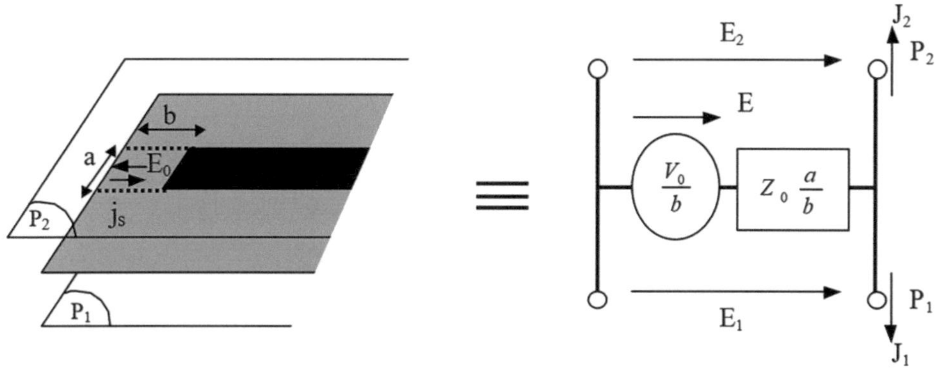

Figure 3.
Simplified representation of a coaxial excitation.

choices do not affect the result of the study of a circuit, i.e., the impedance seen from the source.

Assuming that the input E of the coaxial is connected to an internal impedance generator Z_0 and providing a voltage V_0, we obtain the relation between $\boldsymbol{E}_0$ and $\boldsymbol{J}_s$.

$$\boldsymbol{E}_0 b = V = V_0 - Z_0 I = V_0 + Z_0 \boldsymbol{j}_s a \quad (7)$$

From Eq. (7), we deduce the electric field source $Z_0 a/b$. In the second member of Eq. (7), we notice a plus sign at the orientation change j_s after its simplified representation (**Figure 3**). The scheme in **Figure 3** translates the relation (7) of electric and magnetic fields at the level of the planes P_1 and P_2. As before, we write by convention:

$$\begin{aligned} \boldsymbol{J}_1 &= \boldsymbol{H}_1 \times \hat{n}_1 \\ \boldsymbol{J}_2 &= \boldsymbol{H}_2 \times \hat{n}_2 \end{aligned} \quad (8)$$

$\hat{n}_1$ and $\hat{n}_2$ are the unit normal outgoing ones taking into account the general conditions of continuity:

$$\boldsymbol{H}_1 - \boldsymbol{H}_2 = \boldsymbol{j}_s \times \hat{n}_{21} \quad (9)$$

We deduce the relations between $\boldsymbol{j}_s$ and $\boldsymbol{J}_1, \boldsymbol{J}_2$. Vectorically multiplying Eq. (9) by $\hat{n}_{21} = \hat{n}_1$, it comes:

$$\boldsymbol{J}_1 + \boldsymbol{J}_2 = -\boldsymbol{j}_s \quad (10)$$

The transition relation from Eq. (7), considering Eq. (10), is then written as:

$$\boldsymbol{E}_1 = \boldsymbol{E}_2 = \frac{V_0}{b} - Z_0 \frac{a}{b}(\boldsymbol{J}_1 + \boldsymbol{J}_2) \quad (11)$$

2.1.3 Unilateral excitation of circuit

Previously, excitation by a wire produced a negligible effect. It is interesting to consider the opposite case to compare the results obtained by the two approaches. It is now assumed that a metal strip entirely masks the aperture AB (**Figure 4**) so that the metal strip integrally reflects the incident waves over the source.

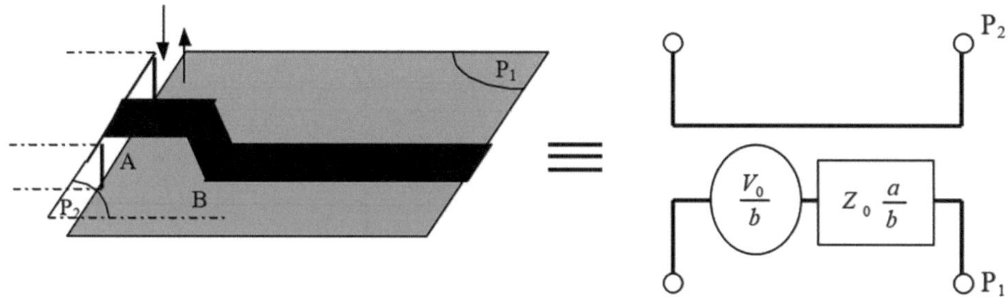

Figure 4.
Excitation of a circuit by a metal tongue.

In the upper part, at the source level, this one being masked, only a short-circuit appears. On the other hand, the schematic of the source with its internal impedance appears in the lower part (**Figure 4**).

Note: The unilateral or bilateral planar source is not identical to the source on the fundamental mode. In fact, everything happens as if a "coupling quadrupole" existed between the source and the fundamental mode line [12]. The determination of this coupling has been the subject of numerous studies [13].

2.2 Auxiliary sources

These sources are used to study two distinct parts of a circuit separately, especially when their respective dimensions are very different from each other.

In the planar circuit approach, it is assumed that the thickness of the "small circuit" is infinitely small.

Therefore, it is equivalent to a possibly nonlinear surface impedance or, as for the main sources, to a field or current source with internal impedance. The simplest circuit that can be studied by auxiliary sources has the main source, S_0 conducting elements, and an impedance surface S_1 (**Figure 5**).

An auxiliary source's analysis consists of simply replacing S_1 by a source, calculating the coupling quadripole Q between S_0 and S_1, then closing the quadrupole by the impedance present in the circuit. We deduce the impedance seen from the main source S_0.

An illustration of the use of a surface impedance or an auxiliary source (defined by a constant imposed electric field $\boldsymbol{E}_0$) to model "localized elements" in planar structures with a method of moments is given in **Figure 6**. **Figure 7** shows the reflection

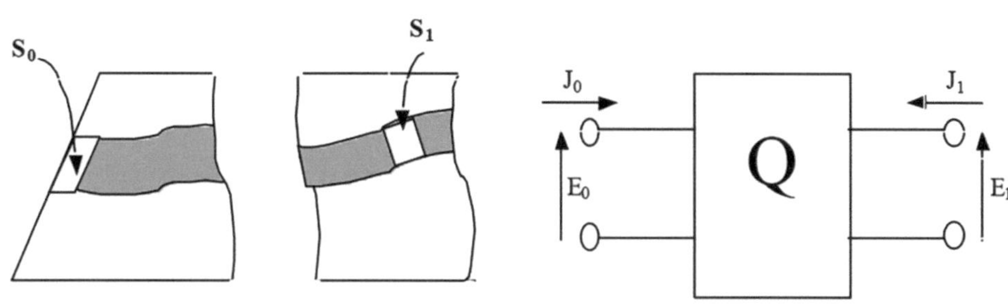

Figure 5.
Circuit comprising main source and an auxiliary source.

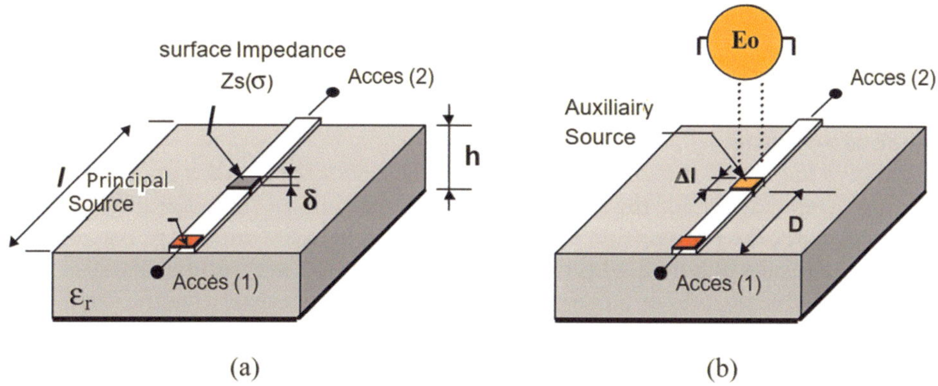

Figure 6.
Microstrip line: With a localized surface impedance (a) or with a localized auxiliary source (b).

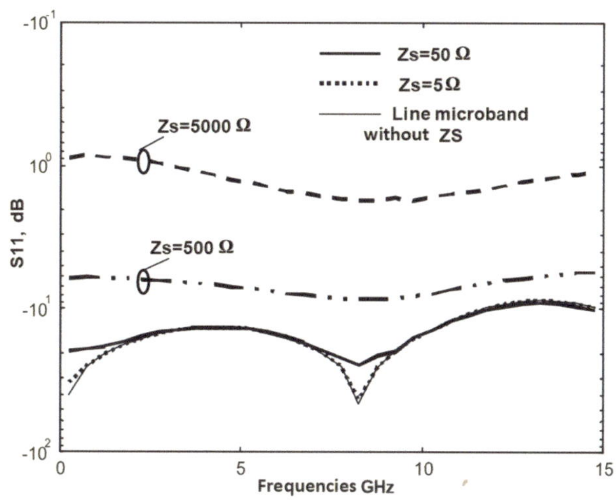

Figure 7.
Effects of localized surface impedance value on the reflection coefficient S_{11} seen from the main excitation source with l = *6 mm, $\varepsilon_r = 9.8$, and* h = *1.35 mm.*

coefficient S_{11} seen from the main excitation source versus frequency for different values of the surface impedance Z_s.

The simulation results corresponding to low values of Z_s (less than 5 Ω, for example) are identical to those obtained for a microstrip line without surface impedance. In this range of Z_s variation, the two approaches by localized surface impedance or auxiliary source are equivalent. As the value of the surface impedance increases, the reflection coefficient increases, inducing a lower transmission between the accesses Eqs. (1) and (2) defined in **Figure 6**.

Note:

The influence of the auxiliary source position in relation to the main excitation source as a function of frequency shows that by placing the auxiliary source at a distance, $D = l/2$ for example, the second resonance frequency of the line disappears, which reappears for a distance $D = l/8$, the width of the auxiliary source is taken equal to the tenth of the line length l.

Thus, except for the differences relative to the previous remark related to the position of the auxiliary source with respect to the main excitation source, the two approaches by localized surface impedance and auxiliary source give identical results as long as the dimensions of the auxiliary source (δ) remain small with respect to the operating wavelength.

Relative to l, these dimensions become of the order of the operating wavelength for very high frequencies. For the auxiliary source dimensions of the order of the operating wavelength, the propagation is no longer negligible in the domain of the auxiliary source. The results obtained with the two approaches are different. In this case, imposing an electric field or a magnetic field does not necessarily give the same results.

If the surface S_1 is large enough, the description of the fields and currents surface at its level must require a set of basis functions, the output E_1, J_1 of Q is then a set of ports (one per basis function), Q is a multipole that must be calculated. The interest in introducing auxiliary sources is not obvious in this case. On the other hand, if S_1 is sufficiently small, then a single function is sufficient. There is a bottleneck at its level, and thus only one electromagnetic calculation (the elements of the Q matrix) for a set of load impedances in S_1..

There is a case where the impedance of S_1 is defined in terms of current and voltage (localized elements). Under this assumption, as for the main sources, quasistatic limit is verified, and distribution of fields and currents for this basis function must satisfy the uniqueness of the definition of voltages and currents:

$$rot(\boldsymbol{E}_1) = 0; div(\boldsymbol{j}_s) = 0 \tag{12}$$

The choice of the auxiliary source is then very small; the knowledge of the limits and the connectors is enough to define it. Returning to the concept of coupling between parts of a circuit, one external Ω_1 and the other internal Ω_2 or localized, the description of an auxiliary source then becomes natural. The two parts of the circuit are connected; two or more metallic parts exist at their borders **Figures 8** and **9**).

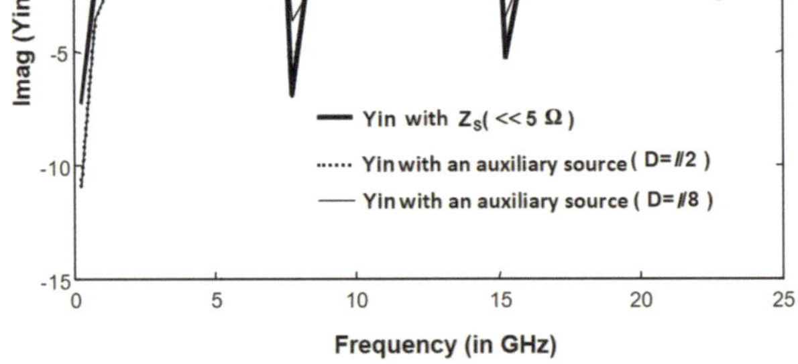

Figure 8.
Position's effects of the auxiliary source described by the reflection coefficient S_{11} seen by the main excitation source that varies against frequency with l = *6 mm,* ε_r = *9.8, and* h = *1.35 mm.*

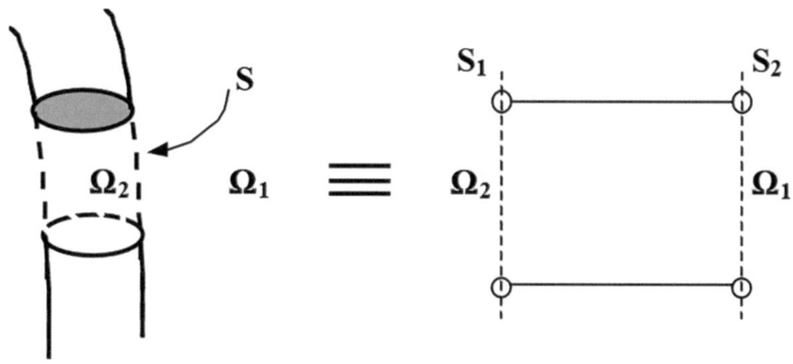

Figure 9.
Decomposition of a circuit into external and internal domains.

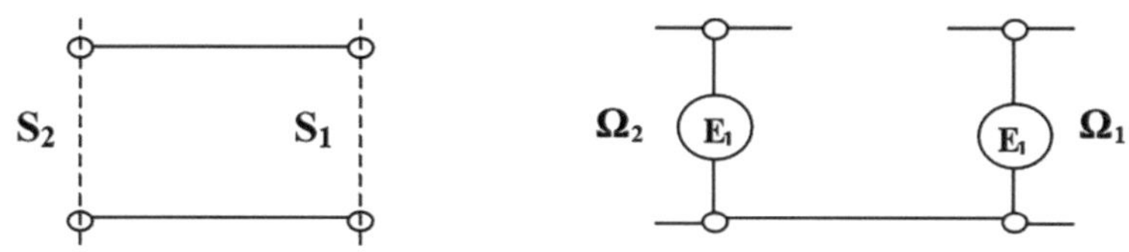

Figure 10.
Placement of auxiliary sources.

In the vicinity of S exist two surfaces S_1, S_2; one in Ω_1 and the other in Ω_2. The disconnection operation between Ω_1 and Ω_2 is done in two steps (**Figure 10**):

- Setting up a magnetic wall in place of S (this is an equivalent schematic of an open circuit).
- Introduction of a source over S_1 for the study of Ω_1 or over S_2 for the study of Ω_2.

For a planar source, the magnetic wall prohibits, for an electric field source, any component normal to the source (**Figure 11**). Thus $\boldsymbol{E}_z = 0$ from which: $\mathrm{div}_s(E) = 0$. So, $(\frac{\partial E_x}{\partial x} + \frac{\partial E_y}{\partial y} = 0)$ and $E = -\mathrm{grad}(v)$ (quasistatic hypothesis).

These two conditions allow us to derive a general formulation for auxiliary sources of any shape [11]. In the case of a rectangular source of dimension a, b, we pose:

- For a field source, $\boldsymbol{E}_0$ is known, and after electromagnetic calculations, the surface current distribution $\boldsymbol{J}_s$ is obtained.

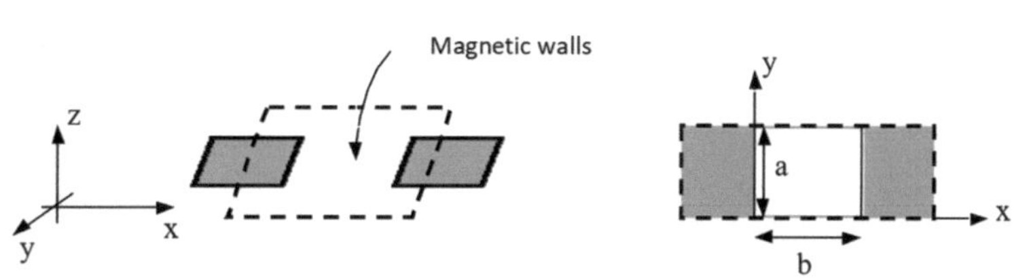

Figure 11.
Auxiliary planar source.

By posing $\boldsymbol{E}_0 = Vf_0$, we deduce:

$$I = \langle f_0 | \boldsymbol{J}_s \rangle \quad (13)$$

where f_0 is a function with two components $(1/b, 0)$. The scalar product is defined as the integral over the whole surface of the source; we obtain the total intensity going through the source by Eq. (13), which gives:

$$\boldsymbol{J} = If_0{}' \text{ and } V = \langle f_0{}' | \boldsymbol{E} \rangle \quad (14)$$

In this case, $f_0{}'$ is equal to $1/a$. This procedure is, of course, valid for the main sources.

Note: An auxiliary source can be considered a mode source for the main sources. This is the case, for example, if we consider a circuit inserted on a ground plane with a periodic motif. The latter can be studied independently employing modal excitations, which coincide with the modes of the case in transverse resonance. The impedance seen by each mode will be introduced in the operators necessary to study the circuit placed above the ground plane. The periodic motif is equivalent to excitation by the fundamental mode of such a guide under a variable incidence. We find in this process the essential quality of the auxiliary source the part not concerned (here the circuit above the periodic motif plane) can change; the impedance seen by each mode in transverse resonance will be the same. Global modeling will not be necessary when only one part of the circuit is variable.

2.3 Auxiliary sources

Virtual sources are well known to users of the method of moments. For example, we give them the current density $\boldsymbol{j}_s$ on a perfect conductor; we deduce the tangential electric field, and, writing that the latter is null on the conductor, we find an equation that allows us to determine the unknown current. If we consider $\boldsymbol{j}_s$ as a source, we can see that it does not deliver any power, neither real nor complex. Hence, the term virtual, as opposed to real, is attributed to this type of source. This property can be expressed more operationally. The definition of a virtual current source is:

A source defined in a domain D is virtual if its dual magnitude is nullified in D.

The dual quantity of $\boldsymbol{j}_s$ is the tangential electric field E and vice versa. The equivalent schema of an interface with a virtual surface current source $\boldsymbol{j}_s$ and the main field source $\boldsymbol{E}_0$ is presented as follows (**Figure 12**): As the value of the virtual source at each point is not known a priori, it is represented as an adjustable current source with an inclined arrow.

The working domain consists of three subdomains: the source domain D_S, the metallic domain D_M, and the dielectric domain D_D (**Figure 13**).

Since the virtual source overlaps the source domain, and the dual quantity $\boldsymbol{j}_s$, i.e., the electric field, is annulled in the latter, we deduce the equivalent schema of the interface in D_S and D_M ($\boldsymbol{E}_0$) is zero on the metal, so in D_M only the short-circuit appears).

Concerning the dielectric, being outside the virtual source, the latter is zero. The equivalent schema shows an open circuit. At each point of D, the boundary conditions are correctly expressed by the schema of **Figure 12**. It is thus sufficient to describe the behavior of the electromagnetic field in the vicinity of D. By expressing the relations between the waves in the upper and lower parts of D using the transverse modes; we

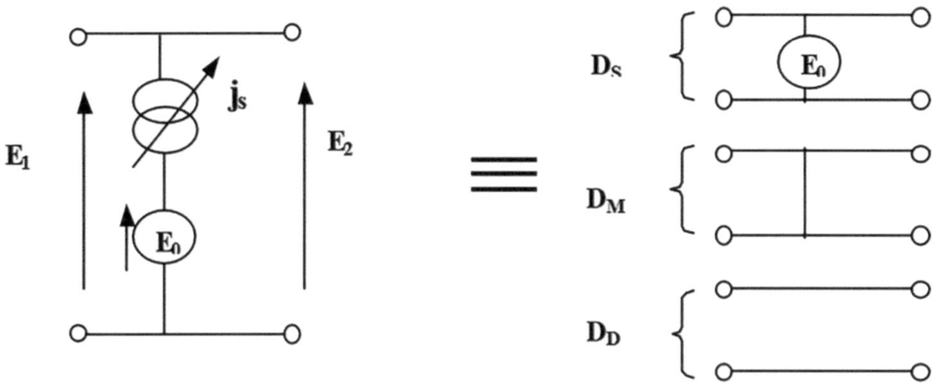

Figure 12.
Representation of an interface with a field source and a perfect conductor.

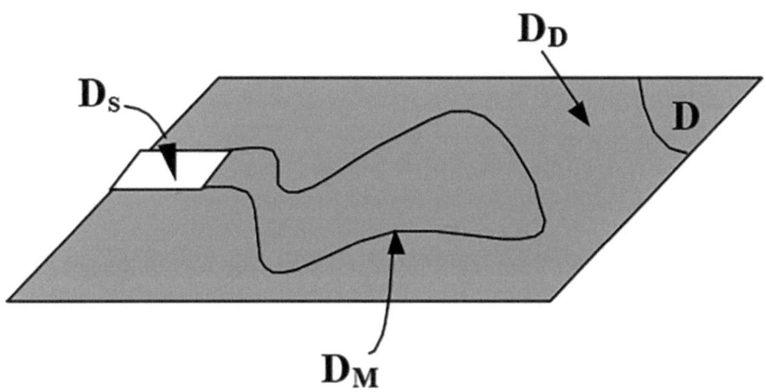

Figure 13.
Interface made of metal, dielectric, and a source.

hold a scheme similar to the one in **Figure 1** described by condition (6) (that assuming $J_0 = 0$). The following paragraph will discuss this process and the resolution of a circuit problem. It will highlight the interest and limitations of auxiliary sources. This will also lead to a rational definition of what we will understand by localized elements in the last part.

3. Use of auxiliary sources in planar circuits

The auxiliary sources allow the passage from a large scale at which the calculation is purely electromagnetic, to a smaller scale that corresponds to the connection of purely electromagnetic, to a smaller one which corresponds to the connection of circuit elements, most often characterized elsewhere.

The main and auxiliary sources are of such dimensions that the definition of voltage and current are intensity are unambiguously defined, as mentioned in the previous paragraph. Taking the general scheme of a planar circuit comprising one or more main sources and auxiliary sources, we can represent the different interactions in the following way (**Figure 14**): Q is the multipole of electromagnetic interactions.

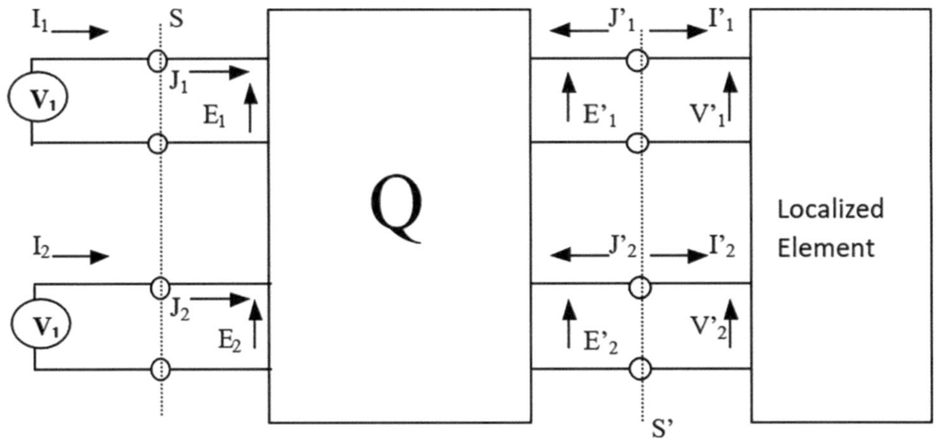

Figure 14.
Interactions between main and auxiliary sources.

At the junction level, surfaces between sources and the surrounding medium, S and S', it must be possible to define at the same time an electric field, a voltage, a current, and a magnetic field; this is why the condition of Eq. (12) has been imposed.

However, this is not without problems because a global calculation (without auxiliary sources) gives a form of the field which does not verify this condition. This is shown by the simple example treated in the following.

This raises the problem of the validity of introducing the auxiliary sources; the analysis of this example gives an order of magnitude of the maximum dimensions to take for these last ones (**Figure 15**).

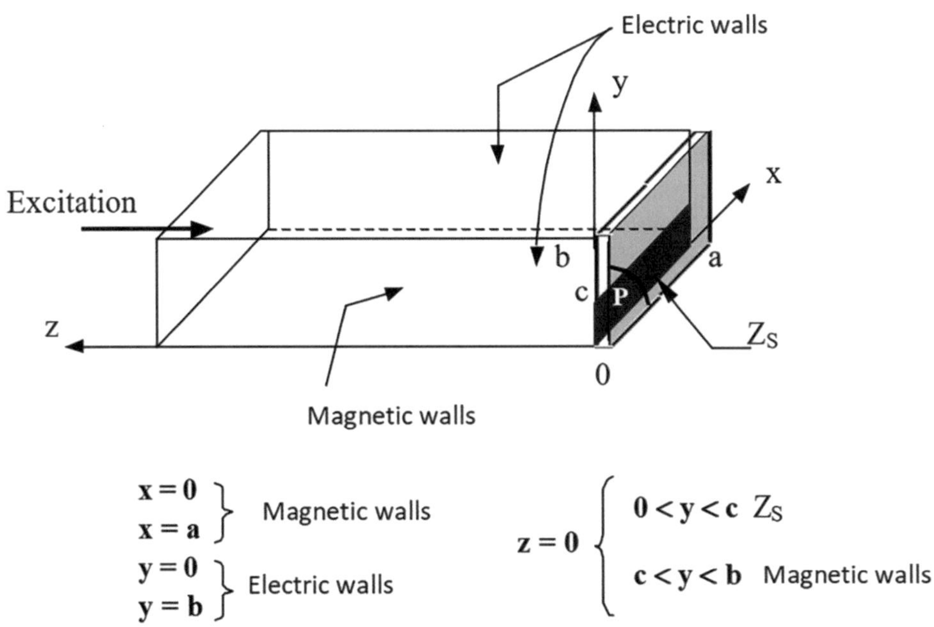

Figure 15.
Microstrip line loaded by a surface impedance with sidewalls magnetic and top and bottom electric.

4. Modeling localized impedance in microstrip circuit

Let us consider a microstrip line loaded by an impedance represented by a surface impedance of value Z_s. The line is assumed to be excited by the TEM mode. The edge effects are neglected and, to simplify the numerical approach, it is assumed that the line is simply bordered by magnetic walls (**Figure 16**). A magnetic wall P also limits the functional area.

This problem can be addressed in two ways. Either the circuit is modeled globally by taking a virtual source (test function) instead of Z_s for $0 < y < c$, or we take an auxiliary source for $0 < y < c$. We calculate the quadrupole Q between this source and the main excitation, then we close this source by impedance Z_s, following the circuit shown in **Figure 14**.

From the calculation developed in the appendix, the following conclusions can be drawn: first, if we want the same result between a direct approach and an indirect approach through auxiliary sources, it is necessary to take as many auxiliary sources as there are test functions in the first approach; it is necessary to take as many sources. This means that if one seeks precision of the results by taking into account, for example, the effects of edges to define the shape of the electric field of a single auxiliary source, other the difficulty that this way of making will present to evaluate the impedance seen of the source (as it was specified in II-2), this selected shape may be inadequate for a different impedance. If the precision is judged insufficient with the only auxiliary source, it will be necessary to introduce others and to study the action of these on the localized element, if possible. The second remark concerns the problem that arises as soon as the dimensions of the subdomain corresponding to a localized element tend to zero. In the example given in an appendix, a capacitance that tends to infinity appears parallel to an impedance whose dimension tends to zero; the capacitive nature comes from the TM modes alone present in the structure studied. For a more realistic impedance placed at the end of a microstrip line, we would find an inductance in series with the impedance tending to infinity with the width of the latter, in addition to the capacitance in parallel.

These remarks raise the difficulty of defining the behavior of a circuit of very small dimensions placed in a box. It is intuitively clear that the details of the internal functioning of this circuit are not affected by the large external circuit; only its electrical characteristics matter, the geometrical parameters do not influence in themselves. Hence, the idea of locating the impedances by admitting that their dimensions are zero. As this approach is not possible, because of the divergences of this discrepancy, we must return to the process of measuring an impedance and see how it can help to define its behavior when it is inserted in any circuit, because the circuit, because the only accessible parameters are those measured, or modeled from a fine description of the element.

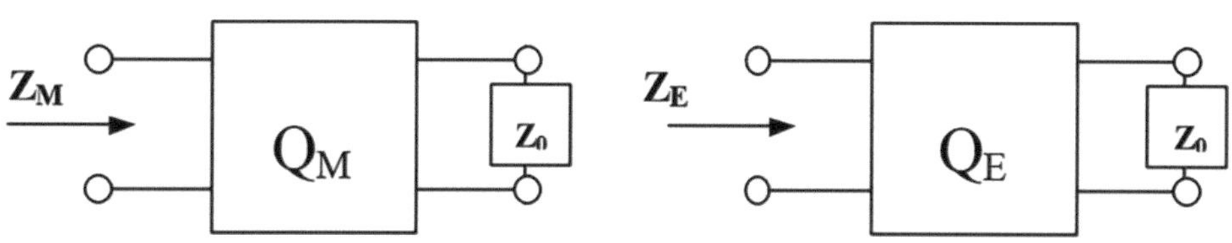

Figure 16.
Relationship between measured impedance input impedance of a device and intrinsic impedance.

5. Equivalent localized elements

To introduce the equivalent impedances, the simplest way is to go back to the previous example, taking into account the fact that the impedance coming from the surface impedance, of value:

$$Z_0 = Z_s \frac{c}{a} \tag{15}$$

which can be called "nude impedance," is not directly accessible.

This impedance is introduced into a measuring device that can be taken similar to the previous one with parameters a_0, b_0, c_0 instead of a, b, c of the device in which impedance is inserted for its practical use. However, in the electromagnetic calculation, the auxiliary source is charged by Z_0, so it is important to know the latter. To do this, a process of "de-embedding" is used.

Knowing the input impedance of the measuring device, we deduce Z_0 by an electromagnetic calculation. Using the results obtained in Eq. (15), we can write the measured impedance Z_M as:

$$Z_M = \frac{1}{j\omega C_0(a_0, b_0, c_0) + \frac{1}{Z_0}} \tag{16}$$

Hence, the expression of Z_0:

$$Z_0 = \frac{1}{-j\omega C_0 + \frac{1}{Z_M}} \tag{17}$$

Placing this expression in a used device, homogenizing the load with a surface impedance,

$$Z_s = Z_0 \frac{a}{c} \tag{18}$$

We deduce the input impedance Z_E

$$Z_E = \frac{1}{j\omega C(a, b, c) + \frac{1}{Z_M}} = \frac{1}{j\omega(C - C_0) + \frac{1}{Z_M}} \tag{19}$$

There is a relationship between the intrinsic impedance Z_0 and the input impedance (or the measured impedance), a homographic relation represented by a coupling quadrupole Q_E or Q_M (**Figure 13**).

Z_0, in the case considered here, is made with a surface impedance. As shown in Eq. (17), several values of Z_s, a, c, can give the same impedance Z_0. Due to the small size of the localized elements, several different devices can have identical electromagnetic behavior, which is reflected here by the same impedance value Z_0. Therefore, the measured impedances will be identical for the same measuring device. By definition, two impedances are "equivalent" when placed in the same device; their measured values are the same. Taking two surface impedances of dimensions a, c and a', c' (**Figure 17**), placed in a device similar to that of **Figure 12**, we see that the surface

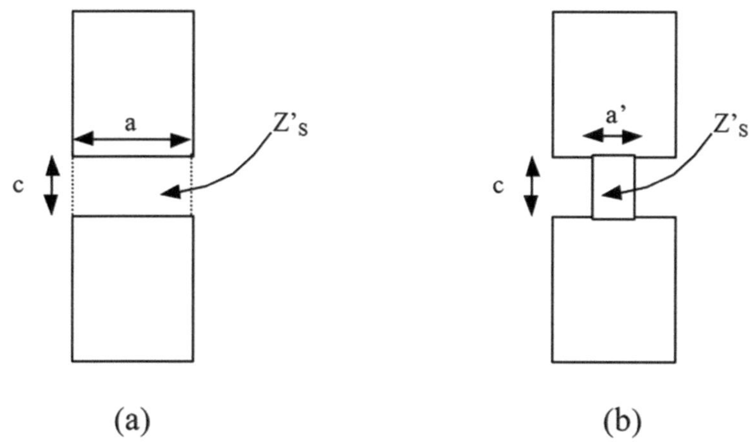

Figure 17.
Equivalent admittances.

impedances Z_s and Z_s'chosen so that the impedances are equivalent and roughly proportional to a anda', according to formula of Eq. (17).

However, the calculation corresponding to **Figure 17b** shows that the TE modes must be taken into account, which leads to the appearance of an inductance; the need to obtain an identical measured impedance in both cases, therefore, requires the surface impedance Z_s' to have a capacitive part to compensate the effect of this inductance.

A common case concerns active or passive dipoles [14–18]. The measuring device is not specified in the manufacturer's data, but an equivalent scheme is usually proposed. This scheme integrates the environment of the dipole in the measurement process. In particular, the capacitance between the terminals is taken into account and should be removed to reach the intrinsic (or nude) impedance (**Figure 18a**): If this precaution is not taken, the capacitance C will be integrated into the localized impedance or in the equivalent medium [19] but, in the calculation of the global device, there will be the presence of external field lines (**Figure 18a**) in the electromagnetic resolution. These field lines contribute to capacity C, and they have already been taken into account to establish the equivalent scheme.

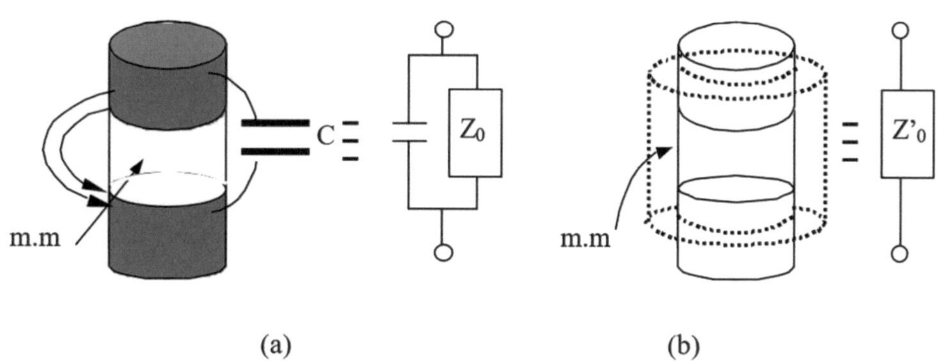

Figure 18.
Dipole and equivalent scheme for two choices of surfaces limits of the active element (magnetic walls: m.m.*).*

In summary, a part of C is counted twice. To avoid this, it is necessary to enclose the "box" in the magnetic wall (surface S' in **Figure 11**) the field lines surrounding the box. Thus, the impedance of the equivalent scheme becomes an intrinsic impedance (**Figure 18b**).

The case of transistors is the case of quadrupoles. Assuming that the dimensions are small enough for the equivalence hypothesis between localized elements to be verified, the shape of the transistor does not have to be specified in detail (exact dimensions of the electrodes, interdigitated or in-line nature, etc.); they are all equivalent to each other by hypothesis. We can choose a simple one, integrated with a classical coplanar measuring device assumed known. This device is shown in **Figure 19a**. Modeling allows deducing the coupling matrix between two auxiliary sources S_1, S_2. Let Z_0 be the impedance matrix of this intrinsic transistor. This matrix will be used in the final device to model the transistor in an arbitrary environment symbolized here by a quadrupole Q connected enclosure to the sources S_1, and S_2 themselves closed on Z_0.

The process is led in three stages:

- Choice of a geometrical shape for the sources and accesses of the G, S, D transistor (**Figure 19a**)

- From the known Z_M, we deduce the intrinsic coupling matrix Z_0 between S_1, S_2.

- We introduce the sources S_1 and S_2 in the circuit to be studied without modifying their dimensions and relative positions. An electromagnetic calculation allows determining the multipole Q_E and the input impedanceZ_E.

Therefore, in introducing active elements in a circuit, there is the necessity to first reach an intrinsic element by an operation of "deshabillage."

This element is then connected to one or more auxiliary sources, linking this element and the circuit studied. This operation is simplified in the case of low frequencies. In the first step, the inductance of the supply wires (**Figure 19a**) and capacitances due to connections can be neglected.

Let a_1, b_1, a_2, b_2 be the dimensions of S_1, S_2. We have successively:

$$V_{GS} = E_1 b_1 \text{ and } V_{DS} = E_1 b_1 + E_1 b_1$$

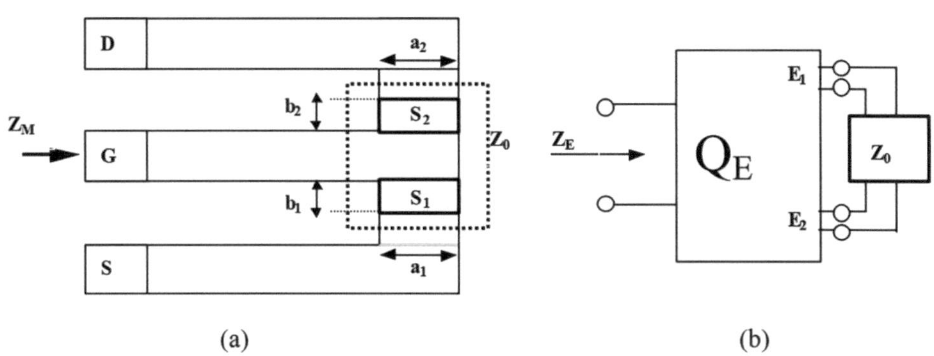

Figure 19.
Transistor in a measuring device (a) and inserted in a circuit (b).

$$I_{DS} = J_2 a_2 \text{ and } I_{GS} + I_{DS} = J_1 a_1. \tag{20}$$

We deduce the relation between Z_M and Z_0:

$$\begin{vmatrix} E_1 \\ E_2 \end{vmatrix} = Z_0 \begin{vmatrix} J_1 \\ J_2 \end{vmatrix}, \begin{vmatrix} V_{GS} \\ V_{DS} \end{vmatrix} = Z_M \begin{vmatrix} I_{GS} \\ I_{DS} \end{vmatrix} \tag{21}$$

$$Z_0 = \begin{vmatrix} \frac{1}{b_1} & 0 \\ -\frac{1}{b_2} & \frac{1}{b_2} \end{vmatrix} Z_0 \begin{vmatrix} a_1 & -a_2 \\ 0 & a_2 \end{vmatrix} \tag{22}$$

Thus, access to Z_0 is simplified and does not require any electromagnetic calculation. This will be the case when the capacities at the terminals of S_1 and S_2 are negligible. It is enough for this that the sources S_1 and S_2 are of dimensions a_1, a_2, sufficiently small.

Therefore, the localized element is an equivalence class, and in the use of localized elements, one arbitrarily chooses one of the elements for electromagnetic modeling. The localized elements are thus accessible by an operation that gives rise to equivalence. This equivalence relation makes it possible to avoid the introduction of zero dimensions that are not possible in electromagnetic calculations.

6. Evaluating surface impedance models for terahertz applications

The design of the terahertz antenna demands particular attention on the choice of materials to fulfill the microfabrication process.

As illustrated in **Figure 20**, a resonant antenna element for the terahertz application proposed here consists of three multilayers, from top to bottom: a gold or graphene dipole, polydimethylsiloxane (PDMS) substrate, and a ground plane. Gold and graphene are good conductive and are not susceptible to oxidation in air, while PDMS has comparatively low losses in the terahertz range.

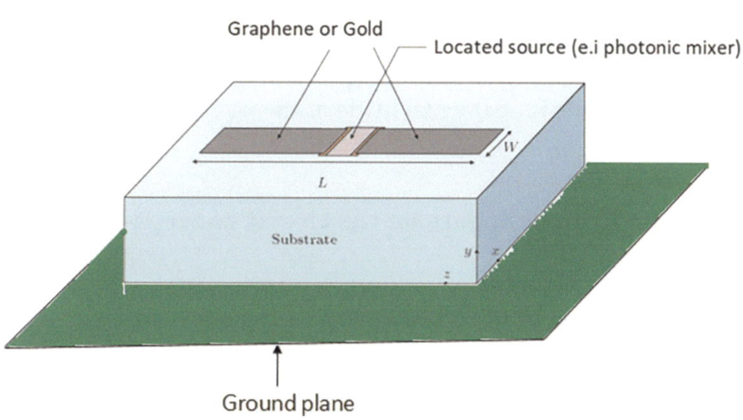

Figure 20.
Dipole-like antenna made of two graphenes (or gold) strips on a substrate and integrated with a located gap source in the center (example of the photonic mixer).

Various metals are selected for the top and bottom layer metallizations due to their selectivity for patterning, as they respond to various etchants. If the same metal is employed, the leakage of the etchant through the PDMS will damage the ground layer during the modeling of the upper metallization.

In the simulations stage, the material properties of the upper metals are obtained from a Drude model to evaluate the surface impedance Z_{SR} [20]. More details are given in [21] knowing that Z_{SR} is expressed by:

$$Z_{SR} = \sqrt{\frac{j\omega\mu_0\mu_r}{\sigma_r + j\omega\varepsilon_0}} \text{ With } \sigma_r = \frac{\sigma_0}{1 + j\omega\tau} \tag{23}$$

where angular frequency $\omega = 2\pi f$, and f = frequency of the driving electric field, μ_0 = permeability of free space, μ_r = relative permeability, ε_0 = permittivity of free, σ_0 = intrinsic bulk conductivity at DC, and τ = phenomenological scattering relaxation time for the free electrons (i.e., mean time between collisions).

Except for the graphene model, as proposed in [22, 23], a single graphene layer can be modeled by a 2D surface conductivity. Other models (with a classical, semi-classical, and quantum mechanical treatment) are proposed to evaluate the surface impedance, as provided in [24]. We utilize the frequency-dependent conductivity of a monolayer of graphene based on Kubo's formula [22, 23].

$$\sigma(\omega, \mu_c, \Gamma, T) \approx -j\frac{q_e k_B T}{\pi\hbar^2(\omega - 2j\Gamma)}\left(\frac{\mu_c}{k_B\Gamma} + 2\ln\left(e^{-\frac{\mu_c}{k_B\Gamma}} + 1\right)\right) \tag{24}$$

Where ω is the angular frequency; μ_c is the chemical potential; $\Gamma = 1/2\tau$; τ is the transport relaxation time; T is the temperature; k_B is the Boltzmann constant, and $\hbar$ is the reduced Planck constant. It is important to note that only the intra-band term is considered in relation to Eq. (24), which gives good and exact results for frequencies limited to a few THz.

The implementation of surface impedance in the formulation of a simplified equivalent circuit is described in Eq. (34) and is detailed in the given articles [23, 25]. The terahertz antenna is studied based on graphene nanoribbon using the MoM-GEC formulation [23, 25]. For this purpose, we propose **Figure 21**, which depicts the radiated electric field for several z planes (near field). For a minimal value of Z, the electric field distribution is perturbed by the evanescence modes in the vicinity of the discontinuity surface, which gives information about the antenna structure. For a distance so far, it gives the fundamental mode of the waveguide.

7. Microwave diode modeling: from the circuit concept (U-I) to the field concept (E-H)

The diode structure can be described as an RCL boundary with two different impedances: field and localized [26, 27]. The definition of the lumped and field impedances are as follows:

$$Z_{\text{Lumped}} = \frac{U}{I}, Z_{\text{Field}} = \frac{E_t}{H_t} \tag{25}$$

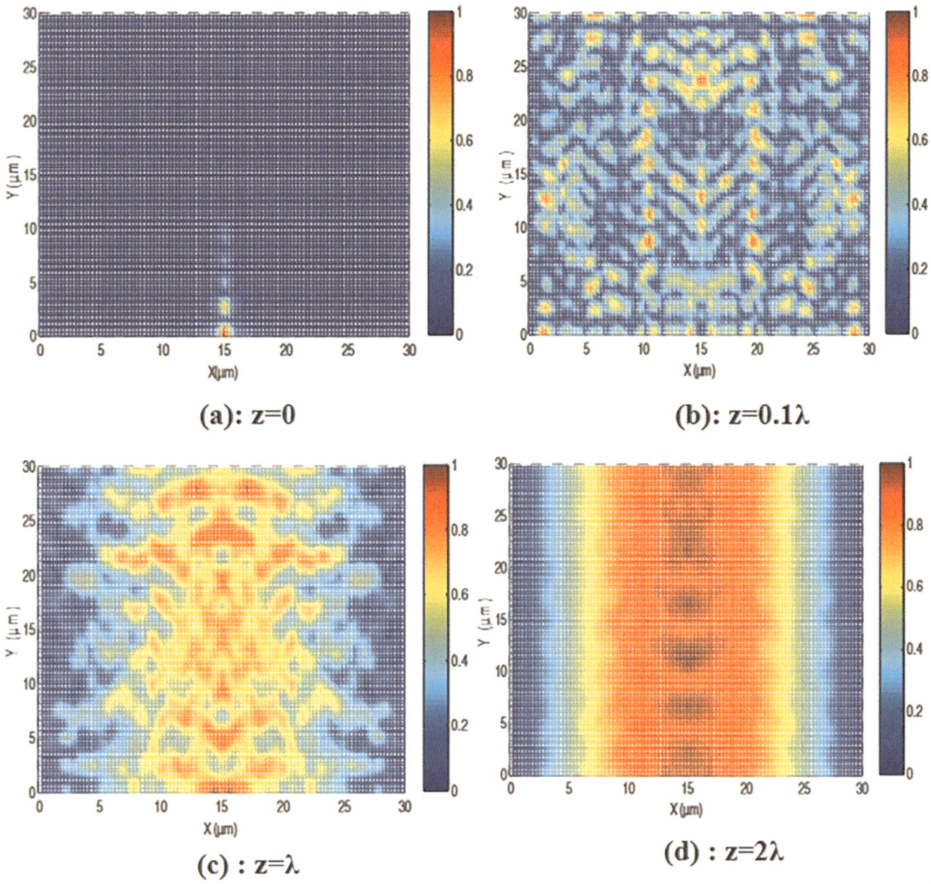

Figure 21.
Normalized radiated electric field of the GNR antenna for different z-planes (planar monopole antenna).

The circuit quantities for U and I may be expressed by the field components as shown below:

$$U = \int_0^l \boldsymbol{E}_t dl \text{ and} \oint \boldsymbol{H}_t dc = \boldsymbol{H}_t w = I \tag{26}$$

By replacing the field quantities in the definition of the impedance, we obtain:

$$Z_{\text{Lumped}} = \frac{U}{I} = \frac{E \times l}{H \times w} = Z_{\text{field}} \frac{l}{w} \Rightarrow Z_{\text{field}} = \frac{w}{l} Z_{\text{Lumped}} \tag{27}$$

Most commercial software do not have a specialized program for the simulation of RLC components. Thus, the simulation is carried out by taking a set of boundaries in the desired impedance values. A component of an arbitrary surface is matched with an equivalent rectangular plate of the identical surface. The width of the equivalent rectangular surface boundary is:

$$w = \frac{S}{l} \tag{28}$$

where S is the sheet area. The relationship between the lumped and field impedances is

$$Z_{\text{lumped}} = Z_{\text{field}} \frac{l^2}{w} \tag{29}$$

The field values are given by $R_{\text{field}} = Rl/w$, $L_{\text{field}} = Ll/w$ and $C_{\text{field}} = Cw/l$ (**Figure 22**).

Generally, in the microwave domain, the particular case of PIN diodes is modeled by an equivalent electric circuit according to its polarization state (ON or OFF). The equivalent models of the PIN diode are shown in **Figure 23**. It is assumed that the frequencies considered are much higher than the transition frequency (in the direct regime) and the relaxation frequency (in the inverse regime).

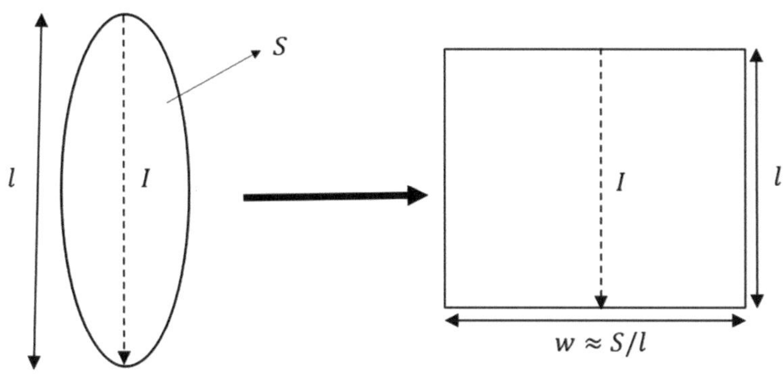

Figure 22.
Transform an arbitrarily shaped surface into an equivalent rectangular-shaped surface.

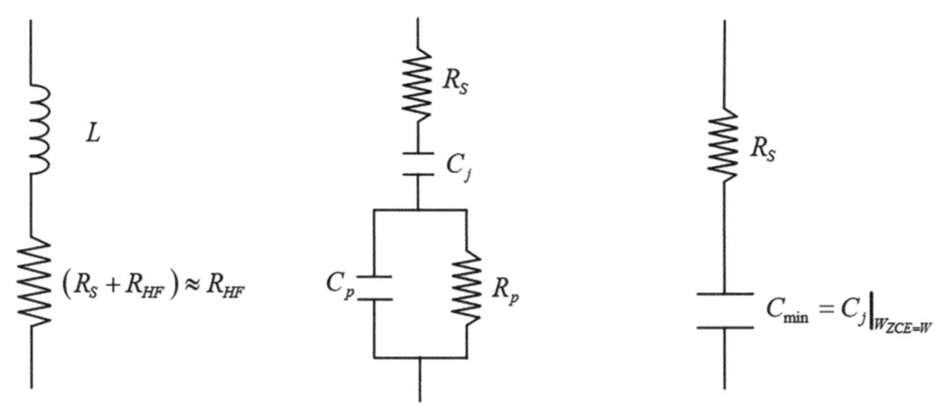

Figure 23.
Equivalent circuit of the PIN diode: (a) direct regime with $f > f_T$ (b) inverse regime with (V = 0), (c) inverse regime $V > V_{PT}$.

Let Z the impedance of the diode in the concept (U-I). It is defined as follows:

$$U = ZI \text{ Such as } Z = \begin{cases} R_s + R_{HF} + jL\omega; \text{direct regime} \\ R_s + \dfrac{-j}{C_{minPT}}; \text{ inverse regime; } (V > V_{PT}) \end{cases} \tag{30}$$

In an attempt to transpose the concept (U-I) to the concept (E-J), the PIN diode was modeled by a surface impedance Z_{diode} of width W_D and height d_D. Let E and J be the field and current relative to Z_{diode}. The relationship between E and J is written as follows:

$$E = Z_{\text{diode}} J \tag{31}$$

Considering the TEM mode, the relationship from concept (U-I) to concept (**E**-**J**) is:

$$\begin{cases} \boldsymbol{E} = \dfrac{U}{d_D} \\ \boldsymbol{J} = \dfrac{I}{w_D} \end{cases} \tag{32}$$

Thus, we have: $\begin{cases} U = ZI \\ E = Z_{\text{diode}} J \end{cases} \Rightarrow \begin{cases} U = ZI \\ \dfrac{U}{d_D} = Z_{\text{diode}} \dfrac{I}{w_D} \end{cases} \Rightarrow Z_{\text{diode}} = \frac{U}{d_D} \frac{w_D}{I}.$

The relation between $Z_{\text{diode}}$and Z is written:

$$Z_{diode} = \frac{w_D}{d_D} Z = \begin{cases} \dfrac{w_D}{d_D} R_s + R_{HF} + jL\omega; directregime \\ \dfrac{w_D}{d_D} R_s + \dfrac{-j}{C_{\min}\omega}; inverseregime (V > V_{PT}) \end{cases} \tag{33}$$

Several examples of diode modeling by an integral method based on the MoM and multiscale methods are presented in the thesis of Sonia Mili [27]. In this context, we propose a simple descriptive structure of an elementary motif in the presence of a diode and its equivalent circuit (CEG), as given in **Figure 24**.

Figure 25 ures represent the distribution of the surface current and the field diffracted by the elementary motif when it is excited by the TEM mode. The current and the field check the boundary conditions on the metal and the dielectric well. The diode is also detected and characterized by a non-zero field and current compared to the metal platelets. The interface between the diode and each of the metal platelets is

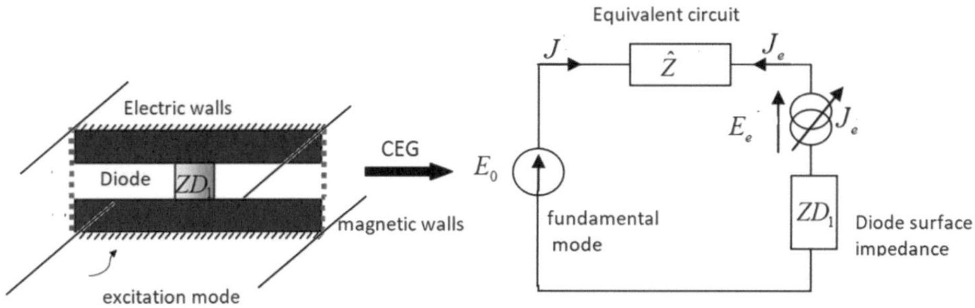

Figure 24.
An elementary motif in the presence of a diode and its equivalent circuit (CEG).

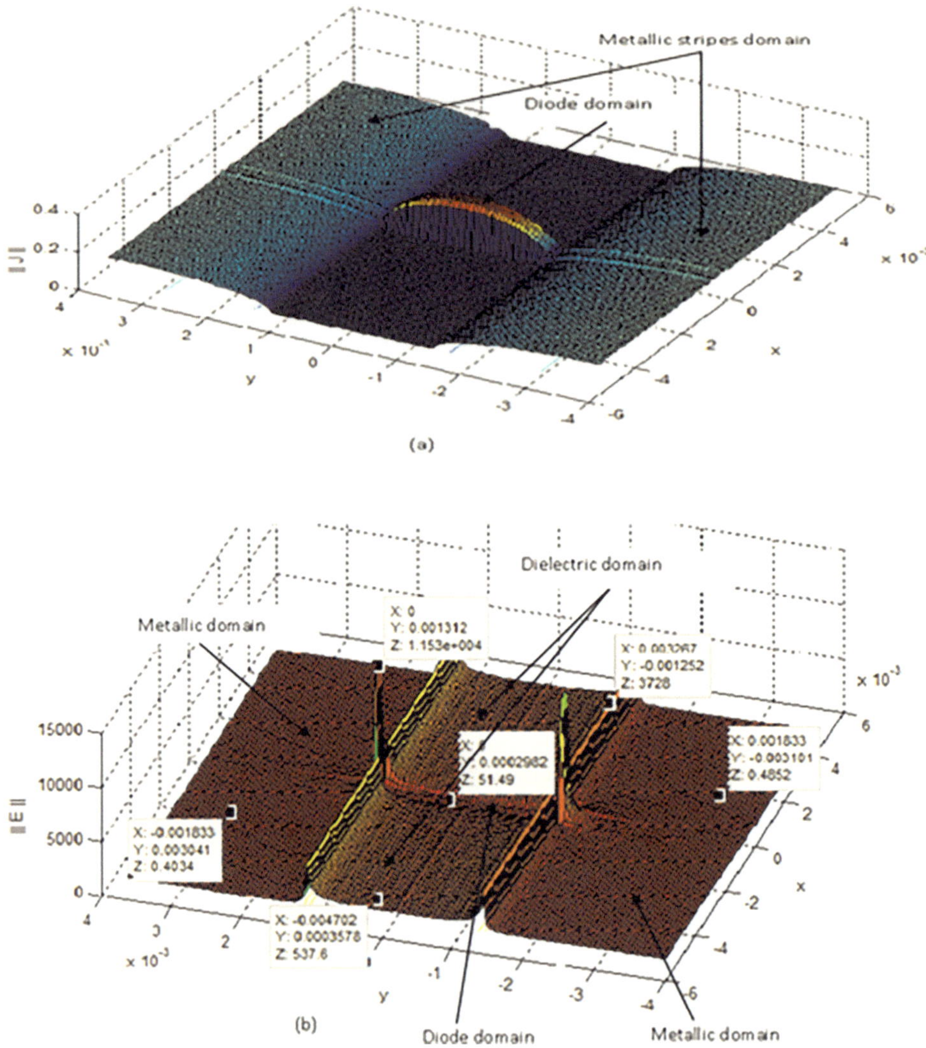

Figure 25.
Distribution of (a) the surface current and (b) the diffracted field on the elementary motif of active array with PIN diodes, f = *2,45 GHz,* a = *10,2 mm,* b = *22,9 mm,* w = *0; 5 mm, diode ON.*

an important discontinuity translated at the level of the diffracted by an abrupt variation of important amplitude.

We also note the existence of the Gibbs effect due to the large values of the field located at the interfaces between the PIN diodes and the surrounding metal platelets. This Gibbs effect reflects the difficulty of approximating such discontinuities by a finite series of continuous modes. Some standard and commercial RF PIN diodes utilized on metasurface designs and programmable antenna arrays are listed in [26].

8. Conclusions

Problems related to scaling in circuits can be solved by introducing auxiliary sources. The coherence of the approach has led to relatively simple models of auxiliary sources. These sources have already been used in the literature, particularly in diffraction, and have been applied here to study the insertion of localized elements in circuits. A simple example has highlighted the limits of the validity of using these sources and has defined a protocol for describing the influence of localized elements in electromagnetism. We have been led to introduce an equivalence relation that allows us to get away from the element's geometry. This last hypothesis, if verified, justifies the intuitive approach, which consists in considering that one must be able to realize an exact electromagnetic calculation by assuming certain parts of the circuit (localized elements) by their "current–voltage" characteristics.

Notes/thanks/other declarations

The author thanks Mr. Henri Baudrand of the N7 Toulouse for his help.

Appendix A. Calculation of a microstrip line impedance

The equivalent scheme of **Figure 12** includes an excitation part using the formula of Eq. (6) (the upper modes are assumed evanescent) and an interface part, which is, in this case, a dipole. The global scheme is thus, after using the one of **Figure 1**, the following is:

By posing $Y_s = Z_s^{-1}$, we obtain the relations (**Figures 26** and 27):

$$\begin{vmatrix} \boldsymbol{E}_0 \\ \boldsymbol{J} \end{vmatrix} = \begin{vmatrix} 0 & 1 \\ 1 & \hat{Y}_M + \hat{Y}_S \end{vmatrix} \begin{vmatrix} \boldsymbol{J}_0 \\ \boldsymbol{E} \end{vmatrix} \tag{34}$$

We can notice that $\hat{Y}_S$ is a diagonal operator because the surface impedance is uniformly defined on the whole surface. Its annulment on the metal subdomain is indeed by the short circuit shown in **Figure 1**.

We define the scalar product as an integral over y between 0 and b of the product of functions:

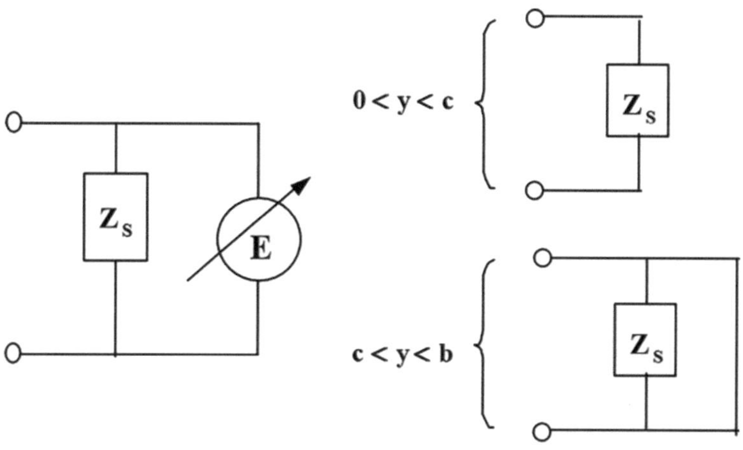

Figure 26.
Equivalent scheme of the closure at z = 0.

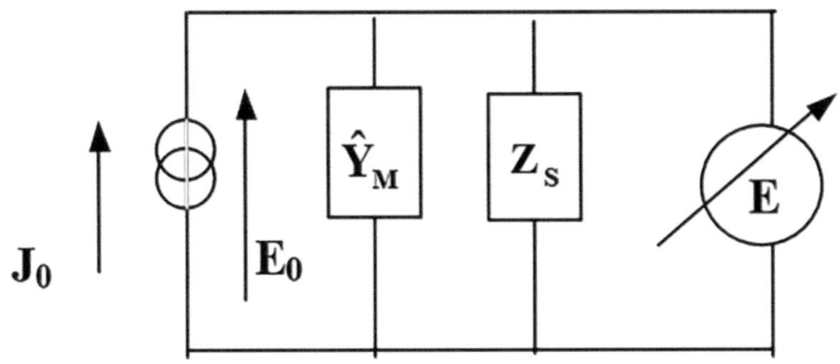

Figure 27.
Equivalent scheme of the structure of ***Figure 12***.

$$\langle f|g\rangle = \int_0^b f_y g_y dy \; (f_x = g_x = 0) \tag{35}$$

In this problem, the E and J fields have no components at x and are independent of x (Invariance by translation along O_x of the structure) so, the integral on the surface can be simplified into a y-integral. The problem is two-dimensional, leading to many simplifications: simple series in the developments, only one mode present, and a scalar formulation. We pose, in a slightly different way than in Eq. (14):

$$J_0 = \frac{1}{a} I f_0' \text{ and } \langle f_0' | E \rangle \tag{36}$$

$f_0' = 1$ in this case, with the definition of the scalar product of Eq. (35). The unknown field in the aperture is expressed on a basis defined from 0 to c. We pose successively:

$$E(y) = \sum_{p=0}^{\infty} x_p g_p(y) \tag{37}$$

with $f_0' = 1, g_0 = \frac{1}{\sqrt{c}}, g_p \neq \sqrt{\frac{2}{c}} \cos\left(\frac{p\pi y}{c}\right)$. The unknowns are g_p; by applying the Galerkin method, the first Equation is scalarly multiplied by f_0' and the second one successively by $g_0, g_1, g_2 \ldots g_p$, p is the order of truncation of the test functions. This gives us:

$$V = \sum_{p=0}^{P} \langle f'_0 | g_p \rangle x_p \tag{38}$$

The other equations have a zero first member because J is zero in the virtual source domain and thus:

$$\langle g_p | f_0' \rangle = 0 = -\langle f_0' | g_p \rangle \frac{I}{a} + \sum_{q=0}^{P} \langle g_p | (\hat{Y}_M + \hat{Y}_s) | g_q \rangle x_q \tag{39}$$

Eliminating the unknown vector of x_p components between Eq. (38) and Eq. (39), we find:

$$\frac{V}{I} = \frac{1}{a} A^t [Y]^{-1} A \tag{40}$$

with, $A = \left[\langle f_0' | \mathbf{g}_1 \rangle, \langle f_0' | \mathbf{g}_2 \rangle, \ \ldots, \langle f_0' | \mathbf{g}_p \rangle \right]$.

$[Y]^{-1}$ is the inverse of the matrix and Y is built on the operator $\hat{Y}_M + \hat{Y}_s$. The general term of Y is written, in this definition:

$$[Y]_{pq} = \left\langle \mathbf{g}_p \middle| (\hat{Y}_M + \hat{Y}_s) \middle| \mathbf{g}_q \right\rangle \tag{41}$$

with as in Eq. (6):

$$\hat{Y}_M = \sum_{n>0} Y_{Mn} \hat{P}_n = \sum_{n>0} \langle f_n | Y_{Mn} | f_n \rangle \tag{42}$$

We can find Eq. (40) with the help of auxiliary sources by representing the problem in **Figure 12** differently: at the level of P, we consider a set of sources for each mode $\mathbf{g}_p$ defined for $y < c$. The multipole Q represents the discontinuity between the guide of height b and that of height c; the sources $\mathbf{g}_p$ will then be closed on the impedance Z_s. A classical modal connection studies the multipole Q. The conditions at $z = 0$ are written:

$$\begin{cases} \mathbf{E} = \mathbf{E}' with 0 < y < b \\ \mathbf{J} + \mathbf{J}' = 0 with 0 < y < c \end{cases} \tag{43}$$

We deduce that, by scalar multiplication by f_n of the first Equation and by $\mathbf{g}_p$ of the second, the generalized transformer relations:

$$\begin{cases} \mathbf{E}_n = \sum_{q=0}^{P} \langle f_n | \mathbf{g}_q \rangle \mathbf{E'}_q \\ \mathbf{J'}_p = - \sum_{m=0}^{\infty} \langle \mathbf{g}_p | f_m \rangle \mathbf{J}_m \end{cases} \tag{44}$$

Using the closure relation on the mode admittances of evanescent modes f_m, $(m > 0)$, the second relation of Eq. (44) combined with the first gives:

$$\begin{cases} \mathbf{J'}_p = \langle \mathbf{g}_p | f_0 \rangle \mathbf{J}_0 + \sum_{m>0}^{\infty} \langle g_p | f_m \rangle Y_{Mn} \langle f_m | \mathbf{g}_p \rangle \mathbf{E}_q{}^P \\ \mathbf{E}_0 = \sum_{q=0}^{P} \langle f_0 | \mathbf{g}_q \rangle \mathbf{E'}_q \end{cases} \tag{45}$$

We notice the similarity between Eq. (44) and Eq. (38), Eq. (41). The difference comes from the definition of I using a function f_0' Eq. (38), whereas here, we use the fundamental mode of the guide.

Now posing the closure relation:

$$J' = -Y_s E'_p \quad (46)$$

We find the expression of the impedance seen from the source:

$$E_0 = B^t \left[\hat{Y}_M + \hat{Y}_S\right] B J_0 \quad (47)$$

With

$$B^t = \left| \langle f_0 | g_1 \rangle, \langle f_0 | g_2 \rangle, \dots \dots \dots, \langle f_0 | g_p \rangle \right| \quad (48)$$

Considering that $f_0 = \frac{1}{\sqrt{ab}}$, (normed fundamental mode), that the scalar product of Eq. (48) concerns the integral in the whole right section, therefore is equal to that Eq. (35) multiplied by a. Finally, $V = E_0 b$ and $I = J_0 a$, we find exactly the expression of Eq. (40). In terms of auxiliary sources, we can say that there is an identity of treatment between the direct approach and the passage through the auxiliary sources provided that these are sufficiently numerous, each having the field distribution corresponding to the functions g_p. The formula of Eq. (40) is simple in the case of a single test function g_0. We can establish the expression of the input admittance and examine to what extent the introduction of a second test function modifies the result.

By taking:

$$g_0 = \frac{1}{\sqrt{c}}; g_p \neq \sqrt{\frac{2}{c}} \cos\left(\frac{p\pi y}{c}\right) \quad (49)$$

$$Y_{Mn} = \frac{j\omega\varepsilon_0}{\gamma_n}; \gamma_n = \sqrt{\frac{n^2\pi^2}{b^2} - k_0^2} \quad (50)$$

The excited evanescent modes are of type TM_{0n}. Hence, the expression of Eq. (44) is as follows.

$$Y = \frac{I}{V} = a \frac{\sum_{n>0} \frac{j\omega\varepsilon}{\gamma_n} \left|\langle g_0 | f_n \rangle\right|^2}{\left|\langle g_0 | f'_a \rangle\right|^2} + Y_s \frac{a}{c} \quad (51)$$

takes the form after development:

$$Y = 2j\omega\varepsilon \frac{a}{b} \sum \frac{1}{\sqrt{\frac{n^2\pi^2}{b^2} - k_0^2}} \left(\frac{\sin\left(\frac{nc}{b}\right)}{\frac{nc}{b}} \right)^2 + Y_s \frac{a}{c} \quad (52)$$

Assuming that the impedance is localized, b representing a quantity of the order of dimension of the box (lower however than the half-wavelength in the vacuum), we can admit that b is very large in front of c. By posing:

$$x = n\frac{c}{b}; dx = \frac{c}{b} \quad (53)$$

x is a practically continuous variable, and the series that appears in Eq. (52) turns into an integral. Using the approximations in Eq. (53), we have:

$$Y = Y_s \frac{a}{c} + 2j\sqrt{\frac{\varepsilon}{\mu}} \int_{\frac{c}{b}}^{\infty} \frac{ak_0}{\sqrt{\pi^2 x^2 - K_0^2 c^2}} \frac{\sin^2(x)}{x^2} dx \tag{54}$$

When c tends to zero, we see that the integral of Eq. (54) diverges as $\log(K_0 c)$, while the purely surface admittance part can be kept constant by making Y_s tend to zero proportionally to c; the capacitive part placed in parallel tends towards infinity, which forbids to consider an impedance of zero dimension.

If we desire to consider the edge effects, it is necessary to introduce other test functions g_p. This gives an idea of the precision of the auxiliary source concept. For example, considering two test functions, we find for the input admittance [according to Eq. (40)]:

$$Y' = a \frac{Y_{11}\left(1 - \frac{Y_{11}^2}{Y_{11} Y}\right)}{\left|\langle g | f'_0 \rangle\right|^2} + Y_s \frac{a}{c} \tag{55}$$

The relative degree of accuracy of the capacitive part is given by the term:

$$\frac{\partial C}{C} = \frac{-Y_{12}^2}{Y_{11} Y_{22}} \text{ With} \begin{cases} Y_{11} = \sum\limits_{n>0} \dfrac{j\omega\varepsilon}{\gamma_n} \left|\langle g_0 | f_n \rangle\right|^2 \\ Y_{22} = \sum\limits_{n>0} \dfrac{j\omega\varepsilon}{\gamma_n} \left|\langle g_1 | f_n \rangle\right|^2 \\ Y_{11} = \sum\limits_{n>0} \dfrac{j\omega\varepsilon}{\gamma_n} \langle g_0 | f_n \rangle \langle f_n | g_0 \rangle \end{cases} \tag{56}$$

From these expressions, we can see that the accuracy depends essentially on the ratio c/b, as a first approximation, the relative error is proportional to $(c/b)^2$; we can therefore admit that the auxiliary sources are a good approximation for dimensions between one-tenth and one-hundredth of the dimensions of the case.

Author details

Bilel Hamdi* and Taoufik Aguili
Communication System Laboratory Sys'Com, National Engineering School of Tunis (ENIT), University of Tunis El Manar, Le Belvédère, Tunis, Tunisia

*Address all correspondence to: hbilel.enit@gmail.com

References

[1] Eleftheriades GV, Mosig JR. On the network characterization of planar passive circuits using the method of moments. IEEE Transactions on Microwave Theory and Techniques. 1996;**44**(3):438-445

[2] Bodereau F, Baillargeat D, Verdeyme S, Aubourg M, Guillon P, Jarthon G, Rogeaux E, Cazaux JL. Global electromagnetic analysis to help the conception of an active module. In: 2000 IEEE MTT-S International Microwave Symposium Digest (Cat. No. 00CH37017); 2000 Jun 11. Vol. 3. London: IEEE; 2020. pp. 1921-1924

[3] Baillargeat D, Larique E, Verdeyme S, Aubourg M, Sommet R, Guillon P. Coupled localized and distributed elements analysis applying an electromagnetic software in the frequency domain. In: 1997 IEEE MTT-S International Microwave Symposium Digest; 1997 Jun 8. Vol. 2. London: IEEE; 1997. pp. 1021-1024

[4] Aubourg M, Madrangeas V, Verdeyme S, Guillon P. Finite element method: References applications to microwave devices. In: Digest of the Fifth Biennial IEEE Conference on Electromagnetic Field Computation; 1992 Aug 3. London: IEEE; 1992. pp. MOF2-MOF2

[5] Kunisch J, Rittweger M, Heinen S, Wolff I. The compression approach: A new technique for the analysis of distributed circuits containing nonlinear elements. In: 1991 21st European Microwave Conference; 1991 Sep 9. Vol. 2. IEEE; 1991. pp. 1296-1301

[6] Guillouard K, Wong MF, Hanna VF, Citerne J. A new global finite element analysis of microwave circuits including lumped elements. IEEE Transactions on Microwave Theory and Techniques. 1996;**44**(12):2587-2594

[7] Harrington R, Mautz J. A generalized network formulation for aperture problems. IEEE Transactions on Antennas and Propagation. 1976;**24**(6): 870-873

[8] Baudrand H, Bajon D. Equivalent circuit representation for integral formulations of electromagnetic problems. International Journal of Numerical Modelling: Electronic Networks, Devices and Fields. 2002;**15**(1):23-57

[9] Baudrand H. Représentation des sources dans les circuits planaires. In: Workshop on Global Simulator of Planar Microwave Circuits. San Francisco: IEEE, MTT; 1996. pp. 17-21

[10] Baudran H. Introduction au calcul des éléments de circuits passifs en hyperfréquences. Toulouse: Cépaduès-éditions; 2001

[11] Pujol S, Baudrand H, Hanna VF, Dong X. A new approach of the source method for characterization of planar structures. In: 1991 21st European Microwave Conference; 1991 Sep 9. Vol. 2. London: IEEE; 1991. pp. 1015-1020

[12] Zhu L, Wu K. Comparative investigation on numerical de-embedding techniques for equivalent circuit modeling of lumped and distributed microstrip circuits. IEEE Microwave and Wireless Components Letters. 2002;**12**(2):51-53

[13] Baudrand H. Representation by equivalent circuit of the integral methods in microwave passive elements. In: 1990 20th European Microwave Conference; 1990 Sep 9. Vol. 2. London: IEEE; 1990. pp. 1359-1364

[14] Baudrand H. Aubert H. Integral equations formulation for active or passive planar devices using equivalent circuit approach. In: Workshop on Global Simulators IEEE-MTT French chapter La Rochelle; 1998

[15] Bajon D, Souny B, Baudrand H. Multisources actives annular slot antenna. In: Annales Des Télécommunications. Vol. 52(3). Berlin/Heidelberg: Springer-Verlag; 1997. pp. 188-201

[16] Cascio L, Tardioli G, Hoefer WJR. Characterization of nonlinear active and passive devices in stub-loaded SCN-TLM. In: Proceedings of 2nd International Workshop on Transmission Line Matrix (TLM) Modelling Theory and Applications. 1997

[17] Lacroux F, Leveque P, Jecko B, Celeste A. Etude théorique d'une rectenna en H pour le transport d'énergie sans fil. In: JNM. Lille. 2003. pp. 21-22

[18] Madrangeas S. Analyse électromagnétique globale de dispositifs microondes intégrant des éléments localisés passifs et actifs. Thèse de doctorat. Limoges; 1996

[19] Mili S, Aguili CL, Aguili T. The renormalization group theory combined to the ms-gec method to study active fractal structures with incorporated pin diodes. Progress in Electromagnetics Research B. 2011;**29**:43-62

[20] Hanson GW. Dyadic Green's functions and guided surface waves for a surface conductivity model of graphene. Journal of Applied Physics. 2008 15; **103**(6):064302

[21] Niu T, Withayachumnankul W, Ung BS, Menekse H, Bhaskaran M, Sriram S, Fumeaux C. Reflectarray Antennas for Terahertz Communications. arXiv preprint arXiv: 1210.0653. 2012 Oct 2

[22] Koohi MZ, Neshat M. Evaluation of graphene-based terahertz photoconductive antennas, 3. Scientia Iranica. Transaction F, Nanotechnology. 2015;**22**:1299

[23] Aidi M, Hajji M, Ben Ammar A, Aguili T. Graphene nanoribbon antenna modeling based on MoM-GEC method for electromagnetic nanocommunications in the terahertz range. Journal of Electromagnetic Waves and Applications. 2016;**30**(8):1032-1048

[24] Lucyszyn S. Evaluating surface impedance models for terahertz frequencies at room temperature

[25] Aidi M, Hajji M, Hamdi B, Aguili T. Graphene nanoribbon modeling based on MoM-GEC method for antenna applications in the terahertz range. In: 2015 World Symposium on Mechatronics Engineering & Applied Physics (WSMEAP); 2015 Jun 11. London: IEEE; 2015. pp. 1-4

[26] Celis Sierra S. Index modulation using reconFigure urable arrays [doctoral dissertation]

[27] Thesis of Dr.Sonia Mili (ask Professor Taoufik AGUILI)

Chapter 4

Waveguide Port Approach in EM Simulation of Microwave Antennas

Faik Bogdanov, Irina Chochia, Lily Svanidze and Roman Jobava

Abstract

This chapter generalizes a recently proposed MoM-based approach to waveguide port excitation (WPE) problems on arbitrary conducting and composite geometries. This approach combines the canonical aperture coupling approach with the EFIE-PMCHWT formulation for composite structures. Each WPE problem in this approach is divided into equivalent sub-problems for internal and external regions, which are solved using the MoM. Internal WPE problems are solved using waveguide modal expansion in the port plane, while external problems are solved using the equivalence principle to reduce these problems to the systems of algebraic equations for unknown electric and magnetic currents. The developed approach is validated on radiation and coupling problems for coaxial ports by comparing simulated results with those obtained by other approaches and measurements. An excellent agreement between the simulated and measured results is demonstrated. Finally, this approach is applied to practical EMC problems for microwave antennas fed by coaxial ports.

Keywords: coupling problem, coaxial port, equivalence principle, method of moments (MoM), waveguide modal expansion, waveguide port

1. Introduction

The excitation problem is of increasing importance at microwave frequencies [1]. Microwave antennas and other microwave devices are often fed by waveguides (rectangular, circular, and coaxial) or transmission lines (such as microstrips). In general, these devices are composite structures consisting of both conductive and dielectric elements. Therefore, the appropriate modeling of the waveguide excitation of such structures is of great interest. Such modeling in numerical methods is usually done by truncating the feed waveguide to create a waveguide port and formulate suitable boundary conditions (BC) imposed on the port. Such BCs should be able to launch an incident wave into the waveguide and absorb the reflected (in active mode) or received (in passive mode) wave without spurious reflections [2].

To date, most approaches to solving the waveguide port excitation (WPE) problem are based on volume discretization methods, such as the finite-element method (FEM) [1, 2], finite difference time domain (FDTD) [3–5], discontinues Galerkin time-domain (DGTD) [6], contour integral method (CIM) [7], etc. Most of these works use various modal absorbing boundary conditions (MABC) [4, 5], developed for time-domain methods as termination conditions imposed on the port.

IntechOpen

At present, many electromagnetic (EM) problems are solved using surface integral equations (SIE) together with the method of moments (MoM) [8]. Within the framework of SIE, the WPE problem was first formulated as an aperture coupling problem for a conducting geometry, and the MoM solution for magnetic currents was obtained in the presence of a short-circuited conductive sheet [9]. This approach was then modified using the pseudo-image method for magnetic currents in addition to electric currents [10]. Further, MoM was applied to the waveguide port problems [11] and antenna radiation problems with aperture port excitation [12]. However, until recently, a MoM-based solution to the WPE problem for arbitrary geometries has been poorly represented in the literature. In our recent works [13–15], such a solution was obtained for radiation and coupling problems for various types of geometries.

This chapter generalizes the recently proposed MoM-based approach to WPE problems [13–15] on arbitrary conducting and composite geometries. The obtained approach combines the canonical aperture coupling approach with the EFIE-PMCHWT formulation for composite structures [16–22]. Each WPE problem in this approach is divided into equivalent sub-problems for internal and external regions, which are solved using the MoM. The internal WPE problems are solved using waveguide modal expansion in the port plane, while the external problems are solved using the equivalence principle to reduce these problems to the systems of algebraic equations for unknown electric and magnetic currents. The obtained solution also considers the problem of material junctions between adjacent surfaces, considered in [19–22].

The developed approach is validated on radiation and coupling problems for coaxial ports by comparing the simulated results with those from other approaches and measurements. In addition, this approach is applied to practical EMC problems for microwave antennas fed by coaxial ports. The MoM calculations were performed using the TriD numerical code incorporated in the EMCoS Studio software package [23].

2. Waveguide port approach for conducting geometry

2.1 Dividing the original problem into equivalent problems

Figure 1a illustrates a canonical waveguide port problem for conducting geometry. This geometry consists of a semi-infinite waveguide *1* with perfect electric conducting (PEC) walls and a microwave structure *2*, which is yet supposed to be conductive. We intend to create port *P* in waveguide *1* to divide the geometry into two regions (*A* and *B*) to truncate the mesh in the region *A* and impose appropriate termination conditions in the port plane.

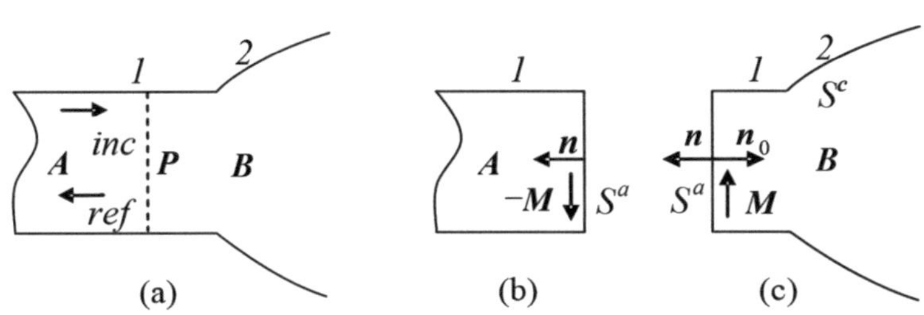

Figure 1.
(a) Waveguide port problem for conducting geometry; (b) Equivalence for the internal region A*; (c) Equivalence for the external region* B.

For this purpose, we follow the classical approach for the aperture problem [9] to divide the original problem into two equivalent problems, as shown in **Figure 1b** and **c**. We introduce a perfectly conducting surface S^a into the port plane P to separate regions A and B and consider two equivalent sub-problems: internal (for region A) and external (for region B). In addition, we introduce equivalent magnetic currents and $\mathbf{M}$ on both sides of S^a to restore the tangential electric fields on the boundary surface S^a. Let us consider these equivalent problems separately.

2.2 Formulation of the internal equivalent problem

Consider an internal equivalent problem for the region A. The total EM field in the region A is composed of the incident field $\mathbf{E}^{inc}, \mathbf{H}^{inc}$ and the reflected field $\mathbf{E}^{ref}, \mathbf{H}^{ref}$ generated by magnetic currents in the presence of a conductor. According to the equivalence principle [24], these currents are related to the total electric field $\mathbf{E}_A^{S^a}$ on the port surface S^a by the relation:

$$-\mathbf{M} = -\mathbf{n} \times \mathbf{E}_A^{S^a} = \mathbf{n}_0 \times \mathbf{E}_A^{S^a} \tag{1}$$

where $\mathbf{n}$ is the internal normal in the region A, and $\mathbf{n}_0 = -\mathbf{n}$ is the propagation direction of the incident wave.

Equation (1) relates the total electric field at the port surface S^a to magnetic currents depending on the geometric and material properties of the external region B. The internal equivalent problem is to find the modal expansion of the total EM field at the port surface S^a through these currents.

2.3 MoM solution of the internal equivalent problem

The total EM field in the region A on the port surface S^a can be generally written as the sum of the incident (+) and reflected (−) TEM (if exists), TE, and TM modes [2, 13]:

$$\mathbf{E}_A^{S^a} = \left(a_0^+ + a_0^-\right) \mathbf{e}_0^{TEM} + \sum_{s=1}^{N^{TE}} \left(a_s^+ + a_s^-\right) \mathbf{e}_s^{TE} + \sum_{s=1}^{N^{TM}} \left(b_s^+ + b_s^-\right) \mathbf{e}_s^{TM} \tag{2}$$

$$\begin{aligned} \mathbf{H}_A^{S^a} = &\frac{1}{Z}\left(a_0^+ - a_0^-\right)\left(\mathbf{n}_0 \times \mathbf{e}_0^{TEM}\right) + \sum_{s=1}^{N^{TE}} \frac{1}{Z_s^{TE}}\left(a_s^+ - a_s^-\right)\left(\mathbf{n}_0 \times \mathbf{e}_s^{TE}\right) \\ &+ \sum_{s=1}^{N^{TM}} \frac{1}{Z_s^{TM}}\left(b_s^+ - b_s^-\right)\left(\mathbf{n}_0 \times \mathbf{e}_s^{TM}\right) \end{aligned} \tag{3}$$

where $a_s^\pm$ and $b_s^\pm$ are the mode amplitudes, $\mathbf{e}_0^{TEM}$, $\mathbf{e}_s^{TE}$ and $\mathbf{e}_s^{TM}$ are the transverse modal functions of TEM, TE and TM waves with wave impedances $Z = \sqrt{\mu/\varepsilon}$, $Z_s^{TE} = \omega\mu/\gamma_s$ and $Z_s^{TM} = \gamma_s/\omega\varepsilon$, respectively, ε and μ are the permittivity and permeability of the medium, Γ_s is the propagation constant, and N^{TE} and N^{TM} are the numbers of accounted TE and TM modes, respectively.

Next, we use the MoM to relate the mode amplitudes of the reflected fields in (2), (3) to the magnetic currents $\mathbf{M}$. The port surface S^a is discretized into planar patches, and the unknown magnetic currents are approximated as

$$\mathbf{M} = \sum_{n=1}^{N^a} M_n \mathbf{f}_n \tag{4}$$

where $\mathbf{f}_n$ are linear independent basis functions (BFs), M_n are unknown expansion current coefficients, and N^a is the number of these BFs on the surface S^a. Substituting now (4) and (2) into (1), multiplying both sides by $\mathbf{n}_0 \times \mathbf{e}_0^{TEM}$, $\mathbf{n}_0 \times \mathbf{e}_{s'}^{TE}$ and $\mathbf{n}_0 \times \mathbf{e}_{s'}^{TM}$, respectively, and integrating over the port surface S^a, we relate the amplitudes of the reflected waves with those of the incident waves and magnetic current coefficients:

$$a_s^- = -a_s^+ - \sum_{n=1}^{N^a} \frac{M_n T_{ns}}{R_s}, \quad b_s^- = -b_s^+ - \sum_{n=1}^{N^a} \frac{M_n T'_{ns}}{R'_s} \quad (s = 0, 1, 2, \dots) \tag{5}$$

where

$$T_{n0} = \int_{S_a} \mathbf{f}_n \cdot \left(\mathbf{n}_0 \times \mathbf{e}_0^{TEM}\right) dS', \quad T_{ns} = \int_{S_a} \mathbf{f}_n \cdot \left(\mathbf{n}_0 \times \mathbf{e}_s^{TE}\right) dS', \quad T'_{ns} = \int_{S_a} \mathbf{f}_n \cdot \left(\mathbf{n}_0 \times \mathbf{e}_s^{TM}\right) dS' \tag{6}$$

$$R_0 = \int_{S_a} \left(\mathbf{n}_0 \times \mathbf{e}_0^{TEM}\right) \left(\mathbf{n}_0 \times \mathbf{e}_0^{TEM}\right) dS', \quad R_s = \int_{S_a} \left(\mathbf{n}_0 \times \mathbf{e}_s^{TE}\right) \left(\mathbf{n}_0 \times \mathbf{e}_s^{TE}\right) dS',$$
$$R'_s = \int_{S_a} \left(\mathbf{n}_0 \times \mathbf{e}_s^{TM}\right) \left(\mathbf{n}_0 \times \mathbf{e}_s^{TM}\right) dS'. \tag{7}$$

Substitution of (5) into (2), (3) determines the total electric and magnetic fields on the port surface in region A through the still unknown magnetic currents.

2.4 Formulation of the external equivalent problem for conducting geometry

Consider now an external equivalent problem for conducting geometry. The scattered EM field in the external region B in **Figure 1c** of the conducting geometry is produced by electric currents $\mathbf{J}$ flowing over surfaces S^a and S^c and equivalent magnetic currents $\mathbf{M}$ at the surface S^a, which can be written as:

$$\mathbf{E}_B^{sc}(\mathbf{J}, \mathbf{M}) = L^{EJ}\mathbf{J} + L^{EM}\mathbf{M} \tag{8}$$

$$\mathbf{H}_B^{sc}(\mathbf{J}, \mathbf{M}) = L^{HJ}\mathbf{J} + L^{HM}\mathbf{M} \tag{9}$$

where L^{EJ}, L^{EM}, L^{HJ} and L^{HM} are the linear integro-differential operators of electric and magnetic fields applied to the electric and magnetic currents, respectively. Applying the boundary conditions for the tangential electric and magnetic fields on the surfaces S^a and S^c, we obtain the following system of integral equations for the unknown electric and magnetic currents $\mathbf{J}$ and $\mathbf{M}$

$$\mathbf{E}_B^{sc}(\mathbf{J}, \mathbf{M})\Big|_{\tan}^{S^a+S^c} = 0 \quad \text{just outside} \quad S^a \tag{10}$$

$$\mathbf{H}_B^{sc}(\mathbf{J}, \mathbf{M})\Big|_{\tan}^{S^a+S^c} = \mathbf{H}_A^{S^a}\Big|_{\tan} \quad \text{just inside} \quad S^a \tag{11}$$

2.5 MoM solution of the external equivalent problem for conducting geometry

To obtain the MoM solution to the BC (10) and (11), we consider, along with Eq. (4), the following expansion for an unknown electric current **J**:

$$\mathbf{J} = \sum_{n=1}^{N^a+N^c} I_n \mathbf{f}_n, \tag{12}$$

where $\mathbf{f}_n$ are the BFs taken the same as for the expansion of magnetic currents in (4), I_n are the unknown expansion current coefficients on the surfaces S^a and S^c, and N^a and N^c are the numbers of these BFs on these surfaces. Substitution of expansions (4) and (12) in (8) and (9) gives the following expressions for the EM field in region *B*:

$$\mathbf{E}_B^{sc}(\mathbf{J}, \mathbf{M}) = \sum_{n=1}^{N^a+N^c} I_n L^{EJ} \mathbf{f}_n + \sum_{n=1}^{N^a} M_n L^{EM} \mathbf{f}_n, \tag{13}$$

$$\mathbf{H}_B^{sc}(\mathbf{J}, \mathbf{M}) = \sum_{n=1}^{N^a+N^c} I_n L^{HJ} \mathbf{f}_n + \sum_{n=1}^{N^a} M_n L^{HM} \mathbf{f}_n. \tag{14}$$

Substituting now (3), (5), (13) and (14) into (10) and (11), introducing the boundary operators $\hat{L}^{JJ} = L^{EJ}\Big|_{S^a+S^c}$, $\hat{L}^{JM} = L^{EM}\Big|_{S^a+S^c}^{\text{outside}}$, $\hat{L}^{MJ} = L^{HJ}\Big|_{S^a}^{\text{inside}}$ and $\hat{L}^{MM} = L^{HM}\Big|_{S^a}$ and testing the resulting equations with appropriate weighting functions $\mathbf{w}_1(\mathbf{r}), \mathbf{w}_2(\mathbf{r}), \dots,$ $\mathbf{w}_m(\mathbf{r})$ leads to the following system of linear algebraic equations

$$\begin{bmatrix} [Z_{mn}^{JJ}] & [Z_{mn}^{JM}] \\ [Z_{mn}^{MJ}] & [Z_{mn}^{MM}] \end{bmatrix} \begin{bmatrix} [I_n] \\ [M_n] \end{bmatrix} = \begin{bmatrix} 0 \\ [V_m^W] \end{bmatrix} \tag{15}$$

with elements defined as:

$$Z_{mn}^{JJ} = -\left\langle \mathbf{w}_m, \hat{L}^{JJ} \mathbf{f}_n \right\rangle, \quad Z_{mn}^{JM} = -\left\langle \mathbf{w}_m, \left(\tilde{L}^{JM} + \frac{1}{2} \mathbf{n} \times \right) \mathbf{f}_n \right\rangle, \tag{16}$$

$$Z_{mn}^{MJ} = -\left\langle \mathbf{w}_m, \left(\tilde{L}^{MJ} + \frac{1}{2} \mathbf{n} \times \right) \mathbf{f}_n \right\rangle, \quad Z_{mn}^{MM} = -\left\langle \mathbf{w}_m, \hat{L}^{MM} \mathbf{f}_n \right\rangle + Q_{mn}^W, \tag{17}$$

$$Q_{mn}^W = \frac{\hat{T}_{m0}}{W} \frac{T_{n0}}{R_0} + \sum_{s=1}^{N^{TE}} \frac{\hat{T}_{ms}}{W_s^{TM}} \frac{T_{ns}}{R_s} + \sum_{s=1}^{N^{TM}} \frac{\hat{T}'_{ms}}{W_s^{TM}} \frac{T'_{ns}}{R'_s}, \tag{18}$$

$$V_m^W = -2 \left[\frac{\hat{T}_{m0}}{W} a_0^+ + \sum_{s=1}^{N^{TE}} \frac{\hat{T}_{ms}}{W_s^{TM}} a_s^+ + \sum_{s=1}^{N^{TM}} \frac{\hat{T}'_{ms}}{W_s^{TM}} b_s^+ \right], \tag{19}$$

where $\tilde{L}^{JM}$ and $\tilde{L}^{MJ}$ are the regular parts of the boundary operators $\hat{L}^{JM}$ and $\hat{L}^{MJ}$, the notation $\langle \mathbf{w}, \mathbf{f} \rangle = \int_S \mathbf{w} \cdot \mathbf{f} dS$ is used for the scalar product, and

$$\hat{T}_{m0} = \int_{S_a} \mathbf{w}_m \cdot (\mathbf{n}_0 \times \mathbf{e}_0^{TEM}) dS', \hat{T}_{ms} = \int_{S_a} \mathbf{w}_m \cdot (\mathbf{n}_0 \times \mathbf{e}_s^{TE}) dS', \hat{T}'_{ms} = \int_{S_a} \mathbf{w}_m \cdot (\mathbf{n}_0 \times \mathbf{e}_s^{TM}) dS' \tag{20}$$

In the case of Galerkin's procedure $\mathbf{w}_m = \mathbf{f}_m$, and coefficients (20) and (6) become the same. The MoM system (15) determines the solution to the waveguide port problem in the conducting geometry.

2.6 Validation of the developed approach for conducting geometry

The developed approach has been validated to simulate the scattering characteristics of a flanged coaxial line as proposed in [25–27]. Such structures are frequently used in biomedical engineering for non-destructive testing of various materials [25–27].

When modeling a coaxial line, it is convenient to choose the port plane at the output of the line to provide fast damping of evanescent waves. In this case, it can be assumed $N^{TE} = N^{TM} = 0$ in (2) and (3) to take taken into account only fundamental, TEM mode with the modal function $\mathbf{e}_0^{TEM} = \mathbf{e}_\rho / [\rho \cdot \ln(D/d)]$, where ρ is the radial distance, $\mathbf{e}_\rho$ is the unit radial vector, and D and d are the outer and inner diameters of the coaxial waveguide.

Figure 2 shows a flanged coaxial line consisting of a coaxial waveguide section with an outer radius $D/2$ = 4.725 mm, an inner radius of $d/2$ = 1.4364 mm, and a length L = 10 mm, ended with a circular disc with a diameter $2R$ = 200 mm. The bottom plane of the waveguide is accepted as a waveguide port, and the structure is excited in this port by TEM mode. To validate the developed approach for conducting geometry, we analyze the case when both the waveguide and outer space have the same permittivity $\varepsilon_r = 2.05$.

Figure 3a and **b** show the magnitude and phase of the reflection coefficient at the end of the coaxial line used as the reference plane. We compare the simulation results obtained using the developed approach, the mode-matching technique [25], and the

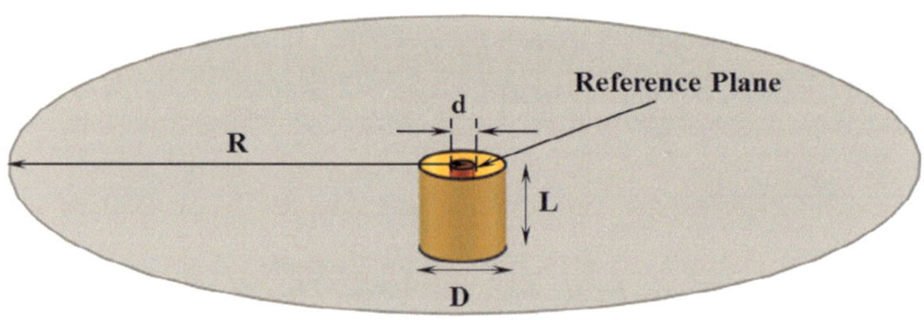

Figure 2.
Geometry of open-ended coaxial line flanged with a circular disc: D/2 = 4.725 *mm,* d/2 = 1.4364 *mm,* L = 10 *mm,* 2R = 200 *mm,* $\varepsilon_r = 2.05$ *inside and outside the line.*

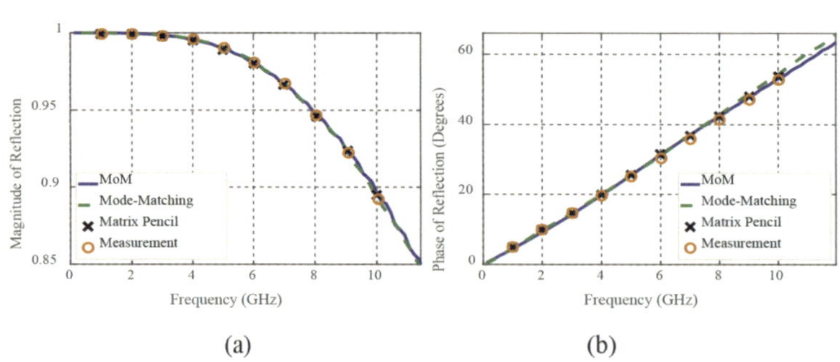

Figure 3.
(a) Magnitude and (b) phase of the reflection coefficient versus the frequency of excitation at the end of the flanged coaxial line, calculated for various approaches.

matrix pencil method [26] with measurement data [27]. Note that infinite flanges are assumed in [25, 26]. Phase data conforms to the time convention $\exp(-i\omega t)$.

Comparison of various results shows excellent agreement between them. However, the phase characteristics obtained by our approach agree somewhat more accurately with the measurement data. Thus, the obtained results validate the developed approach to modeling a coaxial waveguide port for conducting geometries.

3. Waveguide port approach for composite geometry

3.1 Equivalent problems for composite geometry

Figure 4 shows the geometry of the problem, consisting of a composite structure composed of *k-1* homogeneous regions D_i, $i = 1,2,\dots,k-1$, located in the free space region D_0 and exposed to waveguide excitation from the waveguide region B, which will be considered as k-th region of the problem. The region D_k is a finite section of the waveguide, confined by the PEC walls, the port surface S^a, and the dielectric surface S_k^d, through which the structure is fed. The port surface S_a separates the region D_k (B) from the semi-infinite waveguide region A with incident waveguide excitation. In addition, each region D_i is excited, in general, by the incident field E_i^{inc}, H_i^{inc}.

To formulate the waveguide port excitation problem through the port surface S^a, we first consider the aperture coupling problem between the waveguide regions A and B [9]. Thus, we cover the port surface S^a with a PEC sheet and introduce equivalent magnetic currents $-\mathbf{M}$ and $\mathbf{M}$ on both sides of S^a to divide the excitation problem into two different equivalence problems: the internal problem for region A, and the external problem for region B (D_k), as done in Section 2.1. Then, the internal equivalent problem is identical to that formulated in Section 2.2 and solved in Section 2.3. The external equivalent problem requires consideration of equivalent problems for each boundary surface in regions D_i, $i = 1,2,\dots,k$, including the port surface S^a.

3.2 Formulation of the external equivalent problem for composite geometry

An external equivalent problem for composite geometry is reduced to a set of equivalent problems for each conducting and dielectric boundary S_i^c and S_i^d of free space region D_0 $(i = 0)$, composite structure regions D_i $(i = 1,\dots,k\text{-}1)$, and finite

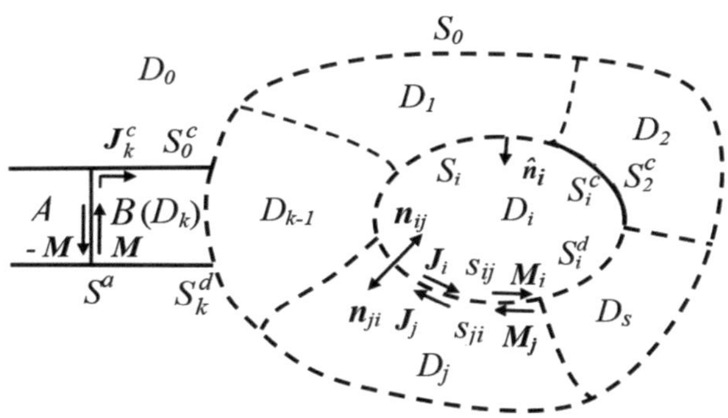

Figure 4.
Waveguide port problem for composite geometry with waveguide excitation.

waveguide region D_k ($i = k$). In turn, each surface S_i^d comprises a set of boundary surfaces $s_{ij} = D_i \cap D_j$ ($i \neq j$), being the interfaces between the regions D_i and D_j.

Per the equivalence principle [24], the total EM field inside the i-th region D_i can be expressed as the sum of the incident field E_i^{inc}, H_i^{inc} and that induced by the total surface currents distributed over its boundary surface S_i and radiating into a homogeneous medium with constitutive parameters ε_i and μ_i of the region D_i. The total electric currents J_i on the boundary surface S_i consist of conducting currents $\mathbf{J}_i^c$, flowing on the inner sides of conducting boundaries S_i^c, and equivalent electric currents $\mathbf{J}_i^d$, flowing on the inner sides of dielectric boundaries S_i^d. Magnetic currents in the region D_i are equivalent currents $\mathbf{M}_i^d$, flowing on dielectric boundaries S_i^d. In addition, in the waveguide region D_k there are equivalent magnetic currents $\mathbf{M}$ on the port surface S^a.

Unknown electric and magnetic currents can be found using the boundary conditions at the conducting boundaries of the composite structure:

$$\left[\mathbf{E}_i^{inc} + \mathbf{E}_i^{sc}\left(\mathbf{J}_i^c, \mathbf{J}_i^d, \mathbf{M}_i^d\right)\right]\Big|_{\tan}^{S_i^c} = 0, \quad i = 0, 1, \ldots, k-1 \tag{21}$$

dielectric boundaries of regions D_i ($i,j = 0, 1, \ldots, k, i \neq j$):

$$\left[\mathbf{E}_i^{inc} + \mathbf{E}_i^{sc}\left(\mathbf{J}_i^c, \mathbf{J}_i^d, \mathbf{M}_i^d, \mathbf{M}\delta_{ik}\right)\right]\Big|_{\tan}^{s_{ij}} = \left[\mathbf{E}_j^{inc} + \mathbf{E}_j^{sc}\left(\mathbf{J}_j^c, \mathbf{J}_j^d, \mathbf{M}_j^d, \mathbf{M}\delta_{jk}\right)\right]\Big|_{\tan}^{s_{ij}}, \tag{22}$$

$$\left[\mathbf{H}_i^{inc} + \mathbf{H}_i^{sc}\left(\mathbf{J}_i^c, \mathbf{J}_i^d, \mathbf{M}_i^d, \mathbf{M}\delta_{ik}\right)\right]\Big|_{\tan}^{s_{ij}} = \left[\mathbf{H}_j^{inc} + \mathbf{H}_j^{sc}\left(\mathbf{J}_j^c, \mathbf{J}_j^d, \mathbf{M}_j^d, \mathbf{M}\delta_{jk}\right)\right]\Big|_{\tan}^{s_{ij}}, \tag{23}$$

and on the port surface S_a and the conducting boundary S_k^c of the k-th region:

$$\left[\mathbf{E}_k^{inc} + \mathbf{E}_k^{sc}\left(\mathbf{J}_k^c, \mathbf{J}_k^d, \mathbf{M}_k^d, \mathbf{M}\right)\right]\Big|_{\tan}^{S^a + S_k^c} = 0 \quad \text{just outside } S^a \tag{24}$$

$$\left[\mathbf{H}_k^{inc} + \mathbf{H}_k^{sc}\left(\mathbf{J}_k^c, \mathbf{J}_k^d, \mathbf{M}_k^d, \mathbf{M}\right)\right]\Big|_{\tan}^{S^a} = \mathbf{H}_A^{S^a}\Big|_{\tan} \quad \text{just inside } S^a, \tag{25}$$

where δ_{ik} is the Kronecker delta, which shows that magnetic currents $\mathbf{M}$ radiate only in a waveguide region D_k. The magnetic field on the right-hand side of (25) is expressed by Eq. (3). The scattered EM fields in (21)–(25) can be expressed in terms of electric and magnetic currents $\mathbf{J}_i$ and $\mathbf{M}_i$ in the dielectric region D_i as

$$\mathbf{E}_i^{sc}(\mathbf{J}_i, \mathbf{M}_i) = -L_i^{EJ}(\mathbf{J}_i) - L_i^{EM}(\mathbf{M}_i) \tag{26}$$

$$\mathbf{H}_i^{sc}(\mathbf{J}_i, \mathbf{M}_i) = -L_i^{HJ}(\mathbf{J}_i) - L_i^{HM}(\mathbf{M}_i) \tag{27}$$

where L_i^{EJ}, L_i^{EM}, L_i^{HJ} and L_i^{HM} are linear integro-differential operators of EM fields applied to currents radiated in the i-th region. It can also be shown [19–22] that the equivalent currents on opposite sides of the dielectric boundaries are related as:

$$\mathbf{J}_i^d = -\mathbf{J}_j^d, \quad \mathbf{M}_i^d = -\mathbf{M}_j^d \quad \text{on } s_{ij} \tag{28}$$

Equation (21)–(25) together with relations (26)–(28) and expansions (3) represent the general (EFIE-PMCHWT) form of integral equations for a composite structure with an arbitrary excitation, including the waveguide port.

3.3 MoM solution of the external equivalent problem for composite geometry

To solve the coupled system of integral Eqs. (21)–(28), we use the MoM to discretize the geometry of all boundary surfaces of the regions D_i $(i=1, \dots, k)$ into the planar patches and to consider the following expansions for the unknown currents:

$$\mathbf{J}_k^c = \sum_{n=1}^{N^a+N_k^c} I_n^c \mathbf{f}_n, \quad \mathbf{M} = \sum_{n=1}^{N_a} M_n, \quad \left[\mathbf{J}_i^c\right]_{i=0}^{k-1} = \sum_{n=1}^{N^C} I_n^C \mathbf{f}_n, \tag{29}$$

$$\left[\mathbf{J}_i^d\right]_{i=0}^{k} = \sum_{n=1}^{N^d} I_n^d \mathbf{f}_n, \quad \left[\mathbf{M}_i^d\right]_{i=0}^{k} = \sum_{n=1}^{N^d} M_n^d \mathbf{f}_n, \tag{30}$$

where $\mathbf{f}_n$ are the suitable BFs, I_n^c, M_n, I_n^C, I_n^d and M_n^d are the unknown expansion current coefficients, and N^a, N_k^c, N^C and N^d are the numbers of BFs on the surfaces S^a, S_k^c, $\left[S_i^c\right]_{i=0}^{k-1}$, if any, and $\left[S_i^d\right]_{i=0}^{k-1}$, respectively. Expansions (29) and (30) take into account relations (28) for unknown equivalent currents on opposite sides of the dielectric boundaries. They also consider the ratios for adjacent currents at material junctions, which are the boundaries between several media [22].

Substituting (29) and (30) into (21)–(25) taking into account (3), (5), (26)–(28) and testing the resulting equations with weighting functions $\mathbf{w}_1(\mathbf{r}), \mathbf{w}_2(\mathbf{r}), \dots, \mathbf{w}_m(\mathbf{r})$, defined in the range of the respective boundary operators, we obtain the following MoM system of linear algebraic equations:

$$\begin{bmatrix} \left[Z_{mn}^{J^cJ^c}\right] & \left[Z_{mn}^{J^cM}\right] & 0 & \left[Z_{mn}^{J^cJ^d}\right] & \left[Z_{mn}^{J^cM^d}\right] \\ \left[Z_{mn}^{MJ^c}\right] & \left[Z_{mn}^{MM}+Q_{mn}^{W}\right] & 0 & \left[Z_{mn}^{MJ^d}\right] & \left[Z_{mn}^{MM^d}\right] \\ 0 & 0 & \left[Z_{mn}^{J^CJ^C}\right] & \left[Z_{mn}^{J^CJ^d}\right] & \left[Z_{mn}^{J^CM^d}\right] \\ \left[Z_{mn}^{J^dJ^c}\right] & \left[Z_{mn}^{J^dM}\right] & \left[Z_{mn}^{J^dJ^C}\right] & \left[Z_{mn}^{J^dJ^d}\right] & \left[Z_{mn}^{J^dM^d}\right] \\ \left[Z_{mn}^{M^dJ^c}\right] & \left[Z_{mn}^{M^dM}\right] & \left[Z_{mn}^{M^dJ^C}\right] & \left[Z_{mn}^{M^dJ^d}\right] & \left[Z_{mn}^{M^dM^d}\right] \end{bmatrix} \begin{bmatrix} \left[I_n^c\right] \\ \left[M_n\right] \\ \left[I_n^C\right] \\ \left[I_n^d\right] \\ \left[M_n^d\right] \end{bmatrix} = \begin{bmatrix} \left[V_m^c\right] \\ \left[V_m^M+V_m^W\right] \\ \left[V_m^C\right] \\ \left[V_m^d\right] \\ \left[V_m^{Hd}\right] \end{bmatrix} \tag{31}$$

where the matrix elements are defined as $Z_{mn}^{\alpha\beta} = -\left\langle \mathbf{w}_m, \hat{L}^{\alpha\beta}\mathbf{f}_n \right\rangle$, $\hat{L}^{\alpha\beta}$ is the respective boundary integral operator, superscripts $\alpha, \beta = \{J^c, M, J^C, J^d, M^d\}$; $V_m^c = \langle \mathbf{w}_m, \mathbf{E}_k^{inc} \rangle$, $V_m^M = \langle \mathbf{w}_m, \mathbf{H}_k^{inc} \rangle$, $\left[V_m^C\right]_i = \langle \mathbf{w}_m, \mathbf{E}_i^{inc} \rangle$, $\left[V_{mn}^d\right]_{ij} = \left\langle \mathbf{w}_m, \mathbf{E}_i^{inc} - \mathbf{E}_j^{inc} \right\rangle$, $\left[V_m^{Hd}\right]_{ij} = \left\langle \mathbf{w}_m, \mathbf{H}_i^{inc} - \mathbf{H}_j^{inc} \right\rangle$ are the voltage elements due to the incident wave in i-th and j-th media, and the elements Q_{mn}^W and V_m^W are the same as those expressed by (18) and (19) and determine the additional inclusions in the matrix and voltage elements due to the waveguide ports.

The MoM system (31) generalizes the solution (15) of the canonical waveguide port problem to the case of composite geometry. In the structure of the MoM matrix of this solution, blocks of waveguide excitation, complex structure, and couplings between these objects through dielectric interfaces are clearly seen.

3.4 Validation of the developed approach for composite geometry

The developed approach has been validated to simulate the scattering characteristics of a single monopole antenna, fed by waveguide excitation from a flanged coaxial line with dielectric filling. **Figure 5a** shows a schematic view of such antenna with a height h_a = 10 mm placed above a square metallic plate of 20 mm × 20 mm, which serves as a reflector. The coaxial line has an outer diameter D = 6.98 mm, an inner diameter d = 2 mm, and a length h_b = 15 mm. The line bottom end is accepted as a waveguide port, and the input impedance of the antenna at this port is simulated for various dielectric fillings of the line.

Figure 6 shows a comparison of the input impedances, calculated by the developed approach for the model of **Figure 5a** with $\varepsilon_r = 1.0001$, by the WPE approach for the conducting model of **Figure 5b**, and by discontinuous Galerkin time-domain (DGTD) method [28]. An excellent agreement between the obtained results is seen, which confirms the equivalence and correctness of both WPE approaches (for conducting and composite geometries) for very low dielectric fillings of coaxial lines.

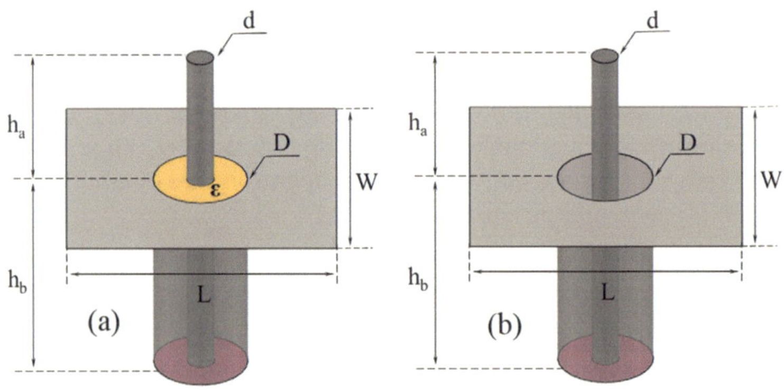

Figure 5.
Single monopole antenna fed by a flanged coaxial line: (a) with dielectric filling; (b) without dielectric filling.

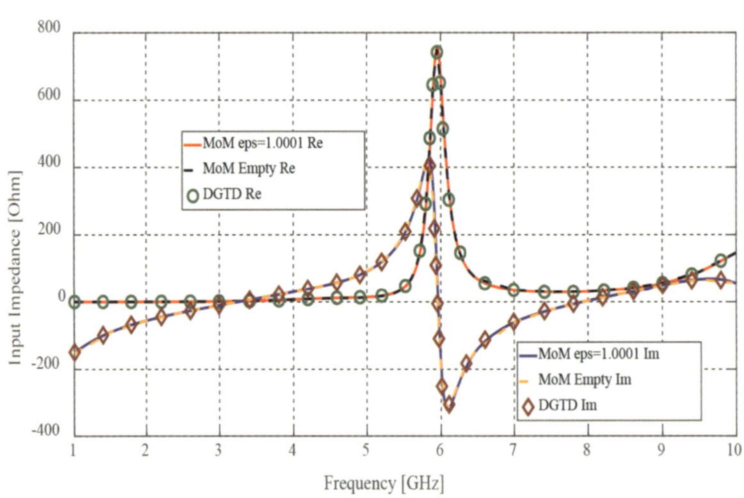

Figure 6.
Comparison of the input impedances of a monopole antenna in the port plane, calculated by the MoM for $\varepsilon_r = 1.0001$ and $\varepsilon_r = 1$ with DGTD method.

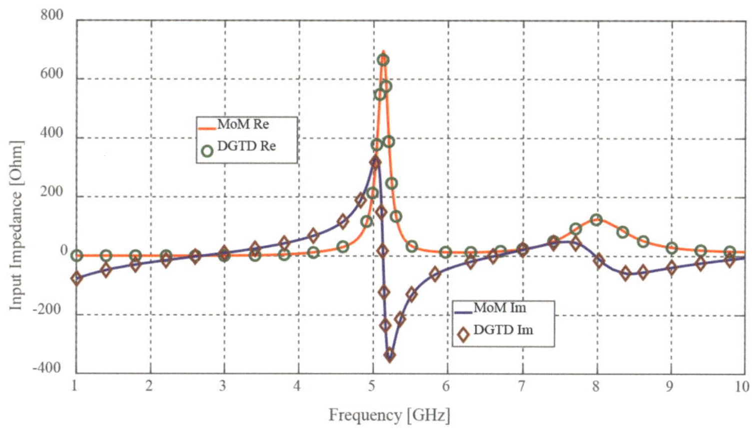

Figure 7.
Comparison of the input impedances of a monopole antenna in the port plane for the dielectric filling of a coaxial line $\varepsilon_r = 2.25$ calculated by MoM and DGTD method.

Figure 7 shows a comparison of the input impedances, calculated for the model of **Figure 5a** with $\varepsilon_r = 2.25$ using the developed approach and DGTD method. An excellent agreement between both results is seen, which validates our approach to treat arbitrary dielectric and geometric parameters of composite structures with waveguide port excitation.

Comparison of **Figures 6** and 7 shows that the use of dielectric filling of the coaxial line shifts the resonances of the input impedance to lower frequencies. In addition, this leads to a change in the line's characteristic impedance from 75 Ω in **Figure 6** to 50 Ω in **Figure** 7. Thus, the developed WPE approach for composite geometries covers a wider area of geometries and provides more control over the characteristics of the analyzed structures.

4. Waveguide port approach in coupling problems

4.1 Problem formulation

Consider the coupling problem between several composite structures fed by waveguide excitations. Although each structure can be formed from an arbitrary number of dielectric regions, for simplicity, we will consider only one-region structures with composite (dielectric and conducting) boundaries. **Figure 8** shows the geometry of the problem consisting of N waveguides W_i radiating into dielectric regions D_i, $i = 1,2, \dots N$, surrounded by closed surfaces S_{D_i} with partially conducting boundaries $S^c_{D_i}$ and inward unit normal $\mathbf{n}_{D_i}$. Waveguides W_i are filled, in general, by dielectrics with permittivities ε_i and permeabilities μ_i, and the regions D_i are filled by dielectrics with parameters ε_{D_i} and μ_{D_i}. An outer space region D_0 is a free space with material parameters ε_0, μ_0.

The waveguide ports P_i in cross-sections S^a_i divide the waveguides W_i into semi-infinite regions A_i and finite regions B_i to truncate the mesh in regions A_i with incident waveguide excitation and act as excitation sources of composite regions D_i through the dielectric boundaries $S^d_{D_iB_i}$ between the regions D_i and B_i. Each region B_i, D_i and D_0 is also excited, for generality, by the impressed EM field $\mathbf{E}^{inc}_\alpha$, $\mathbf{H}^{inc}_\alpha$, $\alpha = B_i, D_i, D_0$.

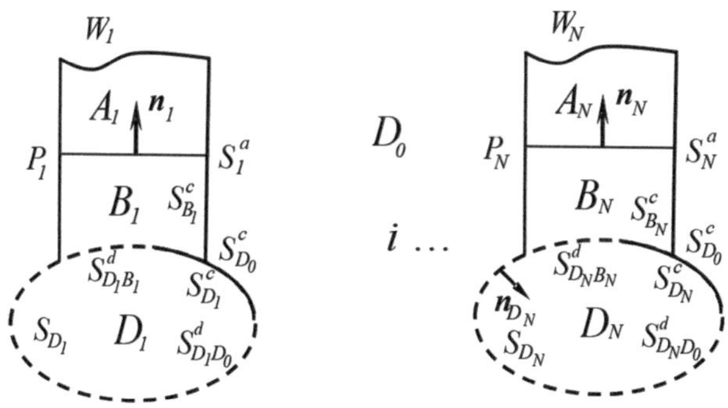

Figure 8.
Geometry of the problem.

To formulate the waveguide port excitation problems through the port surfaces S_i^a, we consider the aperture coupling problems between the regions A_i and B_i to divide an original problem into two sets of equivalence problems: internal problems for regions A_i and external problems for regions B_i, D_i and D_0. For this purpose, we cover the port surfaces S_i^a with PEC sheets and introduce equivalent magnetic currents $-\mathbf{M}_i$ and $\mathbf{M}_i$ on both sides of S_i^a to restore tangential electric fields on the port surfaces S_i^a.

4.2 Solution of the internal equivalent problem

The internal equivalent problems for the considered geometry are similar to those formulated in Section 2.2 and implemented in Section 2.3. According to the equivalence principle [24], the magnetic currents in the regions A_i are related to the total electric field $\mathbf{E}_{A_i}^{S_i^a}$ on the port surface S_i^a by the relation:

$$-\mathbf{M}_i = -\mathbf{n}_i \times \mathbf{E}_{A_i}^{S_i^a} = \mathbf{n}_{0i} \times \mathbf{E}_{A_i}^{S_i^a} \tag{32}$$

where $\mathbf{n}_i$ is an inward normal in the regionA_i, and $\mathbf{n}_{0i} = -\mathbf{n}_i$ is the propagation direction of the incident wave. Thus, the solution of the internal problem is expressed by formulas analogous to those obtained in Section 2.3 with adding the index i, when necessary.

4.3 Formulation of the external equivalent problem

When considering the external equivalent problem, let $S_{B_i}^c$ be the conducting boundary of the region B_i, including the inner sides of the waveguide walls and the conductive part of the boundary surface between the regions B_i and D_i; $S_{D_i}^c$ is the conductive part of the boundary surface S_{D_i}, and $S_{D_0}^c$is the conducting boundary of the region D_0, including the outer sides of the waveguide walls and all conducting boundaries between the regions D_0 and D_i. Further, $S_{D_iB_i}^d$ is the dielectric boundary between the regions D_i and B_i, and $S_{D_iD_0}^d$ is the dielectric boundary between the regions D_i and D_0. Per the equivalence principle [24], the dielectric boundaries between different regions can be replaced by oppositely directed equivalent electric and magnetic currents flowing on both sides of the dielectric interfaces.

The EM field in the waveguide region B_i is created by electric currents $\mathbf{J}_{B_i}^c$ flowing along the port surface S_i^a and conducting surface $S_{B_i}^c$, equivalent electric and magnetic currents $-\mathbf{J}^d_{D_iB_i}$ and $-\mathbf{M}^d_{D_iB_i}$ flowing along the dielectric interfaces $S^d_{D_iB_i}$, and equivalent magnetic currents $\mathbf{M}_i$ flowing along the port surface S^a_i. The EM field in the region D_i is created by electric currents $\mathbf{J}^c_{D_i}$ flowing along the conducting surfaces $S^c_{D_i}$, equivalent currents $\mathbf{J}^d_{D_iB_i}$ and $\mathbf{M}^d_{D_iB_i}$ flowing on dielectric boundaries $S^d_{D_iB_i}$ between the regions D_i and B_i, and equivalent currents $\mathbf{J}^d_{D_iD_0}$ and $\mathbf{M}^d_{D_iD_0}$ flowing on dielectric boundaries $S^d_{D_iD_0}$ between the regions D_i and D_0. The field in the free space region D_0 is created by electric currents $S^c_{D_0}$ flowing along the total conducting boundary of the region D_0, and equivalent currents $-\mathbf{J}^d_{D_iD_0}$ and $-\mathbf{M}^d_{D_iD_0}$ at dielectric boundaries between the regions D_i and D_0.

The unknown currents $\mathbf{J}^c_{B_i}, \mathbf{M}_i, \mathbf{J}^c_{D_i}, \mathbf{J}^d_{D_iB_i}, \mathbf{M}^d_{D_iB_i}, \mathbf{J}^c_{D_0}, \mathbf{J}^d_{D_iD_0}, \mathbf{M}^d_{D_iD_0}$ can be found from the boundary conditions on the port surface and the conducting boundaries of the waveguide region B_i:

$$\left[\mathbf{E}^{inc}_{B_i} + \mathbf{E}^{sc}_{B_i}\left(\mathbf{J}^c_{B_i}, \mathbf{M}_i, -\mathbf{J}^d_{D_iB_i}, -\mathbf{M}^d_{D_iB_i}\right)\right]\Big|^{S^a_i+S^c_i}_{\tan} = 0 \quad \text{just outside } S^a_i \tag{33}$$

$$\left[\mathbf{H}^{inc}_{B_i} + \mathbf{H}^{sc}_{B_i}\left(\mathbf{J}^c_{B_i}, \mathbf{M}_i, -\mathbf{J}^d_{D_iB_i}, -\mathbf{M}^d_{D_iB_i}\right)\right]\Big|^{S^a_i}_{\tan} = \mathbf{H}^{S^a_i}_{A_i}\Big|_{\tan} \quad \text{just inside } S^a_i, \tag{34}$$

and the boundary conditions on the conducting and dielectric boundaries of the regions D_i and D_0:

$$\left[\mathbf{E}^{inc}_{D_i} + \mathbf{E}^{sc}_{D_i}\left(\mathbf{J}^c_{D_i}, \mathbf{J}^d_{D_iB_i}, \mathbf{M}^d_{D_iB_i}, \mathbf{J}^d_{D_iD_0}, \mathbf{M}^d_{D_iD_0}\right)\right]\Big|_{\tan} = 0 \quad \text{on} \quad S^c_{D_i} \tag{35}$$

$$\begin{aligned}&\left[\mathbf{E}^{inc}_{B_i} + \mathbf{E}^{sc}_{B_i}\left(\mathbf{J}^c_{B_i}, \mathbf{M}_i, -\mathbf{J}^d_{D_iB_i}, -\mathbf{M}^d_{D_iB_i}\right)\right]\Big|_{\tan} = \\ &\left[\mathbf{E}^{inc}_{D_i} + \mathbf{E}^{sc}_{D_i}\left(\mathbf{J}^c_{D_i}, \mathbf{J}^d_{D_iB_i}, \mathbf{M}^d_{D_iB_i}, \mathbf{J}^d_{D_iD_0}, \mathbf{M}^d_{D_iD_0}\right)\right]\Big|_{\tan}\end{aligned} \quad \text{on} \quad S^d_{D_iB_i} \tag{36}$$

$$\begin{aligned}&\left[\mathbf{H}^{inc}_{B_i} + \mathbf{H}^{sc}_{B_i}\left(\mathbf{J}^c_{D_i}, \mathbf{M}^{(i)}, -\mathbf{J}^d_{D_iB_i}, -\mathbf{M}^d_{D_iB_i}\right)\right]\Big|_{\tan} = \\ &\left[\mathbf{H}^{inc}_{D_i} + \mathbf{H}^{sc}_{D_i}\left(\mathbf{J}^c_{D_i}, \mathbf{J}^d_{D_iB_i}, \mathbf{M}^d_{D_iB_i}, \mathbf{J}^d_{D_iD_0}, \mathbf{M}^d_{D_iD_0}\right)\right]\Big|_{\tan}\end{aligned} \quad \text{on} \quad S^d_{D_iB_i} \tag{37}$$

$$\left[\mathbf{E}^{inc}_{D_0} + \sum_{i=1}^{N} \mathbf{E}^{sc}_{D_0}\left(\mathbf{J}^c_{D_0}, -\mathbf{J}^d_{D_iD_0}, -\mathbf{M}^d_{D_iD_0}\right)\right]\Bigg|_{\tan} = 0 \quad \text{on} \quad S^c_{D_0} \tag{38}$$

$$\begin{aligned}&\left[\mathbf{E}^{inc}_{D_i} + \mathbf{E}^{sc}_{D_i}\left(\mathbf{J}^c_{D_i}, \mathbf{J}^d_{D_iB_i}, \mathbf{M}^d_{D_iB_i}, \mathbf{J}^d_{D_iD_0}, \mathbf{M}^d_{D_iD_0}\right)\right]\Big|_{\tan} = \\ &\left[\mathbf{E}^{inc}_{D_0} + \mathbf{E}^{sc}_{D_0}\left(\mathbf{J}^c_{D_0}\right) + \sum_{i=1}^{N} \mathbf{E}^{sc}_{D_0}\left(-\mathbf{J}^d_{D_iD_0}, -\mathbf{M}^d_{D_iD_0}\right)\right]\Bigg|_{\tan}\end{aligned} \quad \text{on} \quad S^d_{D_iD_0} \tag{39}$$

$$\begin{aligned}&\left[\mathbf{H}^{inc}_{D_i} + \mathbf{H}^{sc}_{D_i}\left(\mathbf{J}^c_{D_i}, \mathbf{J}^d_{D_iB_i}, \mathbf{M}^d_{D_iB_i}, \mathbf{J}^d_{D_iD_0}, \mathbf{M}^d_{D_iD_0}\right)\right]\Big|_{\tan} = \\ &\left[\mathbf{H}^{inc}_{D_0} + \mathbf{H}^{sc}_{D_0}\left(\mathbf{J}^c_{D_0}\right) + \sum_{i=1}^{N} \mathbf{H}^{sc}_{D_0}\left(-\mathbf{J}^d_{D_iD_0}, -\mathbf{M}^d_{D_iD_0}\right)\right]\Bigg|_{\tan}\end{aligned} \quad \text{on} \quad S^d_{D_iD_0} \tag{40}$$

The scattered EM fields in (33)–(40) are related to the equivalent electric and magnetic currents by Eqs. (26) and (27). After substituting (26) and (27) into (35)–(40), Eqs. (35)–(40) represent a coupled system of integral equations in terms of unknown currents for solving the coupling problem between several composite structures.

4.4 MoM solution of the external equivalent problem

To solve the boundary problem (35)–(40), we use the following MoM expansions for the unknown currents:

$$\left[\mathbf{J}^c_{B_i}\right]^N_{i=1}=\sum_{n=1}^{N^a+N^c_B} I^{c_B}_n\mathbf{f}_n,\ [\mathbf{M}_i]^N_{i=1}=\sum_{n=1}^{N^a} M_n\mathbf{f}_n,\ \left[\mathbf{J}^c_{D_i}\right]^N_{i=1}=\sum_{n=1}^{N^c_D} I^{c_D}_n\mathbf{f}_n, \mathbf{J}^c_{D_0}=\sum_{n=1}^{N^c_{D_0}} I^{c_{D_0}}_n\mathbf{f}_n, \quad (41)$$

$$\left[\mathbf{J}^d_{D_iB_i},\mathbf{J}^d_{D_iD_0}\right]^N_{i=1}=\sum_{n=1}^{N^d} I^d_n\mathbf{f}_n,\ \left[\mathbf{M}^d_{D_iB_i},\mathbf{M}^d_{D_iD_0}\right]^N_{i=1}=\sum_{n=1}^{N^d} M^d_n\mathbf{f}_n,\ [\mathbf{M}_i]^N_{i=1}=\sum_{n=1}^{N^a} M_n\mathbf{f}_n \quad (42)$$

where $\mathbf{f}_n$ are the suitable BFs, $I^{c_B}_n, I^{c_D}_n, I^{c_{D_0}}_n, I^d_n, M^d_n$ and M_n are the unknown expansion current coefficients, and $N^a, N^c_B, N^c_D, N^c_{D_0}$ and N^d are the numbers of these BFs on the surfaces $\left[S^a_i\right]^N_{i=1}$, $\left[S^c_{B_i}\right]^N_{i=1}$, $\left[S^c_{D_i}\right]^N_{i=1}$, $S^c_{D_0}$ and $\left[S^d_{D_iB_i}, S^d_{D_iD_0}\right]^N_{i=1}$, respectively. Substituting now (41) and (42) into (35)–(40) with an accounting of (3), (5), and (26) and (27) for each i-th region and testing the obtained equations with weighting functions $\mathbf{w}_1(\mathbf{r})$, $\mathbf{w}_2(\mathbf{r}), \dots, \mathbf{w}_m(\mathbf{r})$, defined in the range of the respective boundary operators, we obtain the following MoM system of linear algebraic equations:

$$\begin{bmatrix} Z^{J^c_BJ^c_B} & Z^{J^c_BM} & 0 & 0 & Z^{J^c_BJ^d} & Z^{J^c_BM^d} \\ Z^{MJ^c_B} & Z^{MM}+Q^W & 0 & 0 & Z^{MJ^d} & Z^{MM^d} \\ 0 & 0 & Z^{J^c_DJ^c_D} & 0 & Z^{J^c_DJ^d} & Z^{J^c_DM^d} \\ 0 & 0 & 0 & Z^{J^c_{D_0}J^c_{D_0}} & Z^{J^c_{D_0}J^d} & Z^{J^c_{D_0}M^d} \\ Z^{J^dJ^c_B} & Z^{J^dJ^c_D} & Z^{J^dJ^c_D} & Z^{J^dJ^c_{D_0}} & Z^{Z^{J^d}J^d} & Z^{J^dM^d} \\ Z^{M^dJ^c_B} & Z^{M^dM} & Z^{M^dJ^c_D} & Z^{M^dJ^c_{D_0}} & Z^{M^dJ^d} & Z^{M^dM^d} \end{bmatrix} \begin{bmatrix} I^{c_B} \\ M \\ I^{c_D} \\ I^{c_{D_0}} \\ I^d \\ M^d \end{bmatrix} = \begin{bmatrix} V^{c_B} \\ V^M+V^W \\ V^{c_D} \\ V^{c_{D_0}} \\ V^d \\ V^{Hd} \end{bmatrix} \quad (43)$$

where the elements of the block matrices are defined as: $Z^{\alpha\beta}_{mn}=-\left\langle \mathbf{w}_m, \hat{L}^{\alpha\beta}_i\mathbf{f}_n\right\rangle$, $\hat{L}^{\alpha\beta}$ is the respective boundary integral operator, superscripts $\alpha,\beta=\left\{J^c_B, M, J^c_D, J^c_{D_0}, J^d, M^d\right\}$, voltage elements are defined in the same way as in Eq. (31), and the elements of the block matrices Q^W and V^W are expressed by (18) and (19) for each i-th feeding waveguide and determine the additional inclusions in the matrix and voltage elements due to the waveguide ports. The MoM system (43) defines a solution to the coupling problem between several composite geometries. In the structure of the MoM matrix of this solution, blocks of waveguide excitations, complex geometries, and couplings between them are clearly seen.

4.5 Validation of the developed approach for coupling problems

The developed approach has been validated on a two-element antenna array fed by coaxial waveguide ports by comparing the simulation results obtained using the developed MoM approach and the DGTD method [28]. **Figure 9** shows a schematic view of two identical monopole antennas flanged over the PEC plate and fed by coaxial waveguides with generally different diameters and dielectric fillings. The monopoles located at a distance L_a = 40 mm from each other have the same height h_a = 10 mm above the PEC plate with a width W = 40 mm and a length L = 80 mm, which serves as a reflector. Coaxial waveguides have the same inner diameter $d_1 = d_2$ = 2 mm, but generally different outer diameters D_1 and D_2 and relative permittivities ε_1 and ε_2. The depth of each coaxial waveguide under the flange is h_b = 15 mm, and its end is taken as the reference plane of the waveguide port.

Figure 10 shows the real and imaginary parts of the transmission coefficient $S_{21} = a_{02}^{-}/a_{01}^{+}$ between waveguide ports 1 and 2 with the same radii and dielectric fillings: $D_1/2 = D_2/2$ = 6.65 mm, and $\varepsilon_{r1} = \varepsilon_{r2}$ = 5.17, which leads to the same characteristic impedances: $Z_{c1} = Z_{c2}$ = 50 Ω. The developed MoM approach and the DGTD method are compared. The first antenna in these simulations is considered active, and the second is passive. Comparison of these results shows very good agreement between them over a wide frequency range from 1 GHz up to 10 GHz. This validates the developed approach in modeling coupling problems for coaxial waveguide ports with the same characteristic impedance.

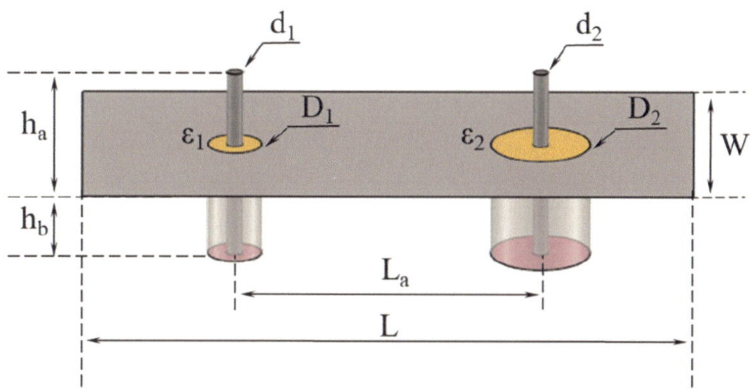

Figure 9.
Schematic view of an array of two identical monopole antennas fed by coaxial waveguides and flanged above the PEC plate.

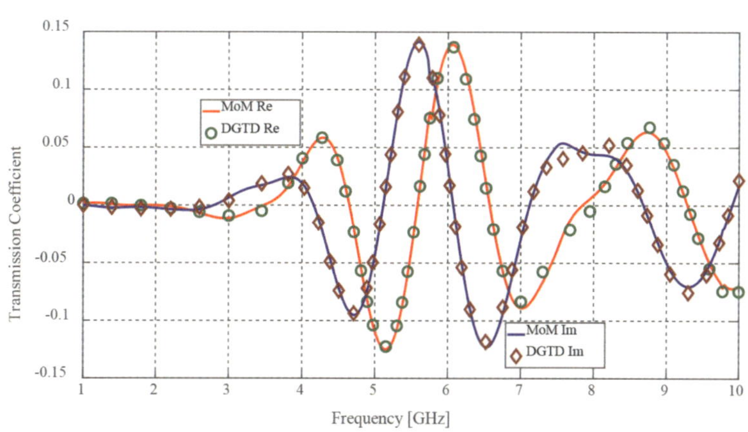

Figure 10.
Transmission coefficient between the of the antenna array waveguide ports with the same parameters of the feeding coaxial waveguides.

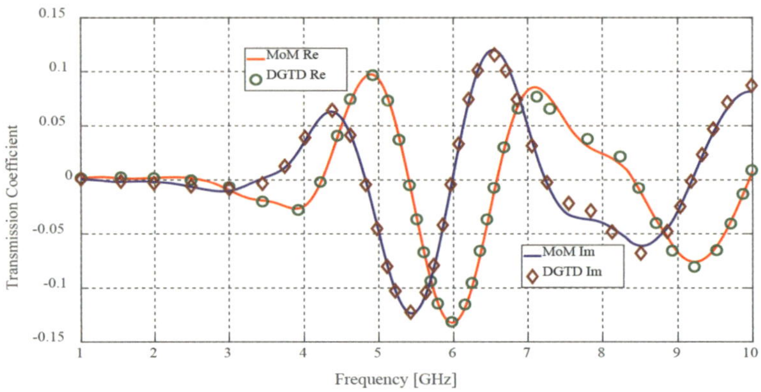

Figure 11.
Transmission coefficient between the waveguide ports of the antenna array for the same permittivities $\varepsilon_{r1} = \varepsilon_{r2} = 2.25$, *but different outer radii:* $D_1/2 = 3.49$ *mm,* $D_2/2 = 6.52$ *mm.*

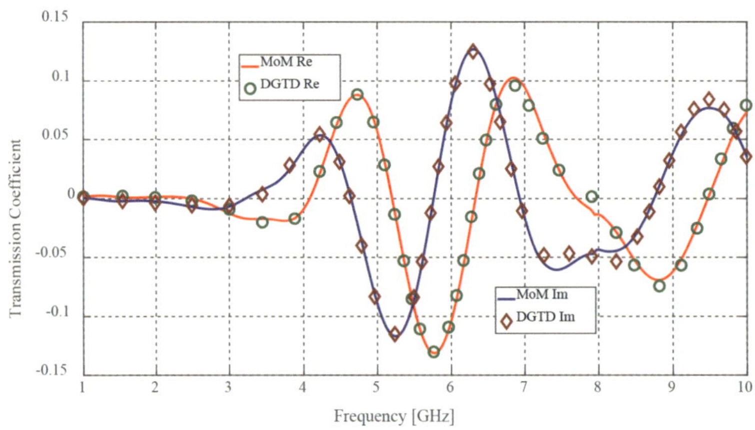

Figure 12.
Transmission coefficient between the waveguide ports of the antenna array for the same outer radii $D_1/2 = D_2/2 = 5.3$ *mm, but different permittivities:* $\varepsilon_{r1} = 4$ *and* $\varepsilon_{r2} = 1.78$.

Figures 11 and **12** show a comparison of the transmission coefficient $S_{21} = a_{02}^{-}/a_{01}^{+}\sqrt{Z_{c1}/Z_{c2}}$ between waveguide ports 1 and 2, calculated by the MoM and DGTD method for different parameters of coaxial waveguides. **Figure 11** is made for the same fillings of waveguides: $\varepsilon_{r1} = \varepsilon_{r2} = 2.25$, but with different outer radii: $D_1/2 = 3.49$ mm and $D_2/2 = 6.52$ mm, while **Figure 12** is performed for different fillings: $\varepsilon_{r1} = 4$ and $\varepsilon_{r2} = 1.78$, but with the same outer radii $D_1/2 = D_2/2 = 5.3$ mm. Both cases result in characteristic impedances of waveguides $Z_{c1} = 50\ \Omega$ and $Z_{c2} = 75\ \Omega$. Comparison of the MoM and DGTD results again shows very good agreement between both simulated results, which validates the developed approach to modeling coupling problems for coaxial waveguide ports with different characteristic impedances.

5. Application of waveguide port approach

The obtained approach has been applied to practical EMC problems for microwave antennas fed by coaxial waveguides. Such waveguides are the most commonly used to excite microwave antennas and electronic devices. This excitation usually uses microwave coaxial connectors, such as BNC and SMA.

5.1 Modeling of two branches feeding large printed UWB antenna

First, based on the measurement data [28], a printed ultra-wideband (UWB) antenna is considered. **Figure 13** shows a schematic view of a large printed UWB antenna with a two-branch-feed, the bottom of which is connected to the core of a 50 Ω SMA connector with waveguide excitation, the covering of which is connected to a metal plate serving as a reflector. The bottom end of the connector is accepted as a waveguide port, and the input impedance of the UWB antenna at the waveguide port is measured and simulated.

The model of a printed UWB antenna is a square metal patch with a length L_a = 40 mm and a width W_a = 40 mm, printed on a dielectric substrate with a length L_b = 43 mm, a width W_b = 47.5 mm, a thickness t = 1.5 mm and material parameters ε_{rd} = 4.4 and $\tan\delta_d$ = 0.02. The antenna is connected to a two-branch-feeding strip with a total width t_1 = 15 mm, a distance between the branches t_2 = 11 mm, and a height of the branches h_1 = 3.5 mm. The UWB antenna is placed at a height h_2 = 3 mm above a metallic plate of a length L = 275 mm and a width W = 207 mm and is connected to the SMA connector. The model of the SMA connector is represented by a coaxial waveguide with an outer radius $D/2$ = 2.125 mm, inner radius $d/2$ = 0.635 mm, and a length L_{con} = 6.8 mm, filled with a polyethylene dielectric with relative permittivity ε_r = 2.24 and loss tangent tan δ = 0.005.

Figure 14 shows a comparison of simulated input impedances of a printed UWB antenna at the waveguide port with measurement results [28]. Comparison of the simulated results with measurement data shows a good agreement between them in a wide frequency range from 1 to 10 GHz. This validates the developed approach to modeling the composite antenna geometries fed by coaxial waveguides with dielectric filling.

5.2 Coupling problem between GPS and SDARS antennas

In conclusion, based on the measurement data [28], the coupling between the GPS and SDARS patch antennas was analyzed in the frequency range from 1 GHz to 3 GHz. **Figure 15** shows the measurement setup (a) and its schematic view (b) for studying

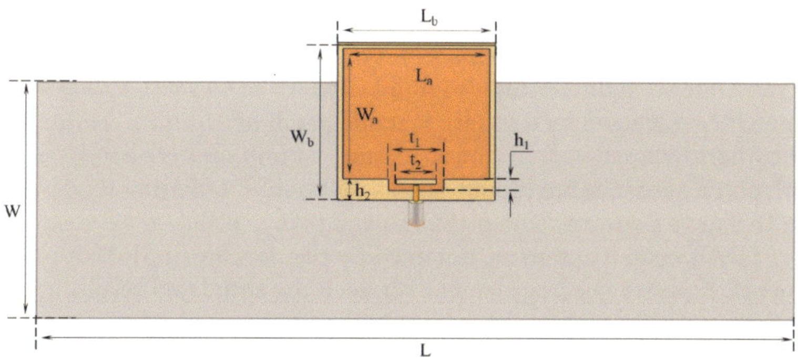

Figure 13.
Schematic view of a large printed UWB antenna with two branch feed connected to a coaxial waveguide port.

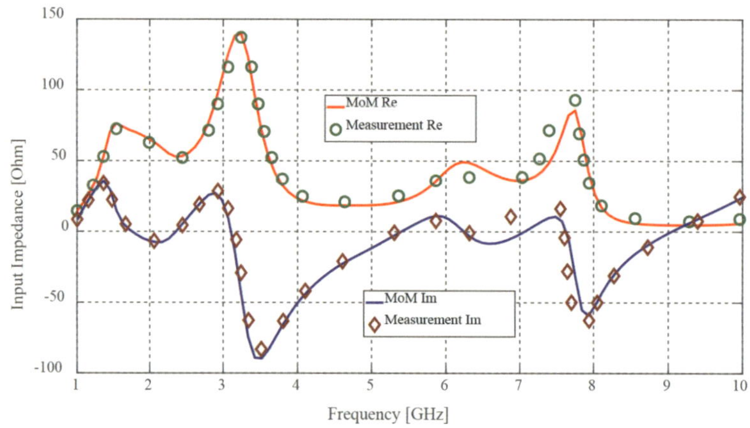

Figure 14.
Comparison of the simulated and measured input impedances of a printed UWB antenna.

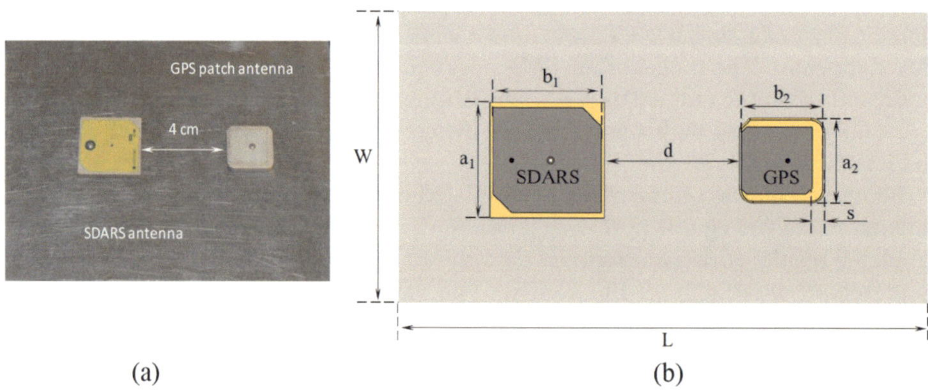

Figure 15.
Measurement setup (a) and its schematic view (b) for the coupled GPS and SDARS patch antennas.

the coupling between active GPS and passive SDARS antennas, separated by a distance of d = 4 cm. Both antennas are fed by 50 Ohm coaxial lines with standard SMA connectors with parameters described in Section 5.1.

The parameters of the setup are the following. The SDARS antenna is a square metallic patch of 32 mm × 32 mm size with two opposite cut corners, printed on a dielectric substrate with dimensions 34 mm × 34 mm × 3.25 mm and ε_r = 4.1. The GPS antenna is constructed by a square metallic patch of 21 mm × 21 mm size with truncated corners, printed on a 25 mm × 25 mm × 4 mm dielectric substrate with ε_{r1} = 20.34. Both patch antennas are mounted on a 190 mm × 145 mm metal plate.

Figure 16 shows a comparison of the transmission coefficient between active GPS and passive SDARS patch antennas, obtained by the developed MoM approach and measurements. A pretty good agreement between the simulated results and measured data in the frequency range of 1–3 GHz is observed. This comparison validates the developed waveguide port approach with measurements to model coupling problems between different composite geometry antennas with coaxial waveguide ports.

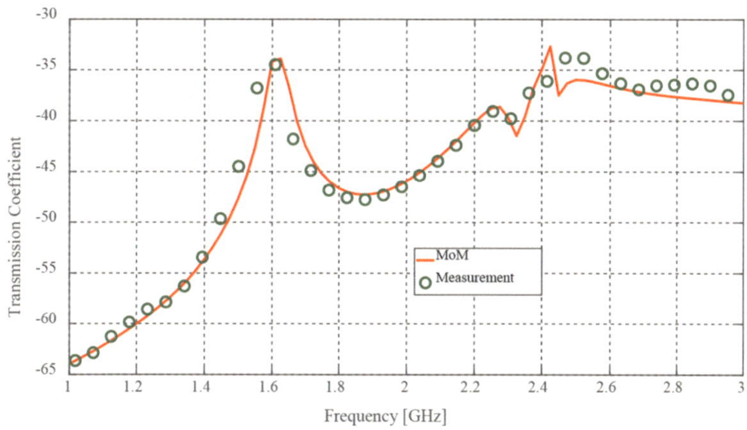

Figure 16.
Transmission coefficient between GPS and SDARS patch antennas.

6. Conclusion

The MoM-based waveguide port approach was developed to model waveguide port excitation problems on arbitrary conducting and composite geometries. The developed approach was validated for modeling radiation and coupling problems for coaxial ports by comparing the simulated results with those obtained by other approaches and measurements. The approach has been applied to practical EMC problems for microwave antennas fed by coaxial connectors. A good agreement between the simulated and measured results has been demonstrated. The efficiency of the developed approach for solving various complex problems with waveguide excitation has been verified.

Author details

Faik Bogdanov*, Irina Chochia, Lily Svanidze and Roman Jobava
EMCoS LLC, Tbilisi, Georgia

*Address all correspondence to: faik.bogdanov@emcos.com

References

[1] Lou Z, Jin JM. An accurate waveguide port boundary condition for the time-domain finite-element method. IEEE Transactions on Microwave Theory and Techniques. 2005;**53**(9):3014-3023. DOI: 10.1109/APS.2005.1551498

[2] Jin JM, Riley DJ. Finite Element Analysis of Antennas and Arrays. Hoboken: Wiley; 2008. 467p

[3] Alimenti F, Mezzanotte P, Roselli L, Sorrentino R. Modal absorption in the FDTD method: A critical review. International Journal of Numerical Model. 1997;**10**:245-264

[4] Alimenti F, Mezzanotte P, Roselli L, Sorrentino R. A revised formulation of modal absorbing and matched modal source boundary conditions for the efficient FDTD analysis of waveguide structures. IEEE Transactions on Microwave Theory and Techniques. 2000;**48**(1):50-59. DOI: 10.1109/22.817471

[5] Flisgen T, Heller J, Rienen U. Time-domain absorbing boundary terminations for waveguide ports based on state-space models. IEEE Transactions on Magnetics. 2014;**50**(2): 145-148. DOI: 10.1109/TMAG.2013.2283065

[6] Li P, Jiang LJ, Bağci H. Transient analysis of dispersive power-ground plate pairs with arbitrarily shaped antipads by the DGTD method with wave port excitation. IEEE Transactions on Electromagnetic Compatibility. 2017; **59**(1):172-183

[7] Štumpf M, Šedĕnka V, Kadlec P. On modeling of excitation ports in the time-domain contour-integral method. In: Proceedings of the International Conference on Electromagnetics in Advanced Applications (ICEAA'17); 11–15 September 2017; Verona, Italy. IEEE; 2017. pp. 1292-1294. DOI: 10.1109/ICEAA.2017.8065509

[8] Harrington RF. Field Computation by Moment Methods. New York: Macmillan; 1968

[9] Harrington RF, Mautz JR. A generalized network formulation for aperture problems. IEEE Transactions on Antennas and Propagation. 1976;**24**(6): 870-873. DOI: 10.1109/TAP.1976.1141420

[10] Wang T, Harrington R, Mautz JR. Electromagnetic scattering from and transmission through arbitrary apertures in conducting bodies. IEEE Transactions on Antennas and Propagation. 1990; **38**(11):1805-1814. DOI: 10.1109/8.102743

[11] Bunger R. Moment-method analysis of arbitrary 3-D metallic N-port waveguide structures. IEEE Transactions on Microwave Theory and Techniques. 2000;**48**(4):531-537

[12] Wen D, Chen H, Zhang J. Radiation Characteristic Analysis of Ridged Horn by Aperture Port Excitation. Xi'an, China: IEEE; 2012. pp. 759-761

[13] Bogdanov F, Chochia I, Svanidze L, Jobava R. Incorporation of MoM-based waveguide port model into the mixed conducting and dielectric geometry. In: Proceedings of the International Symposium on Electromagnetic Compatibility (EMC Europe'17); 4–8 September 2017; Angers, France. IEEE; 2017. pp. 1-6. DOI:10.1109/EMCEurope.2017.8094718

[14] Bogdanov F, Svanidze L, Gheonjian A, Eremyan D, Kutchadze Z, Jobava R.

Application of MoM-based waveguide port approach to the analysis of EM coupling problems related to microwave antennas. In: Proceedings of the European Conference on Antennas and Propagation (EUCAP'18); 9–13 April 2018; London, England. IEEE; 2018. pp. 1-5. DOI:10.1049/cp.2018.1194

[15] Bogdanov F, Chochia I, Svanidze L, Jobava R. Validation of MoM-based solution of waveguide port problem for composite structures applied to microwave antenna and PCB geometries. In: Proceedings of the European Microwave Conference (EuMC'19); 1–3 October 2019; Paris, France. IEEE; 2019. pp. 642-645. DOI:10.23919/EuMC.2019.8910683

[16] Arvas E, Rahhal-Arabi A, Sadigh A, Rao SM. Scattering from multiple conducting and dielectric bodies of arbitrary shape. IEEE Antennas and Propagation Magazine. 1991;**33**(2):29-36. DOI: 10.1109/74.88184

[17] Goggans PM, Kishk AA, Glisson AW. Electromagnetic scattering from objects composed of multiple homogeneous regions using a region-by-region solution. IEEE Transactions on Antennas and Propagation. 1994;**42**(6):865-871. DOI: 10.1109/8.301713

[18] Shin J, Glisson AW, Kishk AA. Analysis of combined conducting and dielectric structures of arbitrary shapes using an E-PMCHW integral equation formulation. In: Proceedings of IEEE-AP-S Symposium (AP-S'00). Vol. 4. 2000. pp. 2282-2285

[19] Kolundzija BM. Electromagnetic modeling of composite metallic and dielectric structures. IEEE Transactions on Microwave Theory and Techniques. 1999;**47**(7):1021-1032

[20] Yiëa-Oijala P, Taskinen M, Sarvas J. Surface integral equation method for general composite metallic and dielectric structures with junctions. Progress In Electromagnetics Research. 2005;**52**: 81-108. DOI: 10.2528/PIER04071301

[21] Carr M, Topsakal E, Volakis JL. A procedure for modeling material junctions in 3-D surface integral equation approaches. IEEE Transactions on Antennas and Propagation. 2004; **52**(5):1374-1379. DOI: 10.1109/TAP.2004.827247

[22] Bogdanov F, Svanidze L, Jobava R. MoM solution to scattering problem on multi-region composite structures with various type material junctions. In: Proceedings of International Seminar/Workshop (DIPED'18); 24–27 September 2018; Tbilisi. pp. 13-18. DOI: 10.1109/DIPED.2018.8543281

[23] EMCoS. EMCoS Studio 2021 [Internet]. Available from: http://www.emcos.com

[24] Harrington RF. Time-Harmonic Electromagnetic Fields. New York: McGraw-Hill; 1961

[25] Cho YH. Analytic and numerically efficient scattering equations for an infinitely flanged coaxial line. PIERS Letters. 2012;**28**:149-158. DOI: 10.2528/PIERL11110901

[26] Tan W, Shen Z. Efficient analysis of open-ended coaxial line using Sommerfeld identity and matrix pencil method. IEEE Microwave and Wireless Components Letters. 2008;**18**(1):7-9. DOI: 10.1109/LMWC.2007.911971

[27] Mosig JR, Besson JE, Gex-Fabry M, Gardiol FE. Reflection of an open-ended coaxial line and application to non-destructive measurement of materials. IEEE Transactions on Instrumentation

and Measurement. 1981;**30**(1):46-51.
DOI: 10.1109/TIM.1981.6312437

[28] Eremyan D, Kutchadze Z, Badzagua I, Gheonjian A, Jobava R. Validation of discontinuous Galerkin method. EMCoS Internal Report. Tbilisi, Georgia; 2013

Chapter 5

One Model of Microwave Heating of Water Drop

Serge Lefeuvre and Olga Gomonova

Abstract

This work deals with the modeling of microwave heating of a water drop. A drop model is reduced to its electric dipoles, masses, and charges are constructed using the associating of COMSOL Multiphysics and Matlab software. The considered model proposes a microscopic point of view on microwave heating, which transforms electrical energy into heat.

Keywords: microwave heating, water drop, electric dipoles, modeling

1. Introduction

Domestic microwave ovens first appeared in American homes at the end of World War II; the American company Raytheon, which manufactured the radars for the Liberty Ships, was redirecting its production towards civilian applications. This is the case in this chapter which deals with microwave heating. The present chapter is interested in a simple water drop reduced to its electric dipoles, masses, and charges. The drop contains 1536 of them in a 3.585 nm edge cube. These dipoles are in permanent interaction and move according to a brownian movement or similar. For these dipoles, receiving an electromagnetic wave means receiving an external force additional to those they receive from the other dipoles of the drop. Their movement generates friction and thus transforms part of the captured energy into heat. We must also keep in mind the adage: "a cold-cooked carrot is no longer raw". Thus, heat produces irreversible changes. However, the cloud of dipoles similar to those of water is not a drop of water. It makes it easy to calculate its collective behavior in a microwave field, displacement, and speed of particles but does not put in thermal memory the sequence of events. Only a human program can transform the kinetic energy acquired by dipoles into thermal energy captured by water and invent a temperature scale that accumulates the sequence of events.

The expression "microwave heating" means that the material itself transforms into heat according to the equality $W = J.Q$, the electromagnetic energy it captures (1 calorie = 4.185 Joule, where W is mechanical energy, J is a universal constant, Q is a heat).

Internal movement of two electrical charges of the dipole also results in a current different from that of the free carriers in a metal or ions in the water itself, i. e., $J = \sigma E$ (where σ is electric conductivity and E is an electric field) is displacement current,

$\partial \boldsymbol{D}/\partial t$ introduced by Maxwell completes Ampère's equation rot $\boldsymbol{H} = \boldsymbol{J}$, which becomes rot $\boldsymbol{H} = \sigma \boldsymbol{E} + \partial \boldsymbol{D}/\partial t$. This current was introduced to satisfy its continuity between capacitor terminals, even in a vacuum. The last equation is used interchangeably in metals and dielectrics to translate any heat production type. Therefore, the expression "microwave heating" is completely different from wood, gas, or coal heating, which burns and radiates the energy produced by combustion or diffuses it by conduction on the material surface to be heated. There are significant differences between the two types of heating:

- during microwave heating, the thermal gradient is oriented from the inside to the outside; the material is not heated by its environment but heats it;

- diversity of behavior of the volume is much greater than that of the surface. Moreover, the electrical and magnetic heterogeneity of the materials contributes to the diversity.

2. Electromagnetic waves in everyday life

2.1 Elementary experiments with induction stove and microwave oven

The active part of an induction hob is a spiral supplied with around 25 kHz and located just below the hob. The energy is not radiated, but it decreases exponentially as it moves away from the spiral. It generates a Foucault current in the surrounding conductors, which tends to move away from the inductor, i.e., from the plate.

To be convinced, one can place a sheet of household aluminum on the plate in operation. If the electrical force is greater than the weight of this aluminum sheet, the sheet will fly away and leaves the field. However, if the experimenter puts the finger on the sheet to hold it, he notices heat release. Heat generation is nature's answer of last resort. It is interesting to tear the sheet and bring the two pieces together: an electric arc appears to ensure the current continuity. This arc sometimes denominates the extra-current of rupture.

To test a microwave oven, the experimenter can heat a soup plate slightly moistened and covered with a well-joined kitchen film. He will quickly see the film inflate like a balloon. Water evaporates and naturally finds with steam a mechanical application. Obviously, one should not use valuable plates or plates decorated with conductive paint.

The experimenter could dry a piece of bread put on the same plate but pre-cooled in a refrigerator and always covered with a kitchen film. He will extract drops of liquid water and then analyze them because they include a lot of salts and other components coming from the bread itself. It is indeed very useful to analyze the evaporated drop to highlight a possible peculiarity of microwave heating and try to find out if it is due to internal friction.

It is also interesting to test the widely marketed metal trays. If the tray is placed on the metal grid of the oven, an arc will appear with the vibrations during operation. The arc can also be maintained by directly capturing microwave energy. One must follow the experiment by standing a few meters from the oven because the door grid is not completely impermeable. However, it is made to see since there is a lamp in the oven!

It is instructive to experiment with metal trays filled with common foods to test the quality of heating. The considered domestic experiments also give an order of magnitude of the masses and volumes, power, and time encountered in microwave heating, for example, one liter of water, one microwave kW, and one minute of heating. These proportions are used in the simulation program presented below. Mass and volume are related to the number of dipoles chosen since 18 cm^3 of water contains N, Avogadro's number of molecules, and weighs 18 g. Time is measured by the period of the incident microwave field since period T is the inverse of the chosen frequency.

2.2 Magnetron and power supply

The magnetron is an electromagnetic wave generator, very similar, in its operating principle, to a 50 Hz alternator but which emits at very high frequencies, for example, 2.45 GHz. The magnetron constitutes a cylindrical vacuum diode formed of an electron-emitting cathode and a stator receiving anode. A continuous magnetic field is applied in the axis to spin the electrons. The rotor appears because the anode generates an alternating circular electric field, sometimes in one direction, sometimes in the other to the group–ungroup the electrons in the beam and thus find the characteristics of a real alternator winding. The electrons finish their route and reach the anode by producing heat which must be eliminated by ventilation or circulation of fluid. The electron beam, in a vacuum, can rotate very quickly (much faster than a rotor) and thus generate a high-frequency current. The anode also has an antenna that collects and emits produced electromagnetic energy.

One can find magnetrons in industrial microwave heating and radars, on roads, and in kitchens. The kitchen magnetron has a special power supply: the first 50 Hz alternation (60 in some countries) charges a capacitor, and this charge will be added to the second. Therefore, the magnetron transmits half-time and even less if the user wishes to reduce the power of his oven. Thus, the heated food makes the average value and leaves a little of the thermal energy of the hot regions to diffuse towards the cold regions. In the elementary experiments proposed above, the observer should detect that the balloon above the plate inflates and deflates in time with the power supply.

3. Construction of the model and numerical results

The proposed model studies a water drop, of a cubic form for simplification, in an electromagnetic field. The water dipoles are replaced by point dipoles carrying the masses and electrical charges of hydrogen H and oxygen O and mobile in the aqua (aqueous liquid), possessing all the other mechanical and thermal properties of water and the vacuum permittivity. Aqua and dipoles have the same density as water.

The model was constructed using the associating of COMSOL and Matlab software [1, 2]. The considered model is a cube with an edge length of $aa = 3.585e - 9$ m containing 1536 electric dipoles. The electric charges of the dipoles are $\pm 2 \cdot 1.606e - 19$ C, and their masses are $2 \cdot 0.16e - 26$ kg for the point labeled "H" and $16 \cdot 0.16e - 26$ kg for the point labeled "O" (**Figure 1**) [3].

The following boundary conditions are chosen for modeling: for the faces, z = 0 and z = aa values of the potential V = 0 and V = Vin are applied, respectively, and for the other faces, the value $\partial V / \partial n = 0$ is applied. The modeling contains three main stages: (1) random placement of the dipoles, (2) calculation of the electric fields by COMSOL Multiphysics and repositioning of the dipoles, and (3) exploitation of the obtained results. The mass center, mc, of each dipole is supposed to be fixed at the mass center of the tetrahedrons of meshing of the cube chosen so that the density of the dipoles is the same as that of water molecules in a liquid drop. The points H and O

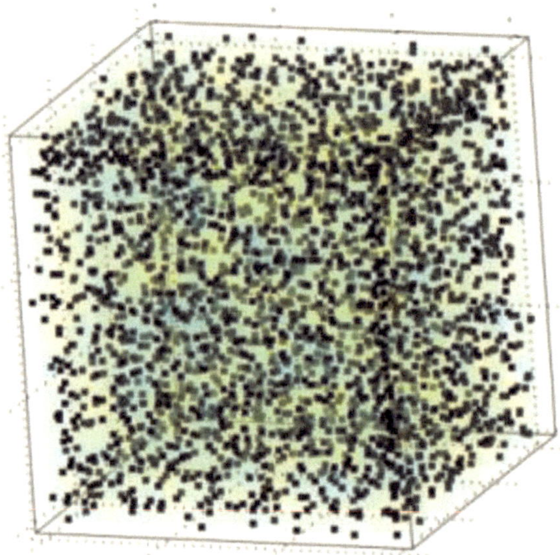

Figure 1.
Model of a water drop reduced to its electric dipoles.

linked to each *mc* are randomly oriented in the local spherical coordinates. The coordinates of all the *mc* are grouped in the permanent matrix *pmc* (size 1536 × 3), generating two random matrices, *pH* and *pO,* of the same size. The formula below represents a part of the first stage of the modeling, i.e., random placement of the 1536 centers of gravity *pmc* in a cube with edge *aa* = 3.585e–9 m:

$$pmc = aa^{*}rand(1536, 3).$$

Taking into consideration length of the dipoles, *lm* = 1.755e-10 m, distances to the point *H* (*lH*) and to the point *O* (*lO*) from the center of gravity respectively:

$$lH = lm^{*}mH/(mH + mO), lO = lm^{*}mO/(mH + mO),$$

the following code was written to determine random positions of oxygen *O* and hydrogen *H* in the cube:

```
for tt = 1:line.
theta = pi*rand; phi = 2*pi*rand;

pH(tt,1) = pmc(tt,1) + lH*sin(theta)*cos(phi);
pH(tt,2) = pmc(tt,2) + lH*sin(theta)*sin(phi);
pH(tt,3) = pmc(tt,3) + lH*cos(theta);

pO(tt,1) = pmc(tt,1) – lO*sin(theta)*cos(phi);

pO(tt,2) = pmc(tt,2) – lO*sin(theta)*sin(phi);
pO(tt,3) = pmc(tt,3) – lO*cos(theta);
end
```

Here, *theta* and *phi* are local spherical coordinates. The initial velocity of the atoms of hydrogen *H* and oxygen *O* is chosen to be zero. These atoms carry mass and electric charges, interact with each other and are under the action of fields received from outside; their centers of gravity play no role.

For calculation, which presents the second stage of the modeling, COMSOL Multiphysics requires classifying the points *H* and *O* according to a precise order: classification of the points according to their coordinates x, y, and z. It is obviously necessary to put markers to reform the dipoles.

The actual calculation requires a certain number of loops "for," which also serve as time markers. The main marker is the period *T* of the wave chosen for heating (note that $T = 1/f$, where $f \approx 2.45$ GHz is frequency). This period is divided into 60 equal parts. The COMSOL calculation is then carried out in each part using static analysis. This heating period is preceded and followed by a period of analysis of the initial situation and final situations.

At the end of each part of the period, obtained results, fields, positions, and speeds replace the initial data, and certain values are stored for further analyses. This is the case with the test dipole numbered (named) "1234," whose position and electric field values are written in an annexed memory. The third stage of modeling, as the main obtained results along with their interpretations, is presented below. **Figure 2** gives values of static potential *V* between two opposite faces of the model cube. The intervals (0, 10) and (70, 80) on the horizontal axis make it possible to analyze the initial and final situations, i.e., the period just before and after applying the electric field. **Figure 3** represents a variation of length (*L*, m) of the dipole "1234" for two different potential amplitudes (which corresponded to the blue and the red line, respectively):

The variation in length around 1.755e–10 m shows that the water molecule undergoes compression and traction actions which are possible sources of internal friction and heat. This variation can also be a source of physicochemical modification analogous to the Kerr effect [4, 5].

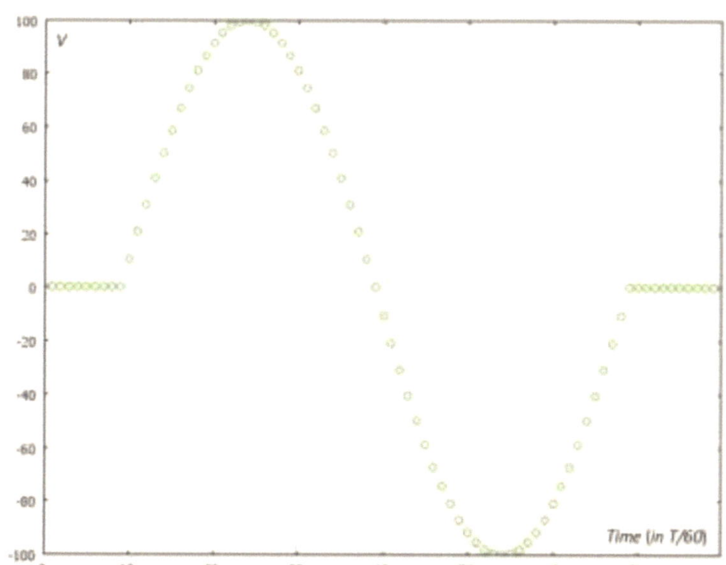

Figure 2.
Values of static potential V between two opposite faces of the cube.

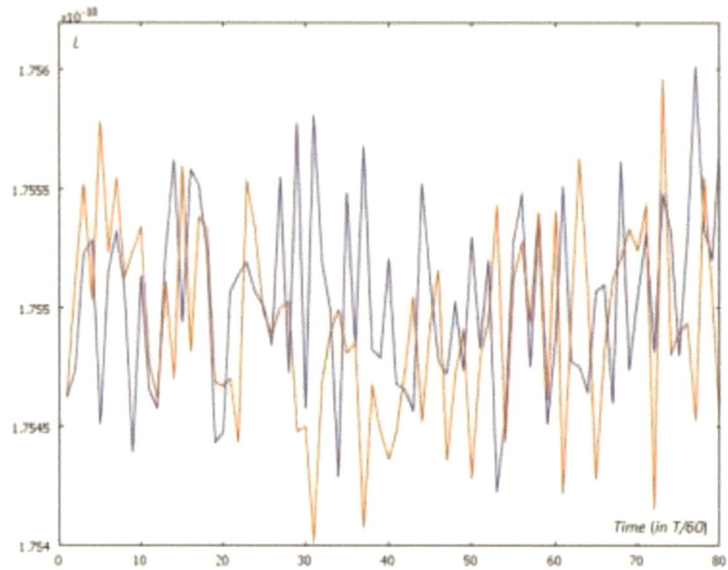

Figure 3.
Variation of dipole length "1234" for two different potential amplitudes.

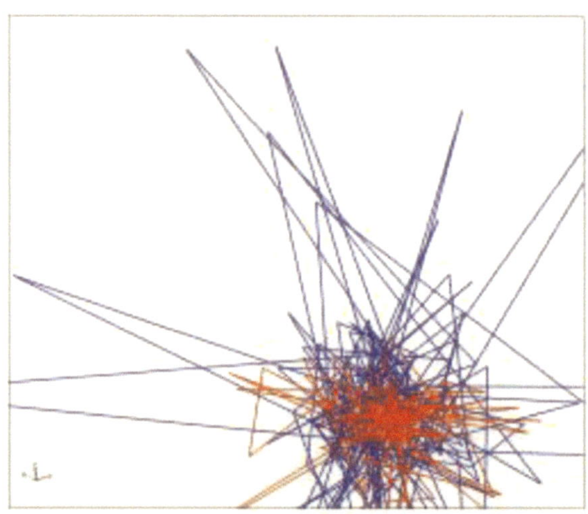

Figure 4.
Electric field applied to dipole "1234" during its Brownian motion in the drop.

Tribology shows that friction is a very general phenomenon that has useful results and others that are less. For example, prehistoric man domesticated fire and created a prototype of the violin, and conversely, the rolling of trucks and cars damages the treads of roads. So it is always prudent to look at the quality of the material heated in the microwave.

Figure 4 presents the electric field applied to dipole "1234" during its Brownian motion in the drop. In the Figure, the red line corresponds to the force received by the oxygen O, and the blue one corresponds to those received by the hydrogen H of the considered dipole.

DOI: http://dx.doi.org/10.5772/intechopen.104949

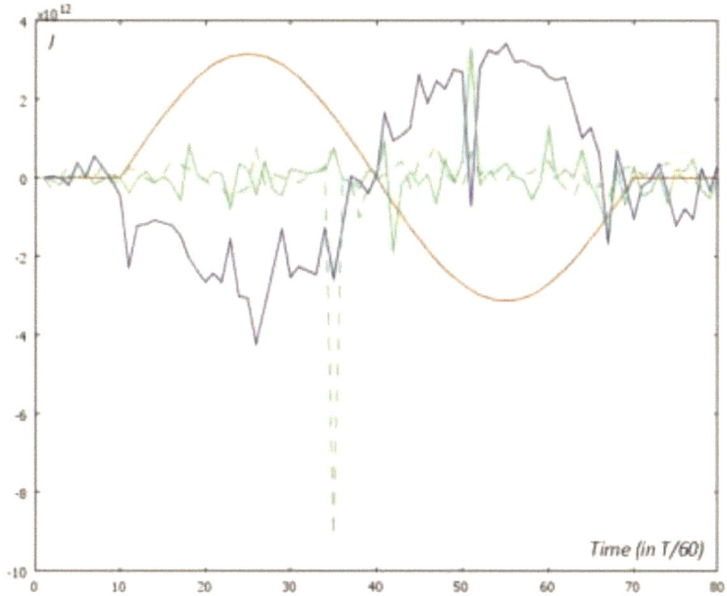

Figure 5.
Heating field and thermal agitation.

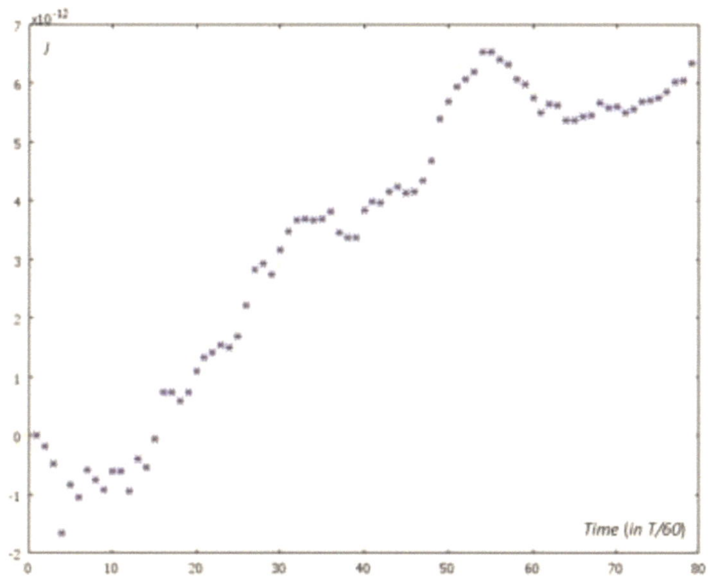

Figure 6.
Accumulation of the kinetic energy of all the dipoles.

In **Figure 5**, the red line corresponds to the impressed heating field. Its coherence is transposed on the blue line presenting the *Ez* field submitted here on hydrogen (dipole “1234”). On the other hand, the two components in *x* and *y* (green line) remain erratic but do not heat up. In addition to heating, these fields could promote

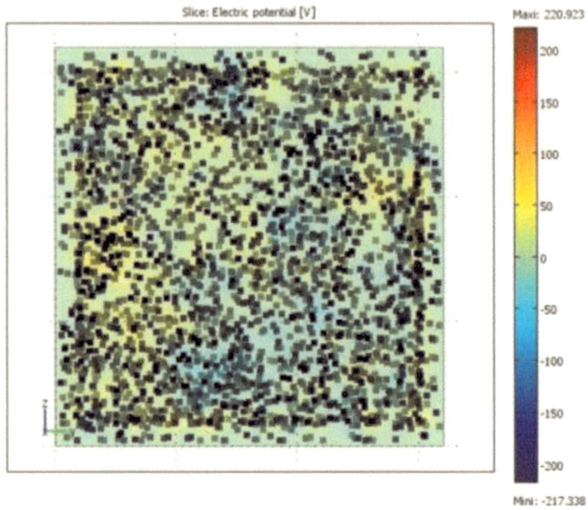

(a) Time = 8*T/60 (on the Figure 2 (b)

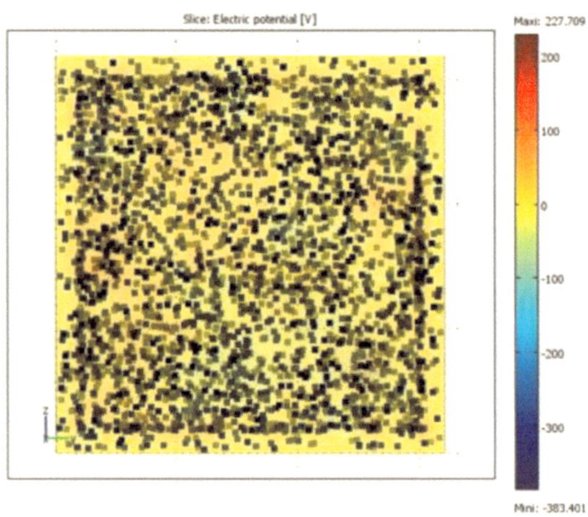

(b) Time = 20*T/60 (on the Figure 2)

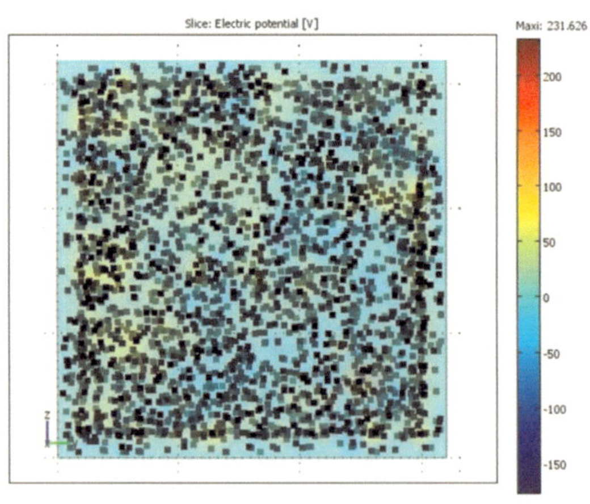

(c) Time = 25*T/60 (on the Figure 2)

Figure 7.
Distribution of electric potential obtained by modeling for different values of calculation time.

other phenomena such as hydrogen bonding. Accumulation of all dipoles' kinetic energy of the considered water drop is shown in **Figure 6**.

The accumulation of kinetic energy denoting the existence of friction induces an accumulation of heat in the cube (the model of considered water drop) because it has no other possible use; the interval (70, 80) of the horizontal axis shows some descent due to the stopping of the heating.

From the macroscopic point of view, this property of heating can be treated as electronic conduction in metals by introducing the imaginary part ε'' added to the real part ε' of the permittivity. This heat accumulation is the consequence of temperature rise that varies according to the different materials and is characterized by its specific heat. The direction of the current does not matter. **Figure** 7(**a**)–(**c**) show the distribution of the electric potential obtained by modeling for various chosen values of calculation time.

Figure 7 (**a**, **b**, and **c**) are produced by COMSOL at the times of heating (8 s, 20 s, and 25 s, respectively) during a period surrounded by two rest phases. These figures are 2D sections perpendicular to the x-axis. The boundary conditions chosen for modeling are: for the face $z = 0$ value of potential $Vin = 0$ and for the face $z = aa -$ $Vin = sin(t/T)$; T is the period. For the faces, $y = 0$ and $y = aa$ value of the applied potential is $\partial V/\partial n = 0$.

The color gives the distribution of electrical energy in the cube. Even in Figure (a), one can distinguish a difference in the distribution of the energy and, therefore, also of the potential.

The dots show the position of oxygen and hydrogen as a function of time. The poles are not always very close; on the sides, there is a string of very tight dipoles, but they are due to a function of Matlab, which is to return the dipoles which had come out of it into the cube.

4. Conclusion

The considered model of the water drop has proposed a microscopic point of view at the microwave heating being a response of the studying material and not a heat transfer. This fact has made it possible to understand the physical origin of the two permittivity components and envisage the production of new materials unimaginable with traditional heating means.

Author details

Serge Lefeuvre[1*] and Olga Gomonova[2]

1 Midival Association, Sainte Foy d'Aigrefeuille, France

2 Reshetnev Siberian State University of Science and Technology, Krasnoyarsk, Russia

*Address all correspondence to: lefeuvre.sae@wanadoo.fr

References

[1] Lefeuvre S, Gomonova O. Modeling at the Nano level: Application to physical processes. In: Musa SM, editor. Computational Finite Element Methods in Nanotechnology. Chapter 16 ed. CRC Press; 2012. pp. 559-584. DOI: 10.1201/b13002. ISBN: 978-1-4398-9323-4

[2] Lefeuvre S. FEM simulation of microwave absorpsion by a water drop. In: 1st URSI Atlantic Radio Science Conference, URSI AT-RASC 2015, 7303048. DOI: 10.1109/URSI-AT-RASC.2015.7303048

[3] Lefeuvre S, Gomonova O. Temperature of a drop of electrical dipoles Δt. In: Proceedings of the 16th International Conference on Microwave and High Frequency Heating AMPERE 18–21 September 2017. Delft, the Netherlands; 2017

[4] Najim M, Matheau JC, Lefeuvre S. Microwave Kerr effect on polar liquids. Applied Physics Letters. 1972;**21**(8): 399–400. DOI: 10.1063/1.1654429

[5] Najim M, Matheau JC, Lefeuvre S. Effet Kerr Ultra-Hertzien du sulfure de carbone. Optic Communications. 1972; **5**(5):416-418

Section 3

Microwave Devices

Chapter 6

Filter Designs Based on Defected Ground Structures

Somdotta Roy Choudhury

Abstract

This chapter concentrates on the filter structures using Defected Ground Structures (DGS). Initially, this chapter discusses the limitations of Electromagnetic Band Gap (EBG) structures and the development of DGS structure from EBG structure. DGS is an area of increasing interest in EBG technology. Here the well-known dumbbell DGS structures' features and physics are also discussed. New investigations are presented on the choice of geometrical shapes for the DGS structure as an element for the proposed filters. All the proposed DGS structures used to implement different types of filters (lowpass, bandpass, and bandstop) are validated.

Keywords: microstrip filter, defected ground Structures, bandstop filter, split ring, resonators

1. Introduction

Electromagnetic band gaps (EBGs) are generally periodic structures that exhibit different bandgaps over specific frequency ranges that prohibit electromagnetic wave propagation [1]. Due to its frequency selective characteristics, EBG structures naturally find applications in microwave filters. The main limitation of EBG structures is their large overall sizes and complex structures, which are not easy to analyse, and in most cases, not feasible to fabricate [2]. Defected Ground Structure (DGS) can overcome large size problems, and the complexity of EBG structures can be overcome by Defected Ground Structure (DGS). DGS structures can produce band rejection in specific frequency bands, and hence, is also known as a kind of EBG structure [3]. A DGS is a non-periodic EBG structure. It is an etched structure in the metallic ground plane of a microstrip structure. D. Ahn et al. first introduced the dumbbell-shaped DGS structure [4]. A variety of geometry etched in the microstrip line ground plane has been implemented to serve as defected ground structure. These structures usually add an extra lumped inductance and capacitance to the microstrip line connected as a parallel resonant circuit in series with transmission lines at both ends. The main features of the DGS structures are as follows in [5–12].

1.1 Some attractive features of DGS structure

Photonic Band Gap (PBG) Structures or Electromagnetic Band Gap (EBG) Structures could be:

- Periodic or non-periodic,
- Symmetric or non-symmetric,
- Easy to be presented as an LC equivalent circuit, and
- Simple and compact size.

1.2 Electrical characteristics of DGS structure

- Disturbs the shielding fields on the ground plane.
- Increases effective permittivity.
- Decrease phase velocity and provide slow-wave characteristics.
- Increase effective capacitance and inductance of the transmission line under which it is placed.
- Provides bandstop characteristics.

1.3 Advantages of using DGS circuits

- The structure is very much simple.
- Wider and deeper stopband characteristics.
- Provide low insertion loss.
- Have extremely small element values.
- Applications of DGS Structures:
- Planar resonator.
- High characteristic impedance transmission lines.
- Microwave circuits such as filters.

Performance improvement of Coupler, Divider/Combiner, Oscillator, Power amplifier and Antenna etc.

2. Different filters using DGS

The filter is one the most important parts of microwave circuit systems, and researchers have proposed different types of filtering circuits using DGS structures in recent years.

2.1 Performance improvement of lowpass filter using hexagonal head dumbbell shaped DGS

The hexagonal head dumbbell-shaped DGS is shown in **Figure 1**. The microstrip line is placed on the top plane and the proposed DGS on the ground plane. The different slot gaps, g = 0.2 mm, transverse slot length, a = 2 mm, b = 3.8 mm, a width of the 50 Ohm microstrip line, W = 1.92 mm.

2.1.1 Simulated results of unit cell

The structure is simulated by IE3D [11, 13] simulation software, and the simulated result shows a one-pole Butterworth prototype response given in **Figure 2**. The cutoff frequency is obtained at 3.3 GHz, and the transmission zero frequency is obtained at 5.1 GHz with attenuation of 23.9 dB. The sharpness factor is 11.6 dB/GHz.

2.1.2 Utility of the hexagonal slot

The advantages of the proposed hexagonal head DGS structure are: (1) for the equal effective area, the hexagonal shape of the slot provides increased path length for the current around the slot as compared to the square-headed slot and (2) for the equal value of resonant frequency the hexagonal slot area is 4.3% more compact than the square slot.

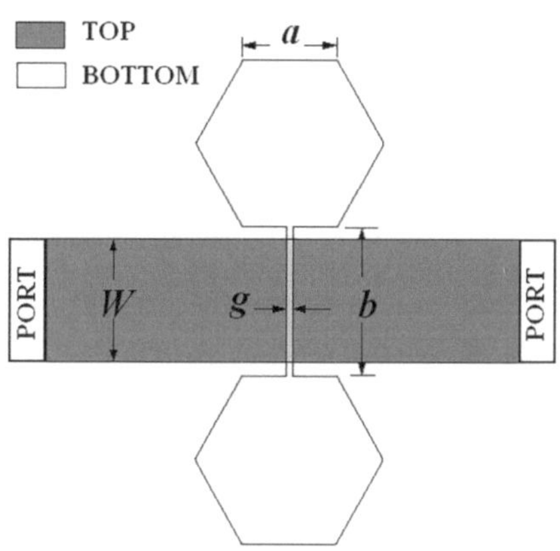

Figure 1.
Schematic diagram of the hexagonal head dumbbell type DGS.

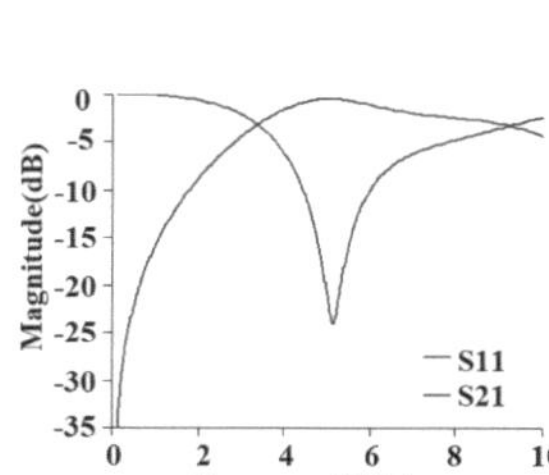

Figure 2.
S-parameter response of DGS structure.

2.1.3 Parametric study of unit cell

Various parametric studies are performed to analyze the proposed DGS structure. Based on the hexagonal-head arm length, the inductance and capacitance of the filter are calculated and listed in **Table 1**. The S-parameter responses for different arm lengths "*a*" of the hexagonal head are shown in **Figure 3**. The variation of the cutoff frequency ("*fc*") and transmission zero frequency ("*fp*") with respect to the arm length is plotted in **Figure 4**.

From **Table 1**, it is observed that the attenuation poles and attenuation zeros decrease with the increase of the arm length (a) of the hexagonal head. With the increase of the arm length, the inductance value increases significantly, and the capacitance value remains almost constant. This is due to the increment in electrical path length around the hexagonal head slot. The change in inductance and capacitance values with different slot head lengths ("*a*") are given in **Figures 5** and **6**, respectively. Another parameter, transverse slot gap "*g*," also greatly affects the filter's S-parameter response. The S-parameter responses for different transverse slots "*g*" of the hexagonal dumbbell DGS are shown in **Figure 7**. The variation of the cutoff frequency (*fc*) and transmission zero frequency (*fp*) with respect to the transverse slot gap is plotted in **Figure 7**. With the increase of this slot, the gap capacitance changes, and correspondingly the current path along the hexagonal head also changes. As a result, the inductance of the structure also changes. This inductance and capacitance change with the structure dimension causes a change in the cutoff and pole frequency of the filter.

The change in the circuit elements are calculated and are tabulated in **Table 2**. It is clear from **Table 3** that the attenuation poles and attenuation zeros decrease with the increase of the transverse slot ("*g*") of the proposed DGS. With the increase of the transverse slot, the inductance value decreases minutely due to the small decrement

Arm length of the Hexagonal Head '*a*' (mm)	3 dB Cutoff frequency 'f_c' (GHz)	Attenuation pole, 'f_p' (GHz)	Capacitance 'C' (pF)	Inductance 'L' (nH)
2	3.3	5.1	0.202	11.5
4	2.5	4.1	0.237	17.0
6	1.7	3.0	0.301	29.1

Table 1.
Circuit parameters of the Hexagonal Head Dumbbell DGS for different hexagonal head lengths "a."

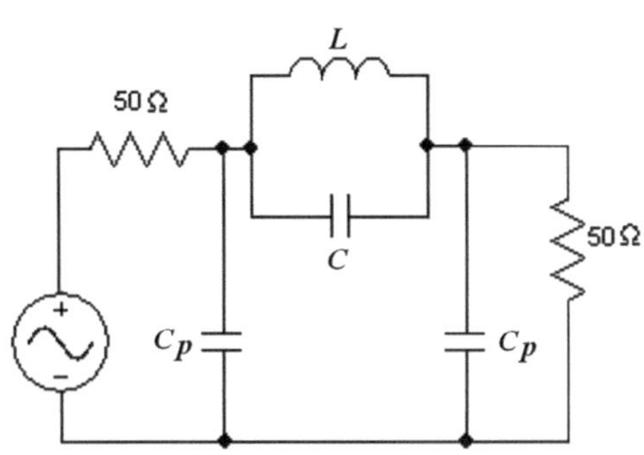

Figure 3.
Equivalent circuit model of the hexagonal head dumbbell type DGS.

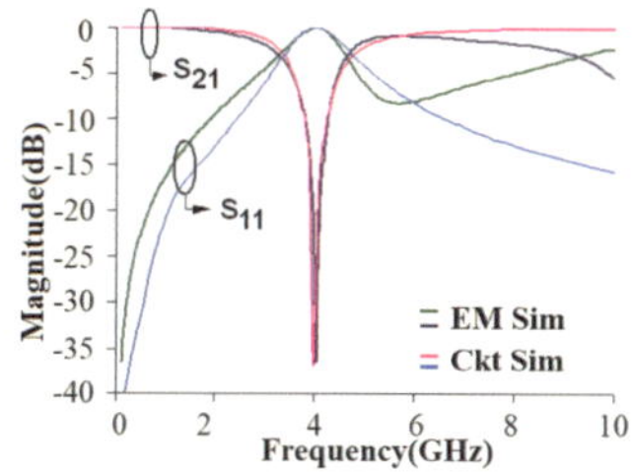

Figure 4.
Comparison of simulated and circuit model responses.

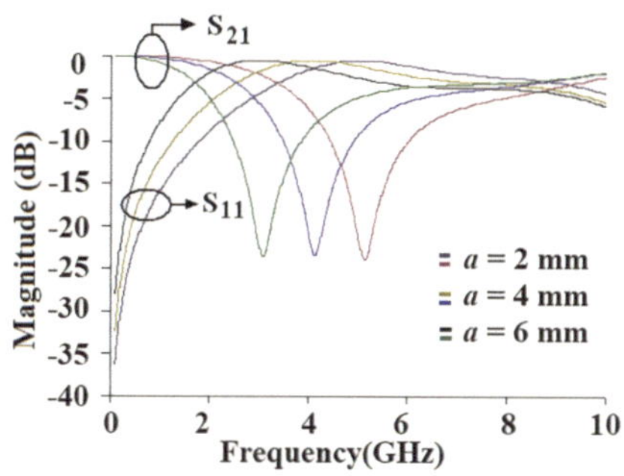

Figure 5.
Variations of simulated s-parameters with hexagonal head arm length "a."

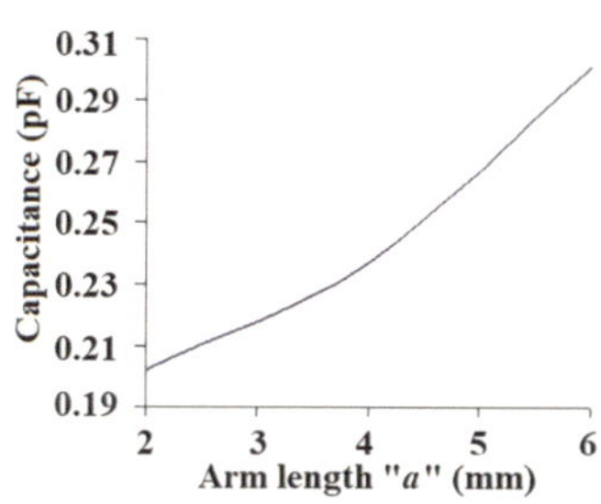

Figure 6.
Variations of equivalent capacitance with hexagonal head arm length "a."

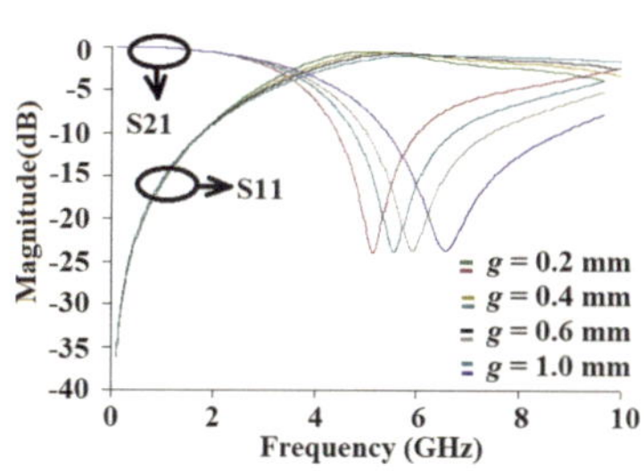

Figure 7.
Variation of simulated S-parameter responses with the change in the transverse slot gap "g."

in the current path length. The capacitance value also decreases with the increase in the transverse slot. The inductance and capacitance values change with different slot gaps ("g").

Transverse slot gap 'g' (mm)	3 dB Cutoff frequency 'f_c' (GHz)	Attenuation pole, 'f_p' (GHz)	Capacitance 'C' (pF)	Inductance 'L'
0.1	1.76	3.7	0.26	7.0
0.5	1.79	4.2	0.19	7.3
1.0	1.83	4.7	0.16	7.4
2.0	1.87	5.3	0.12	7.5

Table 2.
Hexagonal Head Dumbbell DGS characteristics for different transverse slot gap "g."

No. of DGS unit	Pole frequency (GHz)	Cutoff Frequency (GHz)	Sharpness Factor (dB/GHz)	20 dB Attenuation Band (GHz)
Single	5.1	3.3	11.6	0.36
Double	3.8	3.1	33.2	2.10
Triple	2.8	2.5	84.3	4.20

Table 3.
Comparative study among the filter structures using single, double, and-triple element hexagonal head dumbbell DGS.

2.1.4 Improvement of the filter response by dual DGS units

The filtering response of the filter can be improved to higher-order by simply adding another hexagonal head DGS structure of the same dimension side by side with the previous one. The single hexagonal DGS unit provides poor selectivity (11.6 dB/GHz) and small 20 dB rejection bandwidth (360 MHz). Therefore to obtain a good transition between passband and stopband and broader stopband, an array of DGS structures can be used with high and low impedance microstrip lines [14]. The High-low line decreases the insertion loss and provides better matching in the input. Therefore, the main objective is to design an LPF implementing two hexagonal dumbbells DGS units. The schematic diagram of the filter is shown in **Figure 8**. The arm length of the hexagonal head dimension of the individual unit is $a = 2$ mm with a separation of L = 2 mm. The transverse slot dimension of each unit $g = 0.2$ mm, etched out under the compensated transmission line of width, $W_1 = 5$ mm. The transverse slot length, $b = 3.8$ mm.

2.1.5 Simulated response of double unit

The simulated S-parameter response is given in **Figure 9**. From the simulated S-parameter response, the array of two DGS units provides a lowpass filtering response with 3 dB cutoff frequency $fc = 3.1$ GHz, attenuation pole, $fp = 3.8$ GHz, and passband insertion loss of 0.5 dB. This also provides good transition (sharpness factor = 33.2 dB/GHz) between the passband and stopband and wider attenuation characteristics compared to single hexagonal head dumbbell DGS. The 20 dB rejection band is 2.1 GHz.

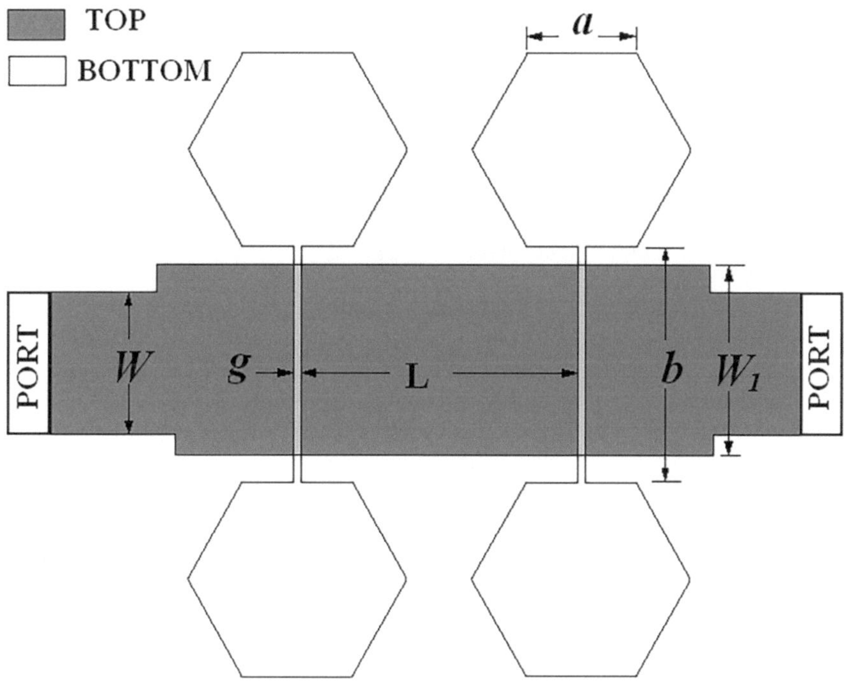

Figure 8.
Simulated S- parameter Response of two DGS units.

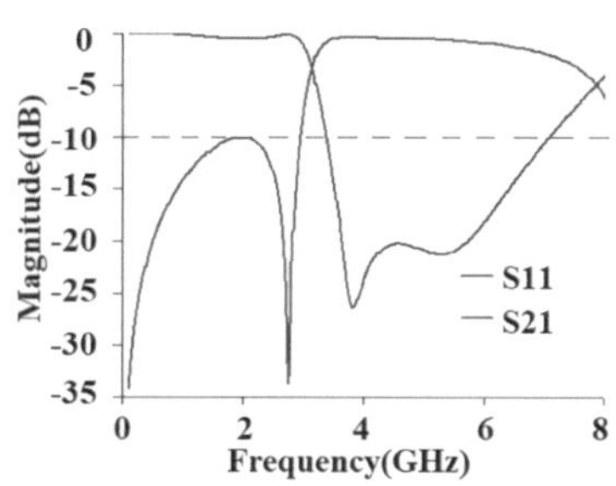

Figure 9.
Layout of the LPF using an array of three DGS units.

2.1.6 Improvement of the filter response by triple DGS units

The performance of the lowpass filter can be improved again by increasing the number of DGS units in the array. Here three DGS units are placed side by side in the ground plane under the high-low line, and all the DGS units have the same dimension. The dimensions of the filter structure are as follows: a = 2 mm, b = 3.8 mm, g = 0.2 mm, W = 1.92 mm, W_1 = 5.6 mm, W_2 = 11.8 mm, L = 2 mm. The layout of the proposed filter using three DGS units is shown in **Figure 10**.

2.1.7 Simulated responses

The S-parameter plot of the proposed filter using three DGS units is shown in **Figure 11**. From the response of the scattering parameters, it is observed that the array

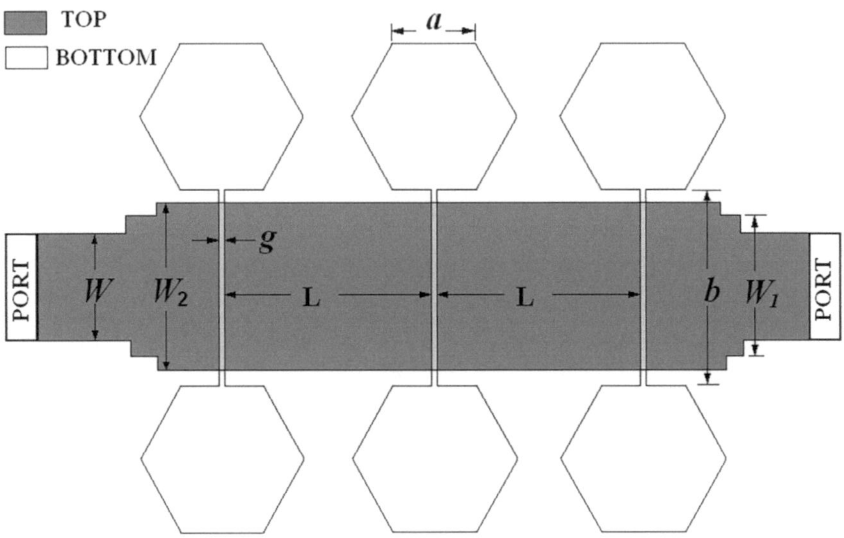

Figure 10.
Layout of the LPF using an array of three DGS units.

of three DGS units provides a lowpass filtering response with 3 dB cutoff frequency, fc = 2.5 GHz, attenuation pole, fp = 2.8 GHz, and passband insertion loss of 0.6 dB. The sharpness of the passband edge of the filter is high, about 84.3 dB/GHz. The 20 dB rejection bandwidth is 4.2 GHz.

2.1.8 Measured results

The proposed structure is fabricated and measured by Vector Network Analyzer (VNA). The critical dimensions of the fabricated structure are a = 2 mm, b = 3.8 mm, g = 0.2 mm, W = 1.92 mm, W_1 = 5.6 mm, W_2 = 11.8 mm, L = 2 mm. i.e., same as that of the simulation procedure. The photographic view of the structure is given in **Figure 12**. The comparison of the measured and simulated responses is shown in **Figure 13**.

The measurement result provides cutoff frequency = 2.6 GHz, Pole frequency = 3.06 GHz, insertion loss = 0.7 dB, 20 dB stop-band = 6.48 GHz, sharpness factor = 60.7 dB/GHz. The measured phase response is given in **Figure 14**. The discrepancy between measured and simulated phase responses is mainly due to the error resulting

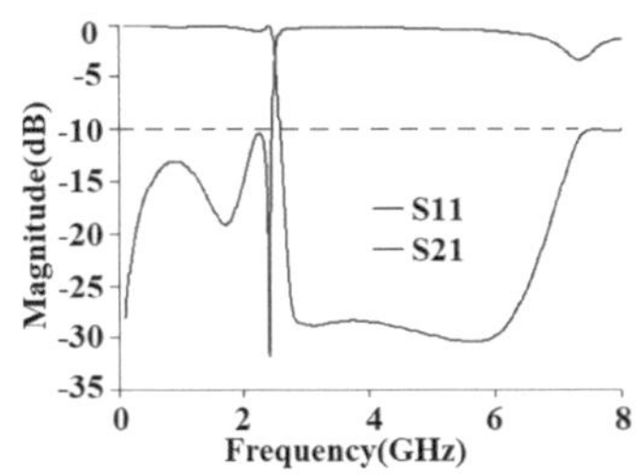

Figure 11.
Simulated S- parameter Response of three DGS units.

Figure 12.
Photographic view array of three DGS units.

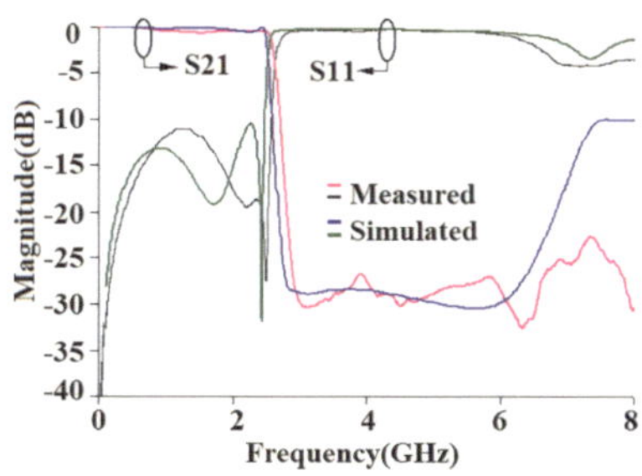

Figure 13.
Comparison of the Simulated and measured s-parameter responses of an array of three hexagonal-head DGS units.

from manual calibration for measuring the phase. So it can be concluded that the present structure provides less insertion loss and better transition between passband and stopband compared to the array of two DGS structures. **Table 3** establishes a comparative study as given below. Thus, from this table, it is clear that the array with higher numbers of DGS units provides a more acceptable response than the array with lower numbers of DGS units. So it can be concluded that, although the big array of DGS units increases the circuit size and complexity, it delivers better filtering performance. **Figure 15** shows a graphical representation of the comparison of the transmission coefficients (S21) performances of the single, dual and triple unit hexagonal head DGSs. Thus, from this table, it is clear that the array with higher numbers of DGS units provides a more acceptable response than the array with lower numbers of DGS units. So it can be concluded that, although the big array

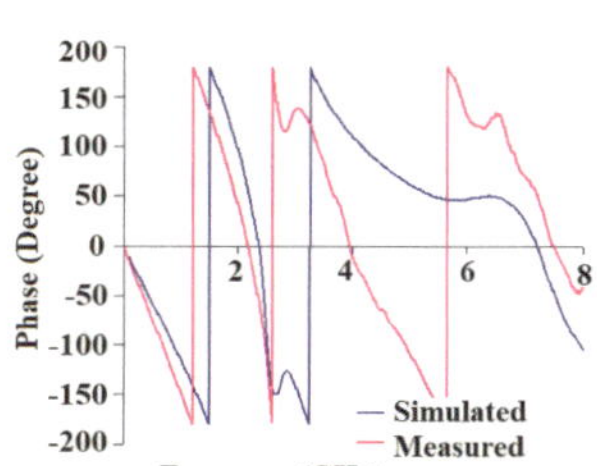

Figure 14.
Comparison of simulated and measured phase responses.

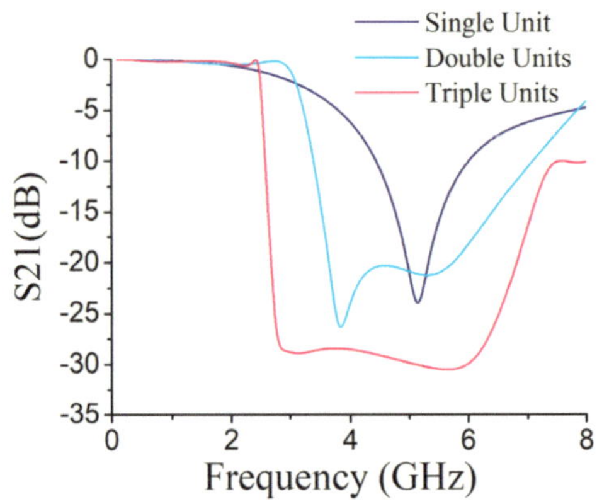

Figure 15.
Comparison of the S-parameters of a single unit, double units, and triple units hexagonal DGSs.

of DGS units increases the circuit size and complexity, it delivers better filtering performance.

2.2 Bandstop filters using circular split-ring type defected ground structure

A compact bandstop filter using circular split-ring type DGS is designed. First, a unit cell circular ring type DGS with a split at the lower portion underneath a microstrip transmission line is analyzed. The microstrip line is placed on the top plane and the DGS on the ground plane, as illustrated in **Figure 16**. The different dimensions of the DGS unit are: - the outer radius of the ring, r_2 = 3.3 mm, the inner radius of the ring, r_1 = 2.9 mm, the width of the ring, g_1 = 0.4 mm, the width of the gap, g_2 = 0.2 mm, a width of the 50 Ohm microstrip line, W = 1.92 mm.

2.2.1 Simulated result

The simulated response shows a one-pole bandstop type response. The attenuation zero frequency is obtained at 4.14 GHz, and attenuation pole frequency is obtained at

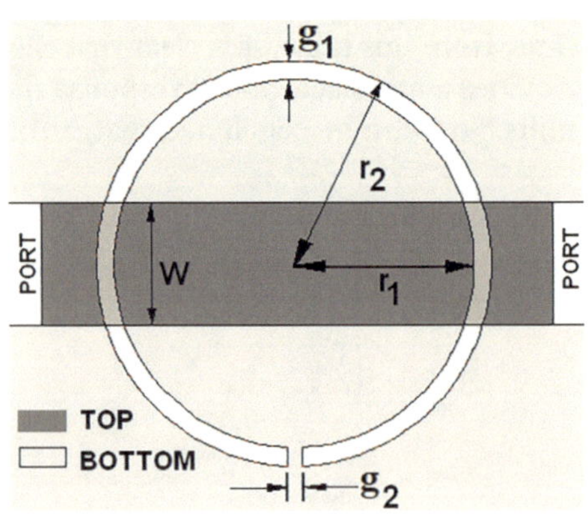

Figure 16.
Schematic diagram of the circular split-ring type DGS.

4.5 GHz with attenuation of 31.5 dB, as shown in **Figure 17**. The sharpness factor at the lower passband edge is 79.17 dB/GHz. Maximum passband insertion loss is –0.07 dB with a return loss greater than −18 dB.

2.2.2 Parametric study

The stopband of the proposed bandstop filter can be tuned by simply changing the inner radius (r_1), keeping the slot width (g_1) and split gap (g_2) of the ring DGS fixed, as shown in **Figure 18**. The ratios of cutoff frequencies and pole frequencies (fc/f0) for different values of the ratios of inner radii and the 50-Ohm line width (r_1/W) are plotted in **Figure 19**. The slot width (g_1) of the split ring DGS also determines the cutoff and the pole frequency, keeping the outer radius (r_2) and split gap (g_2) at constant values as given in **Figure 20**. It is clear from **Figure 21** that changes in the ratios of the slot width to 50-ohm line width (g_1/W) cause variation in the ratios of

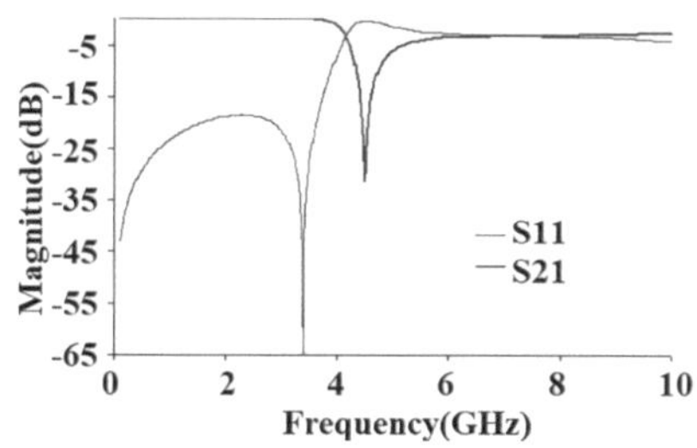

Figure 17.
S-parameters of the circular split-ring type DGS.

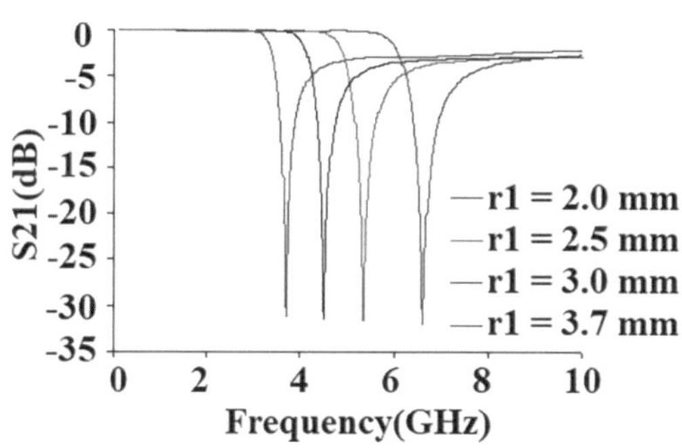

Figure 18.
Tuning of S-parameter responses with the inner radius (r1).

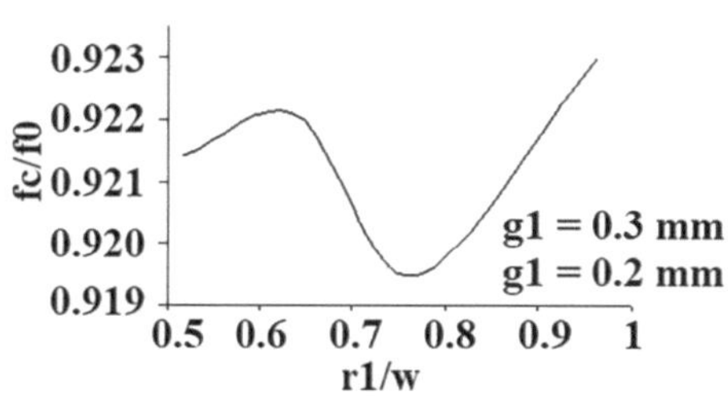

Figure 19.
Tuning of cutoff and pole Frequencies- with the slot width (g1).

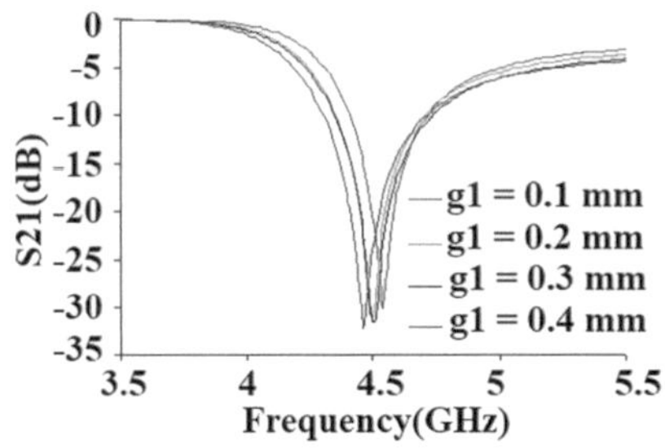

Figure 20.
Tuning of S-parameter responses with the slot width (g1).

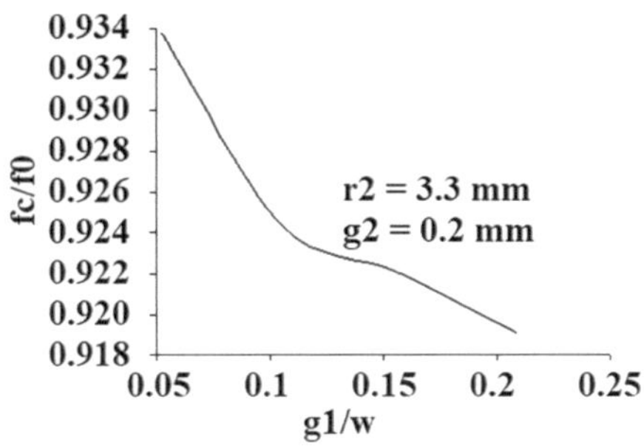

Figure 21.
Variation of the ratios of cutoff and pole frequencies with slot width (g1).

cutoff frequencies and pole frequencies (fc/f0). Again when studying the effect of the parameter g_2 on the frequency response, it is obvious from **Figure 22** that g_2 causes change in the values of cutoff and pole frequencies. The ratios of split gaps and the 50-ohm line widths (g_2/W) are plotted with the ratios of cutoff frequencies and pole frequencies (fc/f0), as shown in **Figure 23**.

2.2.3 Equivalent circuit parameters

Extracted equivalent-circuit parameters (equivalent inductance, L and capacitance, C) of the proposed DGS unit section for different dimensions are calculated by (Eq. (1)) [4] and Eq. (2) [4] and listed in **Tables 4** and **5** given below.

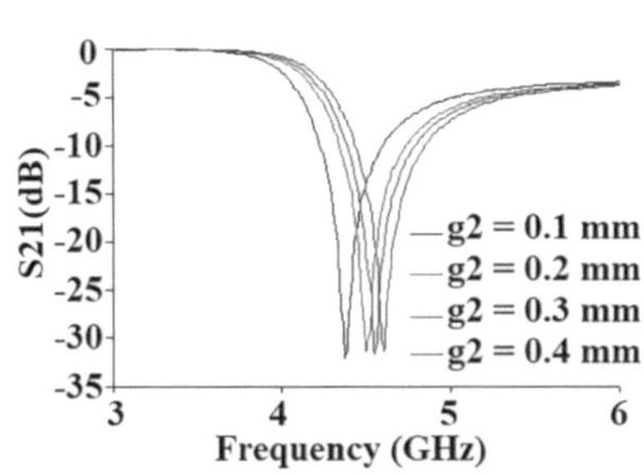

Figure 22.
Tuning of S-parameter of fc and fo with a split gap (g2).

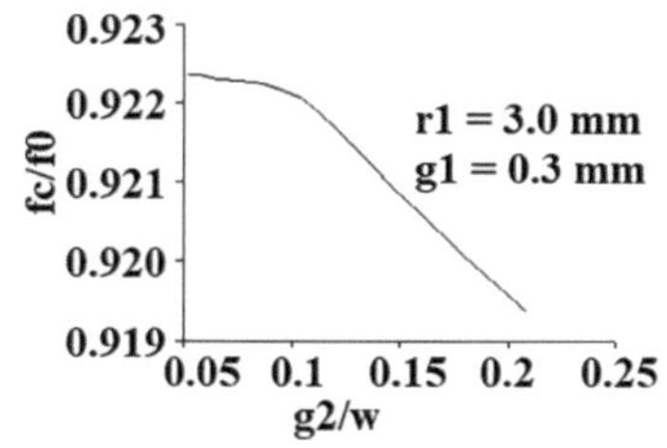

Figure 23.
Variation of the ratios responses with the split gap (g2).

Extracted Parameters & corresponding Frequencies	r_1 = 2.0, r_2 = 2.3 (mm)	r_1 = 2.5, r_2 = 2.8 (mm)	r_1 = 3.0, r_2 = 3.3 (mm)	r_1 = 3.7, r_2 = 4.0 (mm)
Inductance, L (nH)	0.19	0.50	0.57	0.70
Capacitance, C (pF)	1.49	1.77	2.18	2.63
Cutoff Frequency, f_c (GHz)	6.11	4.91	4.14	3.40
Pole Frequency f_0 (GHz)	6.62	5.34	4.49	3.69

Table 4.
Extracted equivalent-circuit parameters of the proposed split ring DGS unit section having dimensions g1 and g2 kept fixed (g1 =0.3 mm; g2 =0.2 mm).

Extracted Parameters & corresponding Frequencies	r_1 = 3.2, g_1 = 0.1 (mm)	r_1 = 3.1, g_1 = 0.2 (mm)	r_1 = 3.0, g_1 = 0.3 (mm)	r_1 = 2.9, g_1 = 0.4 (mm)
Inductance, L (nH)	0.48	0.55	0.57	0.60
Capacitance, C (pF)	2.56	2.24	2.18	2.11
Cutoff Frequency, f_c (GHz)	4.23	4.16	4.14	4.09
Pole Frequency, f_0 (GHz)	4.53	4.50	4.49	4.45

Table 5.
Extracted equivalent-circuit parameters of the proposed split ring DGS unit section having dimensions r2 and g2 kept fixed (r2 = 3.3 mm; g2 = 0.2 mm).

$$C = \frac{\omega_c}{Z_0 g_1 \left(\omega_0^2 - \omega_c^2\right)} \tag{1}$$

$$L = \frac{1}{4\pi^2 f_0^2 C} \tag{2}$$

2.2.4 Improvement of the response

To increase the order of the bandstop filter, two split ring type DGS units are introduced in the ground plane under the 50-Ohm microstrip line, as shown in **Figure 24**. Here in this design, both the DGS units have equal dimensions. The dimensions of the DGS units are: - outer radii of the rings, r_2 = 3.3 mm, inner radii of the rings, r_1 = 3.0 mm, the widths of the rings, g_1 = 0.3 mm, the widths of the gaps, g_2 = 0.2 mm, the length between the two split ring units, L = 17.4 mm, widths of the 50 Ohm microstrip line, W = 1.92 mm.

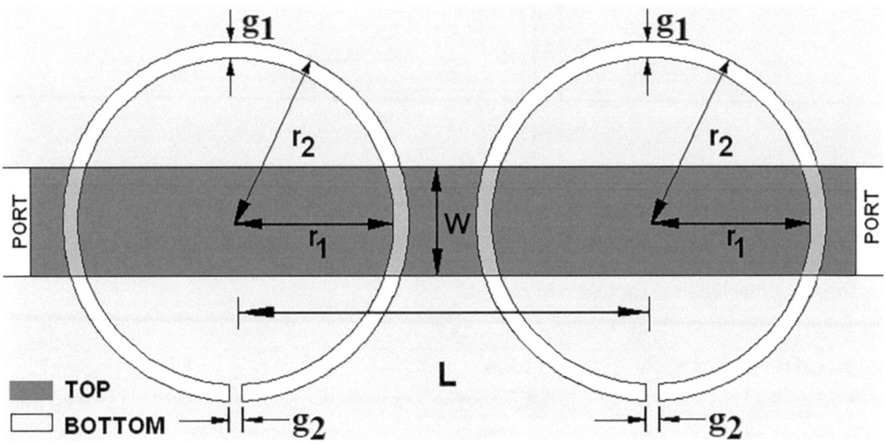

Figure 24.
Schematic diagram of the lowpass filter using double unit split ring DGS.

2.2.5 Simulated result of dual units

The S-parameter response of this structure is given in **Figure 25**. It is clear from the figure that the filter has a cutoff frequency of 3.9 GHz and insertion loss of 0.1 dB. The sharpness of this filter is quite good, about 54 dB/GHz. The main disadvantage of the filter is the poor out-of-band performances, i.e., the stopband outside the passband only ranges from 4.38 GHz to 4.76 GHz at the attenuation level of –20 dB.

2.3 Improvement of out of band performance using open stub

This disadvantage can be overcome by introducing another transmission zero at the region of the stopband [15, 16]. An open circuited stub employed at the top plane creates this additional transmission zero, as shown in **Figure 26**. A compensated high-low impedance line is employed in the microstrip line above the DGS structure for matching purposes. The physical dimensions of the modified structure with open stub are: W_1 = 8.5 mm, W = 1.92 mm, g_1 = 0.3 mm, g_2 = 0.2 mm, L = 9.1 mm, r_1 = 3 mm, r_2 = 3.3 mm. The shunt capacitor due to the stub is given by the following (Eq. (3)) [9].

$$\omega C_2 = \frac{1}{Z_{Oc}} \tan\left(\frac{2\pi W_1}{\lambda_g}\right) \tag{3}$$

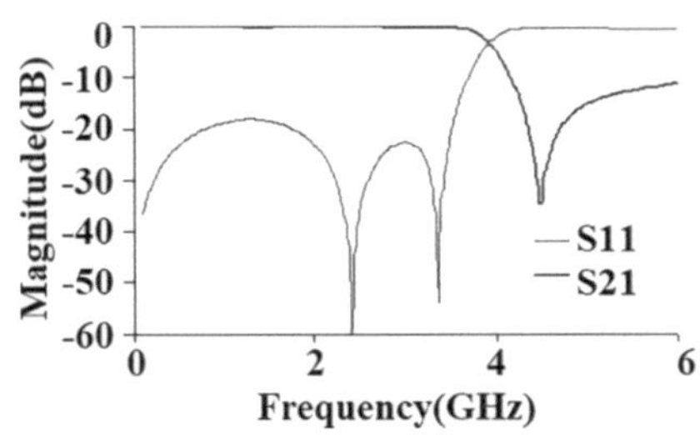

Figure 25.
S-parameter result of the lowpass filter using double unit split ring DGS.

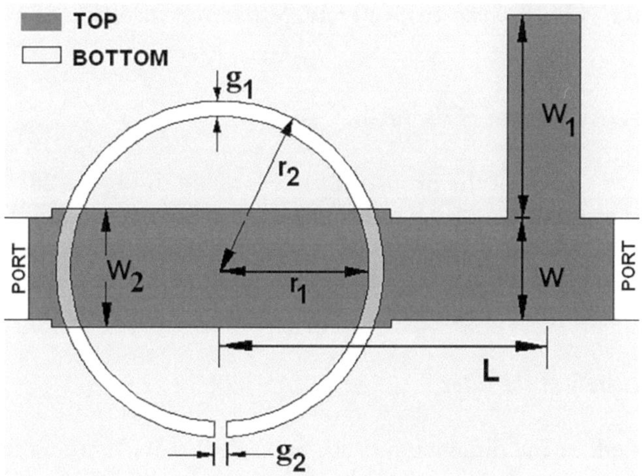

Figure 26.
Schematic diagram of the bandstop filter using a split ring DGS section and open-ended stub.

Here the left side of the equation is the susceptance of the shunt capacitor, and the right side of the equation represents the input susceptance of the open-circuited stub, which has the characteristic impedance "Z_{OC}," and the physical length "W_1" ($W_1 < \lambda g/4$) of the stub. The physical length of the open circuited stub is determined by the (Eq. (4)) below [9].

$$W_1 = \frac{\lambda_{gc}}{2\pi} \tan^{-1} \omega_c C_2 Z_{OC} \tag{4}$$

2.3.1 Simulated response of the structure

The simulated S-parameter response is shown in **Figure 27**. Two prominent resonances are found in the response. The first one is at 4.7 GHz and is responsible for the DGS unit with the same dimensions as given above, and the second pole is at 5.1 GHz and is responsible for the stub. The upper and lower cutoff frequencies are 4.1 GHz and 5.9 GHz, respectively. The filter provides a –20 dB bandwidth of 0.9 GHz and a rejection level of 30 dB. The sharpness of the lower and upper edge of the band are

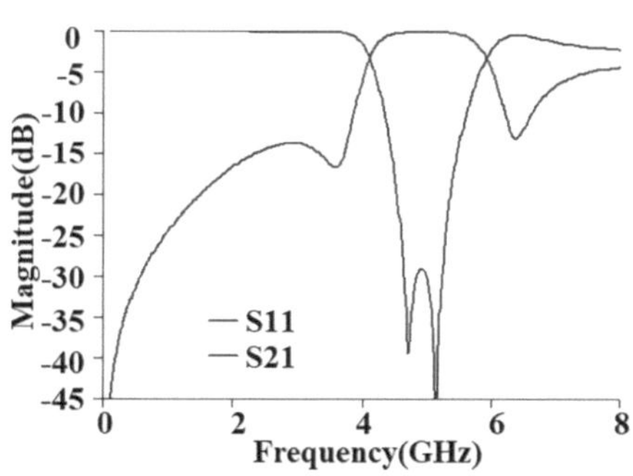

Figure 27.
S-parameter response of the lowpass filter using double unit split ring DGS sections and open-ended stub.

60.1 dB/GHz and 54.8 dB/GHz, respectively. Maximum insertion loss at the passband is 0.2 dB.

2.3.2 Equivalent circuit model of the filter

The equivalent circuit of the proposed filter is given in **Figure 28**, where circular split ring DGS is represented by a parallel resonant circuit connected in series with the source, and an open stub is referred to as a series resonant circuit connected in shunt with the source. The inductance (L2) is obtained due to the high value of Z_{0c}. The comparison of the simulated and circuit response is given in **Figure 29**.

2.3.3 Parametric study of the filter

By changing either the dimension (both inner radius "r_1" and outer radius "r_2") of the split ring DGS or the dimension of the stub (stub length "W_1"), the stopband of the filter can be controlled as shown in **Figures 30** and **31**. The changes of the ring radius and width of the stub with respect to the cutoff and pole frequency are clearly explained by **Figures 32** and **33**, respectively.

The corresponding mathematical expressions for the cutoff and pole frequency of the filter varying with inner radius "r_1" and stub length "W_1" can be obtained from the Eqs. (5)–(8).

$$f_{c1} = -0.72r_1^2 + 3.88r_1 - 0.86 \tag{5}$$

$$f_{p1} = -0.83r_1^2 + 4.35r_1 - 0.8 \tag{6}$$

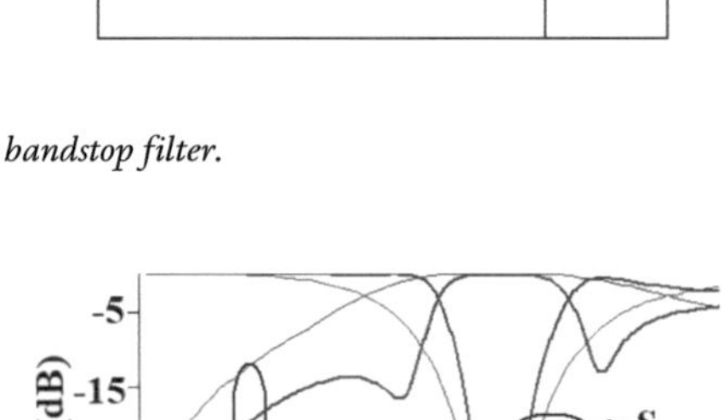

Figure 28.
Equivalent circuit of proposed bandstop filter.

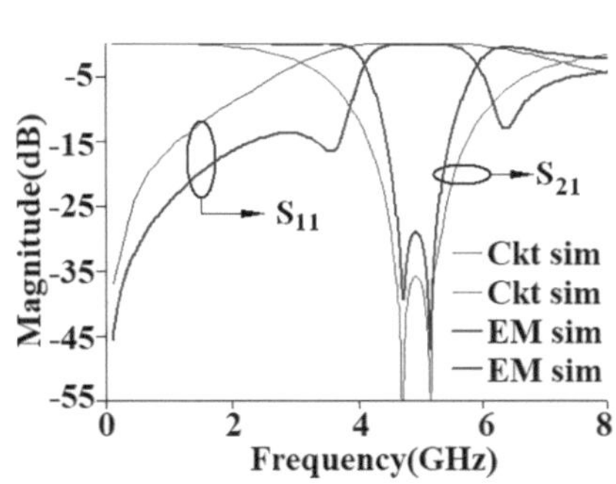

Figure 29.
Comparison of the simulated and circuit response.

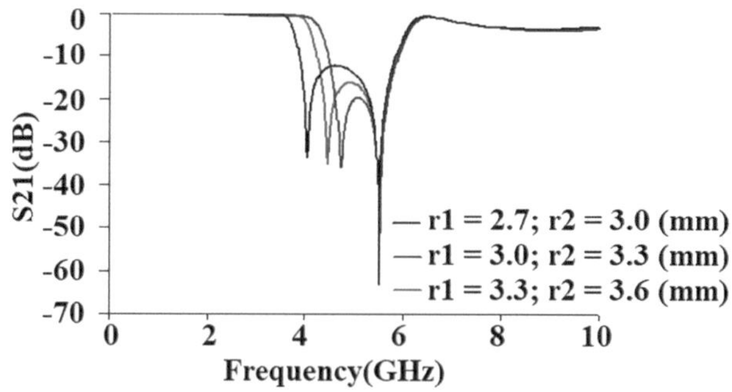

Figure 30.
Variation of S-parameters magnitudes w.r.t outer and inner radii (r2, r1).

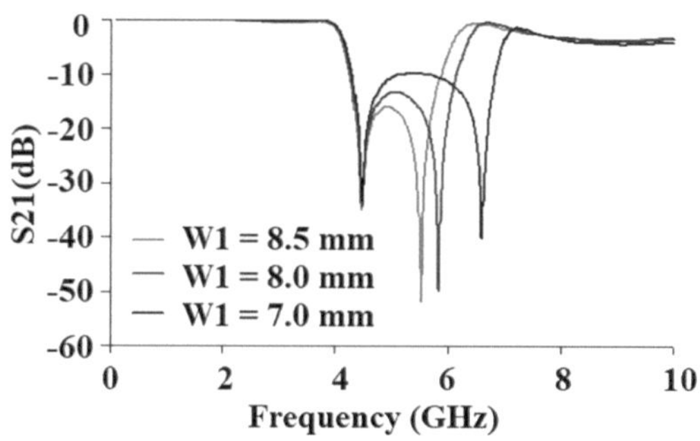

Figure 31.
Variation of the magnitude of S-parameters w.r.t length (W1) of the open-ended stub.

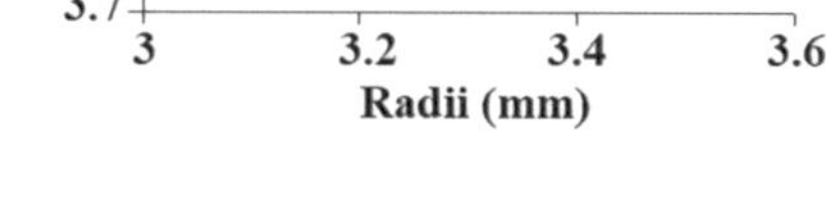

Figure 32.
Change of cutoff and pole frequencies with outer and inner radii (r2, r1).

$$f_{c2} = -0.006W_1^2 + 0.07W_1 + 3.97 \tag{7}$$

$$f_{p2} = 0.08W_1^2 - 2.07W_1 + 16.86 \tag{8}$$

2.3.4 Surface current distribution of the filter

Figures 34 and **35** show the surface current distribution of the filter. At the first resonance, the magnetic current distribution in the split ring DGS is maximum, whereas the electric current density at the stub is negligible.

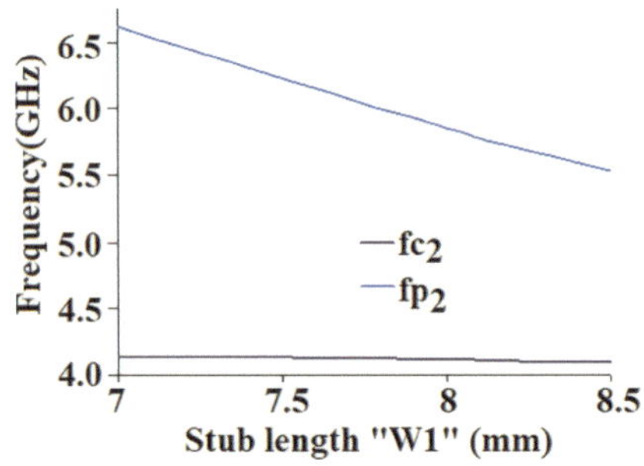

Figure 33.
Change of cutoff and pole frequencies with length (W1) of the open-ended stub.

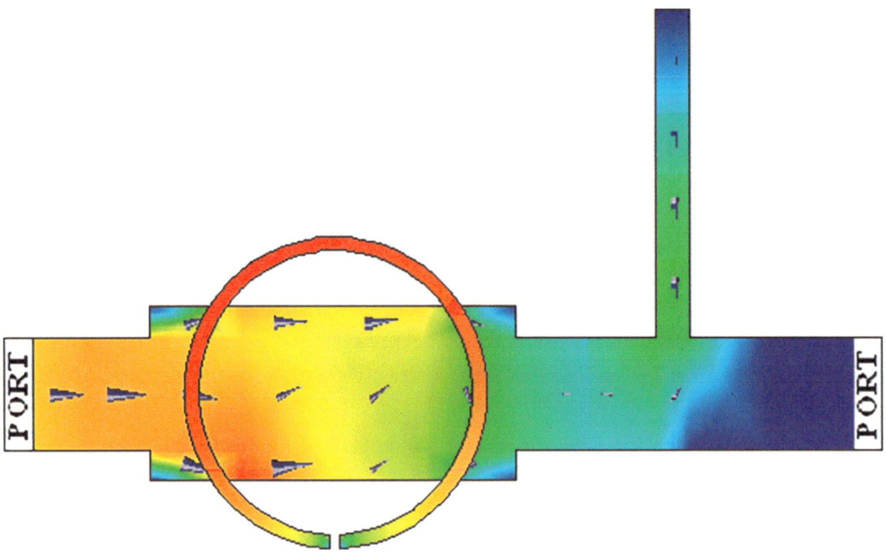

Figure 34.
Surface electric and Magnetic current distribution at 4.71 GHz.

During the second resonance, the split ring DGS's magnetic current is minimal, and the surface electric current distribution is maximum through the stub. The conclusion for the phenomena can be explained, such as that the split ring DGS is responsible for the first resonance whereas the stub provides the second resonance.

2.3.5 Measured result

The prototype fabrication and measurement verify the proposed filter design using the Vector Network Analyzer (VNA). The photographic view of the structure for the top plane and ground plane is illustrated in **Figures 36** and **37**, respectively. It is found in **Figure 38** that simulation and measured S-parameter results are in good agreement. **Figure 39** shows the measured phase response almost linear in the passband region.

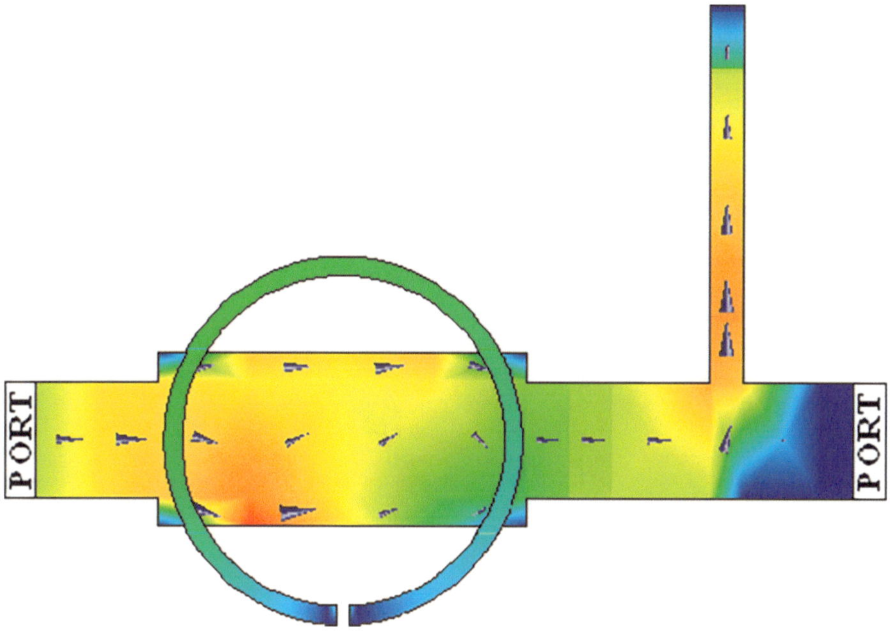

Figure 35.
Surface electric and Magnetic current distribution at 5.16 GHz.

Figure 36.
Photographic view of the Top plane of the Split-ring DGS with Open-stub.

2.4 Increment of bandwidth and roll-off factor of lowpass filter by circular split ring DGS

Another technique is proposed here to increase the stopband, which involves the combination of a third-order lowpass filter (LPF) with bandstop filters. The first step is to design a microstrip elliptic function lowpass filter to develop this filter. The

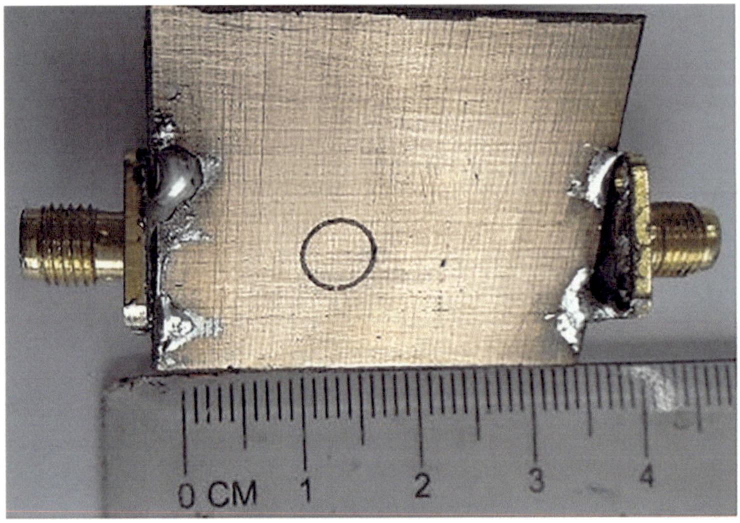

Figure 37.
Photographic view of the bottom plane of the Split-ring DGS with Open-stub.

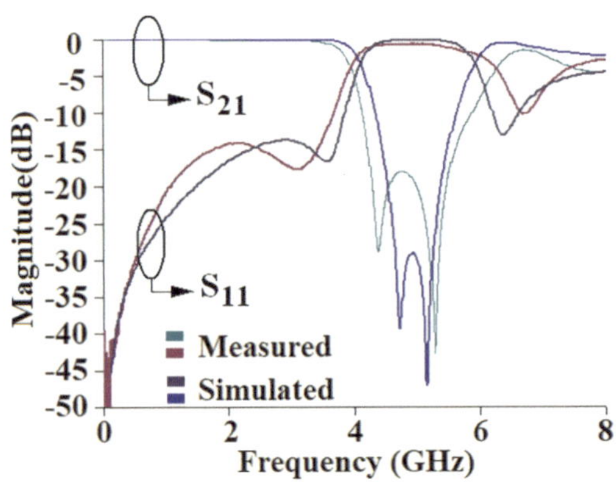

Figure 38.
Comparison of the simulated and measured s-parameter results.

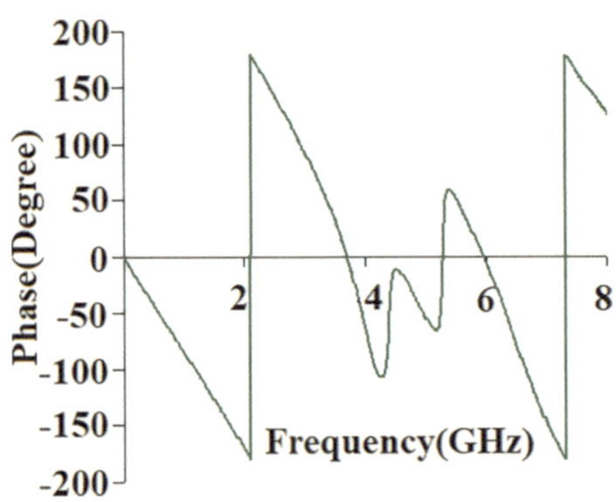

Figure 39.
Measured phase responses.

elliptic filter is chosen due to its sharper cutoff rate for a given number of reactive elements [4]. The bandstop responses are obtained by split ring type DGS units of different sizes. The main purpose of these DGS units is to increase the bandwidth of the rejection band.

2.4.1 Analysis of different LC elements and parameters

The L-C element values, scaled to Z_0 and f_c are determined by the following equations (Eqs. (9) and (10)):

$$L_i = \frac{1}{2\pi f_c} Z_0 g_{Li} \tag{9}$$

$$C_i = \frac{1}{2\pi f_c} \frac{1}{Z_0} g_{Ci} \tag{10}$$

The element values of the third-order elliptic function lowpass prototype are given in **Table 6**, with corresponding L-C values derived from the above equations. All the inductors are realized using high impedance lines with characteristic impedance Z_{OL} = 93 Ohms, whereas all the capacitances are realized using low impedance lines with characteristic impedance Z_{OC} = 14 Ohms. **Table 7** lists all relevant microstrip design parameters calculated using the microstrip design equations [10]. Corresponding design parameters are substituted from **Tables 6** and 7 in the following equations (Eqs. (11) and (12)) to realize the initial physical lengths of the high and low impedance lines and are listed in **Table 8**. The layout of this third-order elliptic function LPF with the design dimension is given in **Figure 40**.

$g_0 = g_4 = 1.0$, Normalized cut off, $\Omega_c = 1$, Passband Ripple, $L_{Ar} = 0.1$ dB.				
Element Values for Elliptic Function LPF prototype	$g_{L1} = g_1 =$ 0.8214	$g_{L2} = g_2 =$ 0.3892	$g_{C2} = g_2 =$ 1.0840	$g_{L3} = g_3 =$ 1.1880
L-C element values	L_1 = 1.64 nH	L_2 = 0.77 nH	C_2 = 0.86 pF	L_3 = 2.36 nH

Table 6.
LPF prototype elements values and L-C element values for third-order elliptic function LPF.

Characteristic impedance (ohms)	Z_{0c} = 14	Z_0 = 50	Z_{OL} = 93
Microstrip line width (mm)	W_C = 10.03	W_0 = 1.92	W_L = 0.59
Guided wavelength at f_c (mm)	λ_{gC} = 44.12	λ_{g0} = 46.97	λ_{gL} = 48.72

Table 7.
Microstrip design parameters for the elliptic function LPF with a dielectric constant of the substrate 3.2 and substrate height = 0.79 mm.

l_{L1} **(mm)**	l_{L2} **(mm)**	l_{C2} **(mm)**	l_{L3} **(mm)**
3.56	1.63	2.16	5.37

Table 8.
Initial physical lengths of the corresponding L-C parameters.

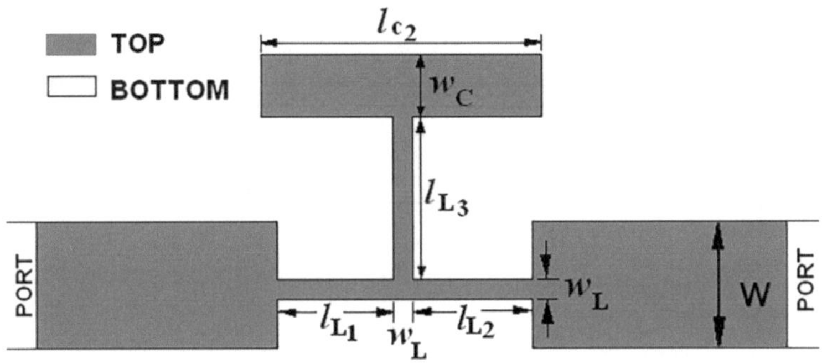

Figure 40.
Schematic diagram of the lowpass filter using open-ended T-shaped stub.

$$l_{Li} = \frac{\lambda_{gL}(f_c)}{2\pi} \sin^{-1}\left(2\pi f_c \frac{L_i}{Z_{0c}}\right) \tag{11}$$

$$l_{Ci} = \frac{\lambda_{gC}(f_c)}{2\pi} \sin^{-1}(2\pi f_c Z_{0c} C_i) \tag{12}$$

2.4.2 Simulated result of the third order elliptic filter

The MoM based EM simulation verifies the design, and the simulated frequency response is illustrated in **Figure 41**. It is clear from the simulation that the upper stopband of the LPF is relatively poor, about 820 MHz at a –15 dB attenuation level. The sharpness factor is 34.7 dB/GHz, with cutoff frequency and pole frequency being 3.3 GHz and 4.5 GHz, respectively.

2.4.3 Bandwidth improvement by DGS units

This third-order elliptic function LPF is then modified by the four split ring-shaped DGS structures. The four DGS units are of different sizes and are distributed in two concentric sets. Each of these sets is constructed by a pair of split ring DGS units and placed on two sides of the LPF. This modified LPF layout is given in **Figure 42**. The dimension of the split ring DGS are:- the inner radius of the smaller split ring DGS placed at the left-hand side, r_1 = 3 mm, the outer radius of the smaller split ring DGS placed at the left-hand side r_1' = 3.3 mm, the inner radius of the bigger split ring

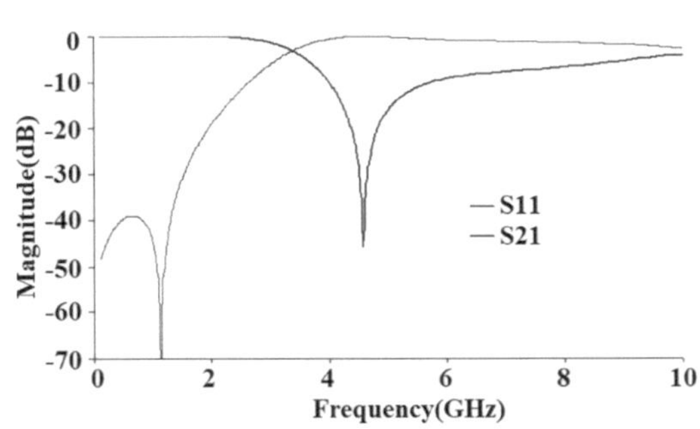

Figure 41.
S-parameter response of the open-ended T-shaped stub.

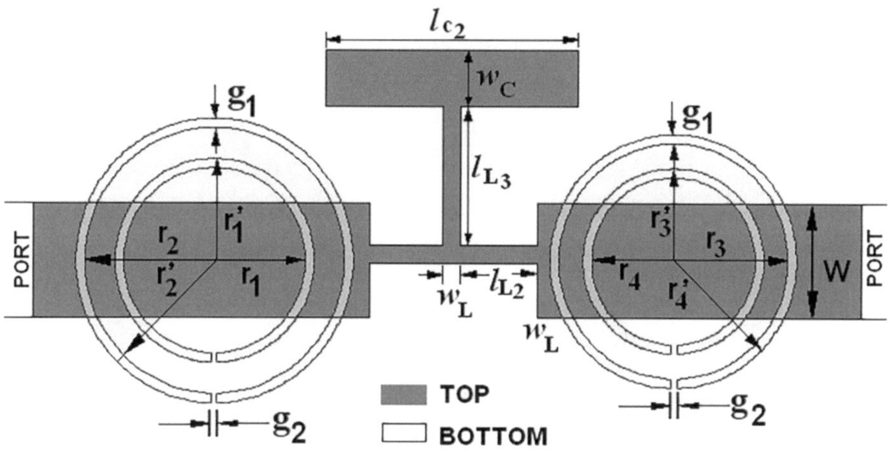

Figure 42.
Schematic diagram of the lowpass filter using open-ended split ring DGS sections and T-shaped stub.

DGS placed at the left-hand side r_2 = 4.3 mm, the outer radius of the smaller split ring DGS placed at the left-hand side r_2' = 4.6 mm, the inner radius of the smaller split ring DGS placed at the right-hand side r_3 = 2.7 mm, the outer radius of the smaller split ring DGS placed at the right-hand side r_3'= 3 mm, the inner radius of the bigger split ring DGS placed at the right-hand side r_4 = 3.7 mm, the outer radius of the bigger split ring DGS placed at the right-hand side r_4' = 4.1 mm, the distance between two sets of DGS units, L = 15 mm, the slot widths of the four DGS units are same and are given by, g_1= 0.3 mm, the split gaps of the DGS units are also alike and are referred to as, g_2 = 0.2 mm.

2.4.4 Simulated result of the modified third order elliptic filter with DGSs

The design is simulated, and the corresponding simulated response is shown in **Figure 43**. It is evident from the response that the filter's stopband is increased, and –15 dB bandwidth is obtained as 2.7 GHz. The proposed filter exhibit four transmission zeros responsible for four DGS units. These pole frequencies occur at 3.1 GHz, 3.9

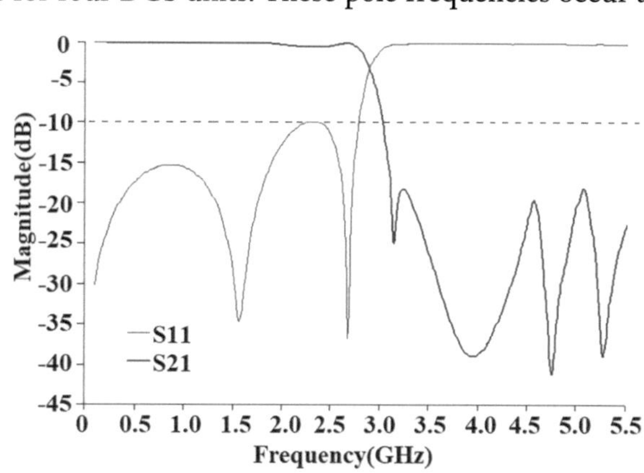

Figure 43.
S-parameter response of the lowpass filter using double unit split ring DGS sections and Open-ended T-shaped stub.

Figure 44.
Photographic view of the fabricated prototype top plane.

Figure 45.
Photographic view of the fabricated prototype bottom plane.

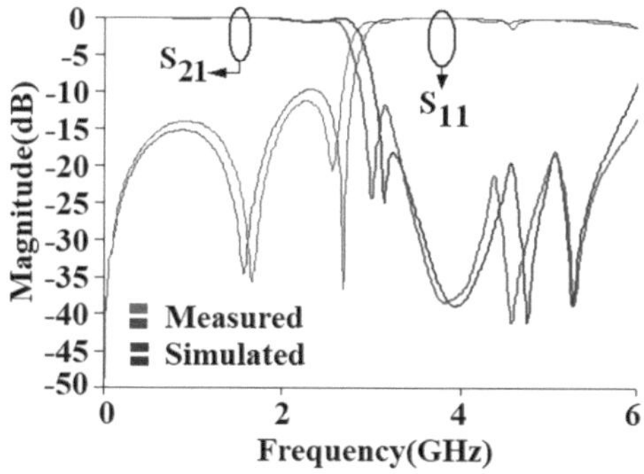

Figure 46.
Comparison of the simulated and measured S-parameter result.

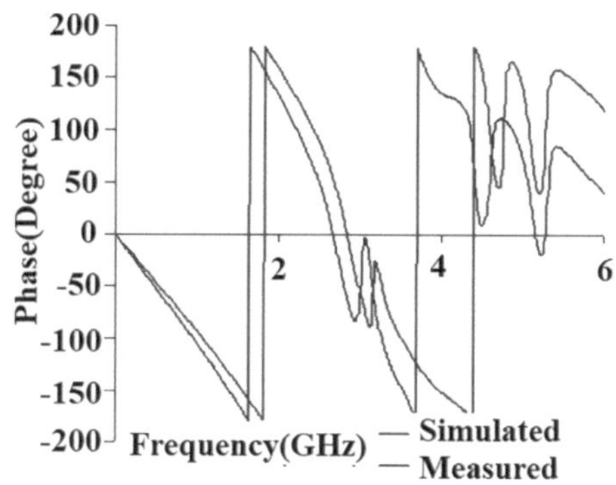

Figure 47.
Comparison of the simulated and measured phase response.

GHz, 4.7 GHz, and 5.2 GHz. The filter's cutoff frequency is obtained at 2.89 GHz with an insertion loss of less than 0.5 dB. The maximum rejection level of the stop-band is 16.3 dB, and the sharpness factor is increased to 100.9 dB/GHz. Therefore the proposed filter provides greater bandwidth and sharper roll-off factor by applying the four split ring-shaped DGS units.

2.4.5 Measured result of the modified third order elliptic filter with DGSs

The fabricated prototype verifies the simulated response, and a photograph of the fabricated layout is shown in **Figures 44** and **45**. There is good agreement between simulated and measured results, plotted in **Figures 46** and **47**, showing the variation of simulated and measured phase response with the frequency.

3. Conclusions

Here in this chapter, different DGS structures and their characteristics are studied. The hexagonal head dumbbell DGS provides stopband performance with a 4.3% reduced size than the conventional square head dumbbell DGS. The circular split ring DGS is very compact and provides a bandstop response. This type of structure has a very narrow slot gap and a very small cross-sectional area of $0.07\,\lambda_0 \times 0.07\,\lambda_0$. This

References	Compactness $\lambda_0 \times \lambda_0$	Sharpness factor at lower side (dB/GHz)	Insertion Loss at passband (dB)
[4]	0.13 × 0.33	5.1	—
[17]	0.30 × 0.06	6	—
[5]	0.03 × 0.08	51.1	−1.3
[9]	0.07 × 0.11	45	−0.43
Proposed unit DGS structures	Compactness $\lambda_0 \times \lambda_0$	Sharpness factor at lower side (dB/GHz)	Insertion Loss at passband (dB)
Hexagonal shape	0.04 × 0.12	11.6	−0.03
Circular Split ring shape	0.07 × 0.07	79.2	−0.07

Table 9.
Some advantages of the proposed DGS structures as compared to state of the art.

DGS structure also provides a very sharp roll-off factor of 79.2 dB/GHz. The above table (**Table 9**) provides some advantages of the proposed DGS structures compared to some unit structures available in the literature. Since here, the main intention is to compare the qualitative analysis of the proposed DGS units to that of the DGS units available in the literature. Due to the non-availability of data, some insertion loss boxes in the above table are kept blank.

Author details

Somdotta Roy Choudhury
Ramrao Adik Institute of Technology, D.Y. Patil Deemed to be University, Navi Mumbai, India

*Address all correspondence to: somdottaroychoudhury@gmail.com

References

[1] Radisic V, Qian Y, Coccioli R, Itoh T. Novel 2-D photonic bandgap structures for microstrip line. IEEE Microwave Guided Wave Letters. 1998;**8**:69-71. DOI: 10.1109/75.658644

[2] Yang F-R, Ma K-P, Qian Y, Itoh T. A uniplanar compact photonic-band gap (UC-PBG) structure and its applications for microwave circuits. IEEE Transactions on Microwave Theory and Techniques. 1999;**47**:1509-1514. DOI: 10.1109/22.780402

[3] Kim C-S, Park J-S, Ahn D, Lim J-B. A novel 1-D periodic defected ground structure for planar circuits. IEEE Microwave Guided Wave Letters. 2000; **10**:131-133. DOI: 10.1109/75.846922

[4] Ahn D, Park J-S, Kim C-S, Kim J, Qian Y, Itoh T. A design of the low-pass filter using the novel microstrip defected ground structure. IEEE Transactions on Microwave Theory and Techniques. 2001;**49**:86-93. DOI: 10.1109/22.899965

[5] Rahman A-B-A, Verma A-K, Boutejdar A, Omar A-S. Control of bandstop response of Hi-Lo microstrip low-pass filter using slot in ground plane. IEEE Transactions on Microwave Theory and Techniques. 2004;**52**: 1008-1013. DOI: 10.1109/TMTT.2004.823587

[6] Man M-K, Sanyal S. A novel defected ground structure for planar circuits. Microwave & Wireless Component Letters. 2006;**16**:93-95. DOI: 10.1109/LMWC.2005.863192

[7] Karmakar N-C, Roy S-M. Quasi-static modelling of defected structure. IEEE Transactions on Microwave Theory and Techniques. 2006;**54**:2160-2168

[8] Weng L-H, Guo Y-C, Shi X-W, Chen X-Q. An overview on defected ground structure. Progress in Electromagnetics Research B. 2008;**7**:173-189. DOI: 10.2528/PIERB08031401

[9] Huang S-Y, Lee Y-H. A compact E-shaped patterned ground structure and its applications to tunable bandstop resonator. IEEE Transactions on Microwave Theory and Techniques. 2009;**57**:657-666. DOI: 10.1109/TMTT.2009.2013313

[10] Parui S-K, Das S. Modeling of modified split-ring type defected ground structure and its application as bandstop filter. Radio Engineering. 2009;**18**: 149-154

[11] Biswas R-N, Kar A. A novel PSO-IE3D based design and optimization of a low profile Dual Slot Microstrip Patch Antenna. IEEE Region 10 International Conference TENCON (TENCON '08); 19–21 November 2008; Hyderabad, India; 2009. p. 1-4

[12] Pozar D-M. Microwave Engineering. 4th ed. USA: John Wiley & Sons; 2017. p. 399

[13] Meyers R-G, Ye Q. Incorporation of Zeland's IE3D in the microwave and RF classroom. In: IEEE Antennas and Propagation Society International Symposium (APS '02). San Antonio, TX; 2002. pp. 688-691

[14] Park J-S, Kim J-H, Lee J-H. A novel equivalent circuit and modeling method for defected ground structure and its application to optimization of a DGS lowpass filter. IEEE MTT-S International Microwave Symposium Digest. 2002;**1**: 417-420. DOI: 10.1109/MWSYM.2002.1011644

[15] Park J-S, Yun J-S, Ahn D. A design of the novel coupled-line bandpass filter using defected ground structure with wide stopband performance. IEEE Transactions on Microwave Theory and Techniques. 2002;**50**:2037-2043. DOI: 10.1109/TMTT.2002.802313

[16] Lim J-S, Kim C-S, Lee Y-T, Ahn D, Nam S. Design of lowpass filters using defected ground structure and compensated microstrip line. Electronics Letters. 2002;**38**:1357-1358. DOI: 10.1049/el:20020889

[17] Kim C-S, Lim J-S, Nam S, Kang K-Y, Ahn D. Equivalent circuit modelling of spiral defected ground structure for microstrip line. Electronics Letters. 2002;**38**:1109-1110. DOI: 10.1049/el: 20020742

Chapter 7

Tunable Filter

Di Lu

Abstract

Tunable filters enable dynamic spectrum access for the wireless systems, and the tunable bandpass filters with constant bandwidth (BW) are most favorable for practical applications. This chapter investigates the synthesis and realization techniques for the tunable filters using the coupling matrix with variable entries synthesizes the tunable filter and guides the filter design. The synthesis method and the matrix extraction procedures for the constant-bandwidth bandpass filter are included, and the typical numerical examples are given. This chapter also discusses the relationship between the theoretical matrix and the physical circuits, and then a planar tunable filter design is presented to verify this relationship. Furthermore, the general approach to designing the constant-bandwidth filters using the element variable coupling matrix is concluded. The planar circuit, as well as the 3D structure realizations, are offered to practically demonstrate the synthesis design approach.

Keywords: tunable filter, constant bandwidth, constant shape, synthesis of tunable filter, element variable coupling matrix

1. Introduction

The tunable filter is generally employed as a switched filter bank with a substantially reduced size. It self-adaptive selects the frequency spectrum and filters out undesired signals to meet the need for tunability/reconfiguration in wireless systems. As exemplified in **Figure 1**, the core part of a typical mobile transceiver module is realized by a transceiver IC with many blocks of filters (or multiplexers), switch matrices, power amplifiers, and two antennas. The filters andswitch matrices constitute the unintegrable switched filter banks that are used to select the signal dynamically. It is predictable that if the compact and high-performance tunable filters replace the bulky filter banks, this part of the circuit will consist of only a transceiver IC, two tunable filters, a wideband power amplifier, and two antennas (right side of **Figure 1**). Nowadays, with a more complex wireless electromagnetic environment, the frequency spectrum is more crowded, and thus it is even more significant to facilitate efficient utilization of the available frequency spectrum. The tunable filter, which plays a crucial role in utilizing the frequency spectrum, has become the hotspot not only in the research area but also for industrial applications.

The filter with tunability and without too much Q degradation known as the high-Q tunable filter has been widely used in the industry. Magnetically tunable filters or yttrium iron garnet (YIG) filter, which provides the tunable band with the high Q factor and multi-octave tuning range, is the essential part in front-ends of the microwave test and

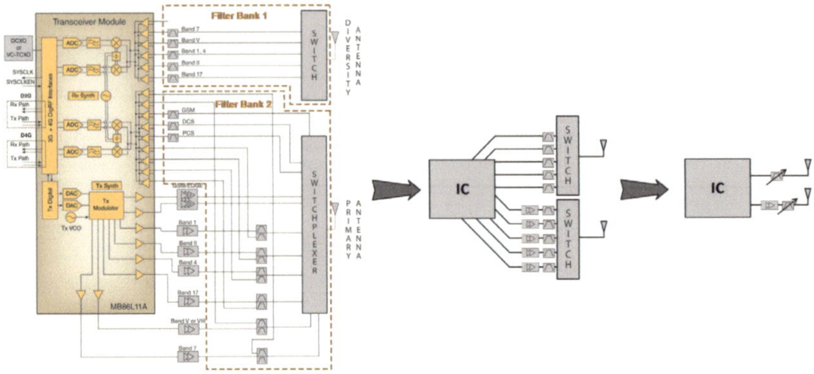

Figure 1.
Tunable filters replace switched filter banks with a substantially reduced size [1].

measurement instruments [2, 3]. This type of filter can be tunable by magnetically adjusting the ferrimagnetic resonance of the crystal YIG spheres, thus resulting in the filter frequency adjustment. The mechanically tunable filter is another type of high-Q filter that can provide the high-Q tunable passband with a relatively compact size (compared with the YIG filter). This type of filter's frequency or response shape is reconfigured by changing the physical dimensions of the filter structures or disturbing the electromagnetic field in the resonators. The mechanically tunable filter has been commonly employed for tunable wireless infrastructure equipment or reconfigurable communication satellite operators to extend their service life and functionality [4–6]. A great deal of high-Q mechanically tunable filters, including coaxial, waveguide, or dielectric resonator structures, have been exploited. For example, mechanically adjusting the end-loading capacitors (or equivalent capacitors) of each coaxial resonator, the coaxial filter can be tunable with a wide tuning range [7–10]. The reconfigurability of the waveguide filter is enabled by reshaping the cavity dimension of the resonator or moving the perturbations inserted in the waveguide cavities [11–15]. For the dielectric resonator filter, moving the movable disks above the dielectric resonators can tune their resonant frequencies resulting in filter passband tunability [16–18].

The ever-increasing demand for the miniaturized and highly integrated wireless system requires the future tunable filter with a more compact size and fully electrical control. The giant tuning mechanisms inevitably make YIG filters and mechanically tunable filters oversize. With this regard, the electrically tunable filter with the semiconductor tuning element has been drawing a lot of attention and getting extensively exploited because of its very compact size, fast tuning speed, and straightforward control mechanism, even though the semiconductor tuning element loaded on the resonator will dramatically deteriorate the filter Q factor. Planar tunable filter is a popular research topic because of its easy integration with semiconductor tuning elements. For example, the planar $\lambda/4$, $\lambda/2$, and multimode resonators loaded by tunable varactors or PIN diodes or both are employed to construct the tunable filters with frequency and bandwidth (BW) control [19–23]. In addition, emerging tuning semiconductor devices such as Radio Frequency Microelectromechanical Systems (RF-MEMS) and ferroelectric devices are also used as the variable capacitor in the planar tunable filters to alleviate the Q factor deterioration [20, 24–26]. Aside from the planar filter, the high-Q tunable three-dimensional (3D) filter with tuning semiconductor elements is also a research hotspot because of its low loss, good power handling, and high selectivity. For example, coaxial filters or quasi-coaxial filters [7, 27–31], dielectric resonator filters [32, 33], and waveguide filters [34] are loaded by the various variable capacitors or switchable devices, thus constructing the tunable/reconfigurable 3D high-Q filters.

Among a large variety of tunable filters, the bandpass filter with the tunable center frequency (CF) is attractive for its widespread application. It is preferable to use the minimum number of tuning elements to control the frequency of the bandpass filter and realize the constant absolute bandwidth (BW). This realization will minimize the Q fact degradation introduced by the loaded tuning elements and maintain the filter response shape as the frequency is tuned. It is also the simplest tunable filter with the most straightforward control mechanism. Therefore, the tunable filter with constant bandwidth has been one of the emerging trends in filter design. For example, the planar filters [35–37], waveguide filters [15], coaxial filters [7], etc., have been all investigated to approach the constant-BW tunable filters.

In general, since the tunable filters have not large-scale replaced the filter banks, the research and development of the tunable filter is still indispensable, especially for the frequency tunable bandpass filter with constant bandwidth. This chapter will mainly deal with the tunable bandpass filter and offer the general synthesis-based approach. The synthesis method, tuning behavior, and physical realization techniques are included and discussed. The synthesis is based on the coupling matrix where the elements are variable, and the coupling matrix with the variable elements can represent the tunable filter response as well as the tuning behavior. The direct relationship between the matrix and the filter realization will be established, and thus various physical structures can be employed to realize the tunable filter accordingly. Furthermore, the constant-BW tunable filter will also be investigated, and the synthesis with its design approach will be included. A planar and a 3D tunable filter design examples will be offered to realize the theory.

2. Element variable coupling matrix (EVCM)

The coupling matrix is the most commonly used technique to synthesize the fixed filter. One can prescribe or optimize the filter function in terms of the fixed filter specification, and then the coupling matrix can be directly extracted from the filter function. After a few iterations of mathematical manipulation/optimization, the coupling matrix can physically correspond to the filter structures. This is the synthesis process for the fixed filter. However, for the tunable filter, the frequency of the resonator is variable. When the filter is tuning (tunable frequency or tunable BW), there are numerous filter prototypes (filter matrices) corresponding to the same filter response. As illustrated in **Figure 2**, one can extract three different coupling matrices according to the same filter response using different tuning frequencies (f_d) and different tuning bandwidths (BWs). This means three coupling matrices can be corresponding to the same filter response. Obviously, the conventional coupling matrix is not appropriate to synthesize the tunable filter, and therefore the element variable coupling matrix (EVCM) is introduced.

The EVCM is derived from the conventional coupling matrix and can be treated as the tunable version of the coupling matrix. First of all, the conventional coupling matrix can be extracted using the frequency-fixed filter synthesis (e.g. using the matrix manipulation method [5] or optimization process [38]) based on the prescribed frequency response. For the lossless $N \times N$ coupling matrix with the termination impedance R_1, the response is calculated based on the current loop model and voltage node model as:

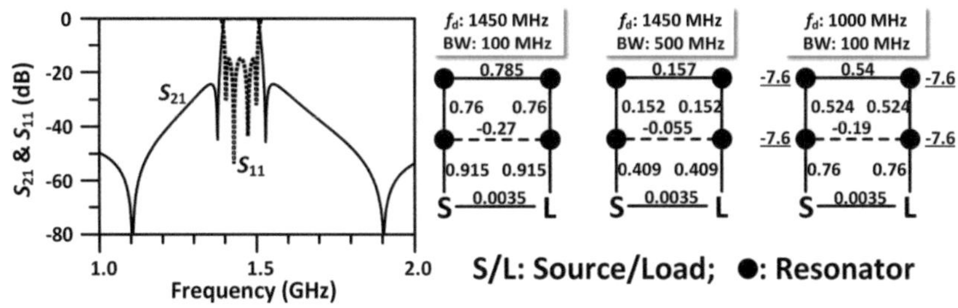

Figure 2.
Same response of the three different classical coupling matrices.

$$\begin{cases} S_{21} = \dfrac{2R_1}{w_0 L \times FBW_d}[A]^{-1}{}_{N,1} = \dfrac{2}{w_0 C R_1 \times FBW_d}[A]^{-1}{}_{N,1} \\ S_{11} = 1 - \dfrac{2R_1}{w_0 L \times FBW_d}[A]^{-1}{}_{1,1} = 1 - \dfrac{2}{w_0 C R_1 \times FBW_d}[A]^{-1}{}_{1,1} \end{cases} \tag{1}$$

where

$$[A] = \begin{bmatrix} 1/(Q_e \times FBW_d) + \Lambda_d + jM_{11} & jM_{12} & \cdots & jM_{1n} \\ jM_{21} & \Lambda_d + jM_{22} & \cdots & jM_{2n} \\ \vdots & \vdots & \vdots & \vdots \\ jM_{n1} & jM_{n2} & \cdots & 1/(Q_e \times FBW_d) + \Lambda_d + jM_{kk} \end{bmatrix} \tag{2}$$

$$\begin{cases} \dfrac{1}{Q_e} = \dfrac{R_1}{w_0 L} = \dfrac{1}{R_1 w_0 C} \\ \Lambda_d = j\left(\dfrac{f}{f_d} - \dfrac{f_d}{f}\right)/FBW_d \\ M_{kk} = \left(\dfrac{f_d}{f_{kk}} - \dfrac{f_{kk}}{f_d}\right)/FBW_d \end{cases} \tag{3}$$

Λ_d and M_{ij} denote normalized frequency, and element of the coupling coefficient between resonator i and j. M_{kk} is the self-coupling coefficient or immittance, representing the corresponding resonator's frequency shift. f_{kk} is the resonant frequency of each resonator (or resonator k). As implied by (1–3), the passband response is calculated from a coupling matrix when the CF (scaling frequency f_d) and fractional bandwidth (*FBW*) are given. However, for the tunable filter, both BW (or*FBW*) and CF are variable. Thus, it is difficult to form a tunable passband by a coupling matrix and then realize the filter accordingly (as demonstrated in **Figure 2**). The tuning range and behavior are also important since the passband is tunable, but the conventional matrix cannot work. Therefore, the coupling matrix is reformulated as:

$$\begin{cases} S_{21} = \dfrac{2}{Q_e}[B]^{-1}{}_{N,1} \\ S_{11} = 1 - \dfrac{2}{Q_e}[B]^{-1}{}_{1,1} \end{cases} \tag{4}$$

$$[B] = \begin{bmatrix} \frac{1}{Q_e + \lambda_d + jm_{11}} & jm_{12} & \cdots & jm_{1n} \\ jm_{21} & \lambda_d + jm_{22} & \cdots & jm_{2n} \\ \vdots & \vdots & \vdots & \vdots \\ jm_{n1} & jm_{n2} & \cdots & \frac{1}{Q_e + \lambda_d + jm_{kk}} \end{bmatrix} = ([Q_e] + \lambda_d[W] + j[m_\Delta]) \tag{5}$$

$$\begin{cases} \lambda_d = j\left(\frac{f}{f_d} - \frac{f_d}{f}\right) \\ m_{kk} = \left(\frac{f_d}{f_{kk}} - \frac{f_{kk}}{f_d}\right) \end{cases} \tag{6}$$

$$[Q_e] = \begin{bmatrix} \frac{1}{Q_e} & 0 & \cdots & 0 \\ 0 & 0 & \cdots & 0 \\ \vdots & \vdots & \vdots & \vdots \\ 0 & 0 & \cdots & \frac{1}{Q_e} \end{bmatrix} \tag{7}$$

Here, the denormalized matrix $[m_\triangle]$ has an interesting feature. For demonstration, a dual-mode resonator filter (or two-order transversal filter) is taken as an example (as shown in **Figure 3**). The matrix $[m_\triangle]$ of this filter can be extracted according to the aforementioned techniques as:

$$[m_\Delta] = \begin{bmatrix} m_{ee} & m_{eo} \\ m_{oe} & m_{oo} \end{bmatrix} = \begin{bmatrix} x & 0 \\ 0 & y \end{bmatrix} \tag{8}$$

For simplification, the external Q factors $Q_{exe/exo}$ for two resonant modes are roughly approximate to Q_e as:

$$Q_e = \frac{(Q_{exe} + Q_{exo})}{2} \tag{9}$$

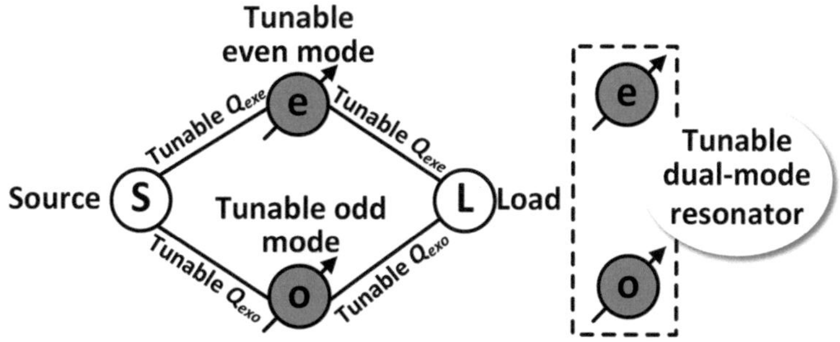

Figure 3.
Architecture of the dual-mode resonator tunable filter.

It can be derived that

$$\begin{cases} x = f_d/f_e - f_e/f_d \\ y = f_d/f_o - f_o/f_d \\ Q_{exe/exo} = \dfrac{f_{exe/exo}}{\Delta f_{\pm 90}} \approx \dfrac{1}{FBW_{exe/exo}} = \dfrac{f_{exe/exo}}{BW_{exe/exo}} \end{cases} \tag{10}$$

Now, it can be found that all elements of matrix $[m_{\triangle}]$ are directly related to the architecture. As long as the parameters of the architecture are defined with the given tuning ranges and the scaling frequency f_d, the responses of the tunable filter (e.g. tuning ranges of CF, BW, return loss (RL)) can be completely determined. The second-order filter example is given for demonstration, as shown in **Figure 4**, where the matrix is denoted as x, y, and Q_e. **Figure 5** presents the responses of the matrix when the x is varied. It is observed that only the even-mode resonant frequency is tuned, implying even-mode resonant frequency is independently controlled by x. Similarly, **Figures 6** and **7** show that the odd-mode frequency and RL independently varied by changing y and Q_e, which suggests that the odd-mode resonant frequency and the RL are independently controlled by y and Q_e.

With this interesting feature, a one-to-one correspondence between the tunable filter responses and this matrix can be established. As a result, the matrix can uniquely represent tunable filter responses. **Figures 8–10** presents the typical responses by purposely varying the matrix elements $[m_{\triangle}]$. As shown in **Figure 8**, the matrix elements x, y, and Q_e are changed from 0.058, 0.034, and 18.765 to 0.028, 0.044, and 27.777, thus forming a 0.8 GHz to 1.2 GHz CF-tunable filter response with a constant 130 MHz BW. Hence, this tunable filter can be synthesized by such a coupling matrix

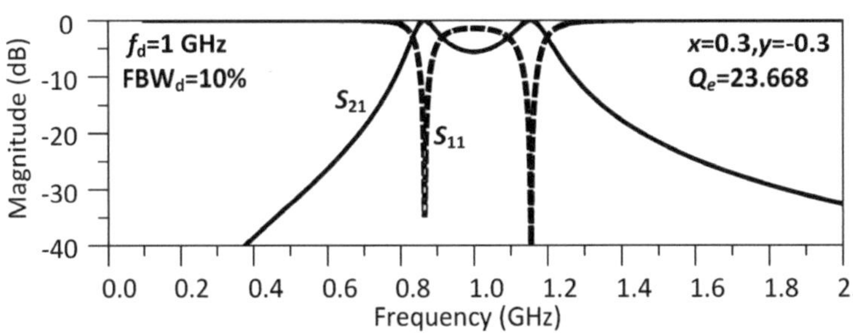

Figure 4.
Theoretical response of a fixed filter example.

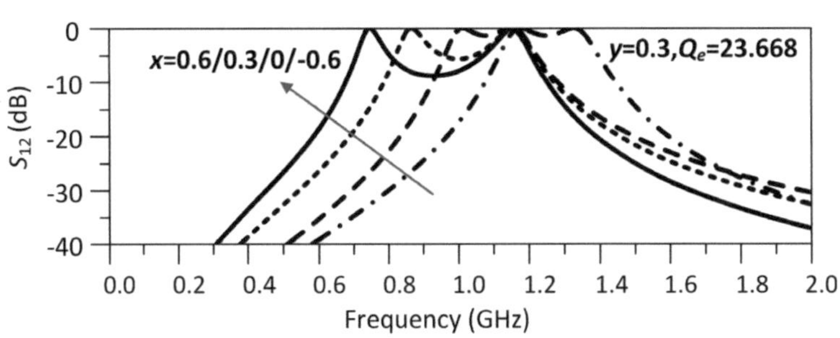

Figure 5.
Theoretical curves for varying x.

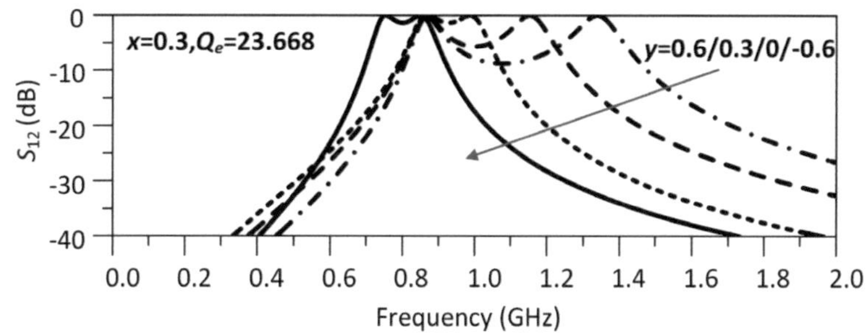

Figure 6.
Theoretical curves for varying y.

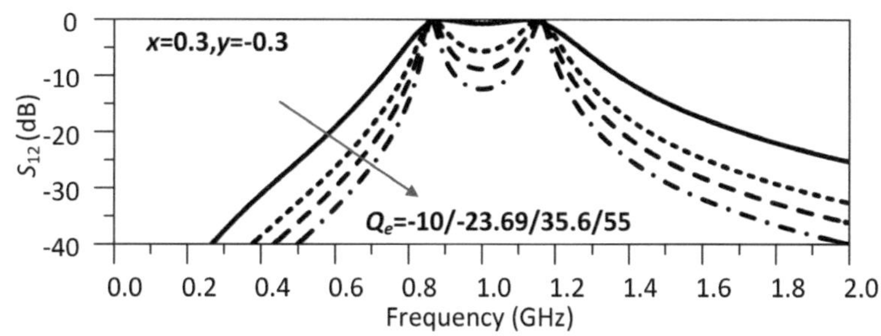

Figure 7.
Theoretical curves for varying Q_e.

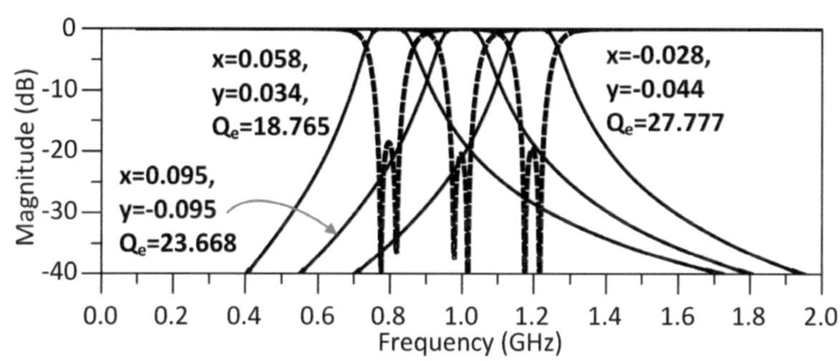

Figure 8.
Theoretical curves for tuning the CF with 130-MHz 3-dB BW and 20-dB RL.

with these variable elements. Similarly, 130 MHz to 300 MHz BW-tunable filter responses with the constant 1 GHz CF (as shown in **Figure 9**) and RL-tunable filter responses with 1 GHz CF and130 MHz BW (as shown in **Figure 10**) are formed by specifying the elements of the matrix. So, the element variable coupling matrix (EVCM) can synthesize these tunable filters.

The physical structure is presented to realize the EVCM, as shown in **Figure 11**. The given circuit is controlled by loaded varactor diodes, and each control element can be used to tune the entry of the EVCM separately. For example, the variable capacitor C_o and C_e, respectively, control the odd-mode and even-mode resonant frequencies, and adjusting C_m will tune the external quality factor Q_e. Therefore, this structure is a fully tunable filter. Since the multimode structures are challenging to be coupled with equal energy (Q_{exo} = Q_{exe}) for every resonant mode, the added circuit configuration

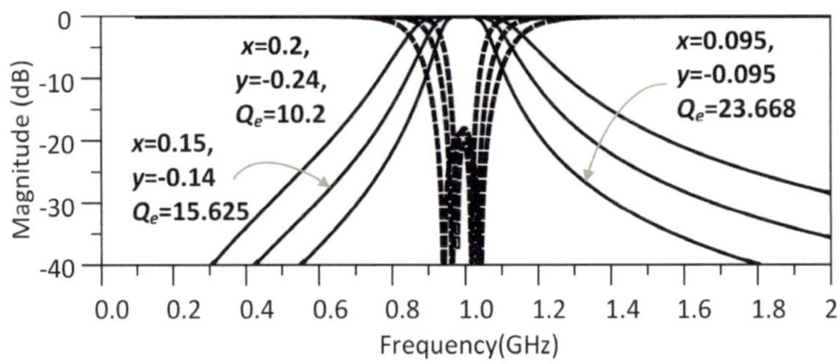

Figure 9.
Theoretical curves for tuning the BW with 1 GHz CF and 20-dB RL.

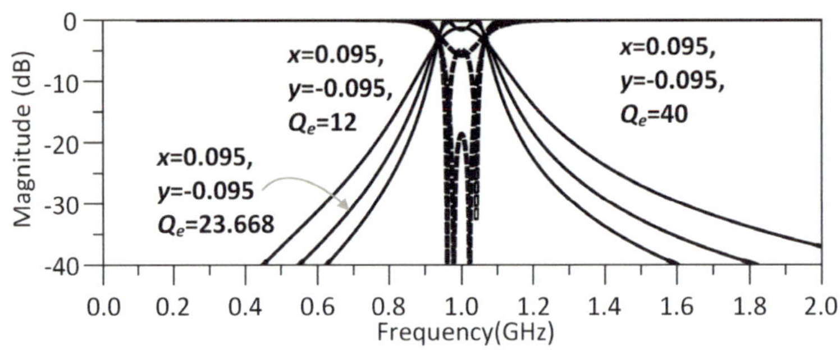

Figure 10.
Theoretical curves for the RL reconfiguration with 1 GHz CF and 130 MHz BW_{3db}.

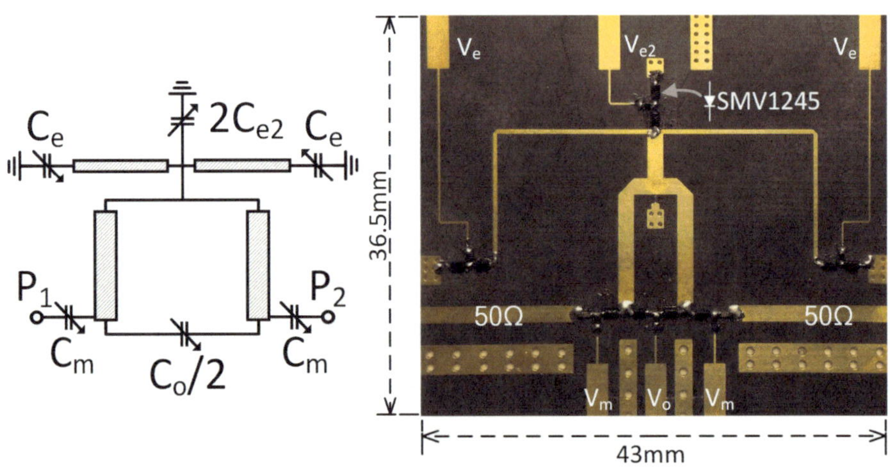

Figure 11.
Transmission line model and physical circuit.

with C_{e2} is employed. Properly adjusting C_{e2} will enforce $Q_{exo} = Q_{exe}$ and fully implement the EVCM.

Figures 12–14 presents the measurement results of the physical circuit corresponding to the calculated results of EVCM. As can be seen, the in-band responses of the filter are all very close to the given EVCM. The CF frequency of the filter can be tuned from 0.8 GHz to 1.2 GHz, and a 130-MHz 3-dB BW is kept nearly

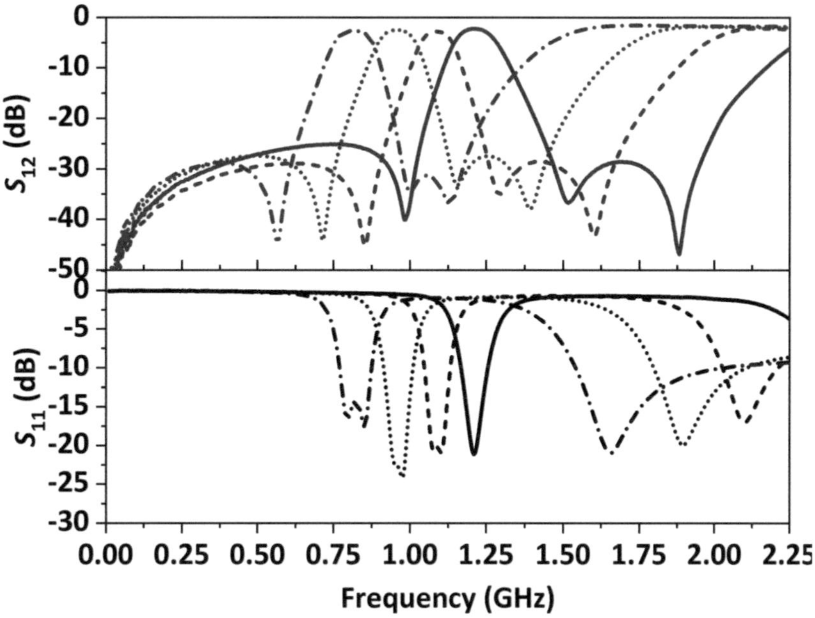

Figure 12.
Measurement curves for tuning the CF with constant BW.

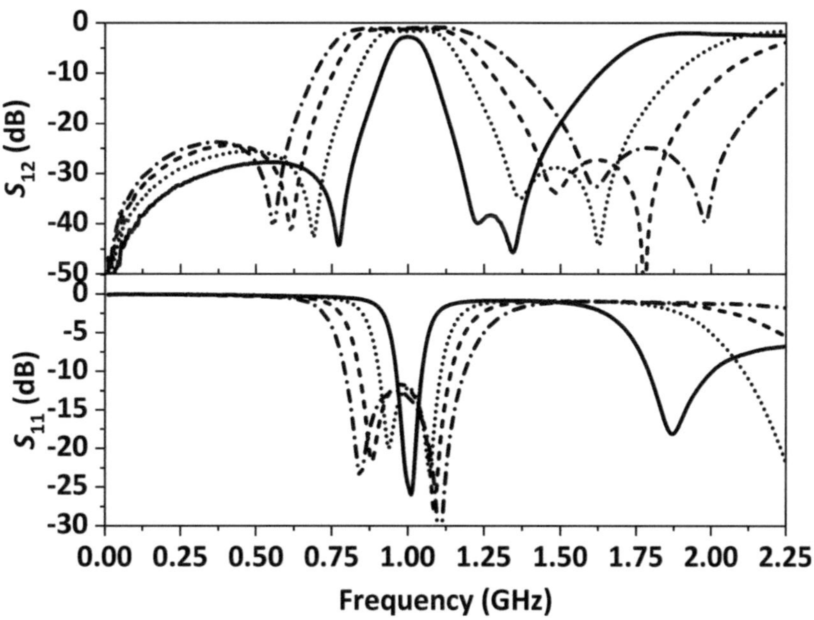

Figure 13.
Measurement curves for tuning the BW with constant CF.

constant, as implied in **Figure 12**. These responses closely agree with the EVCM results as shown in **Figure 8**. The BW varying from 50 to 400 MHz is measured as shown in **Figure 13** when the circuit is fully tuned, which corresponds to the EVCM results as

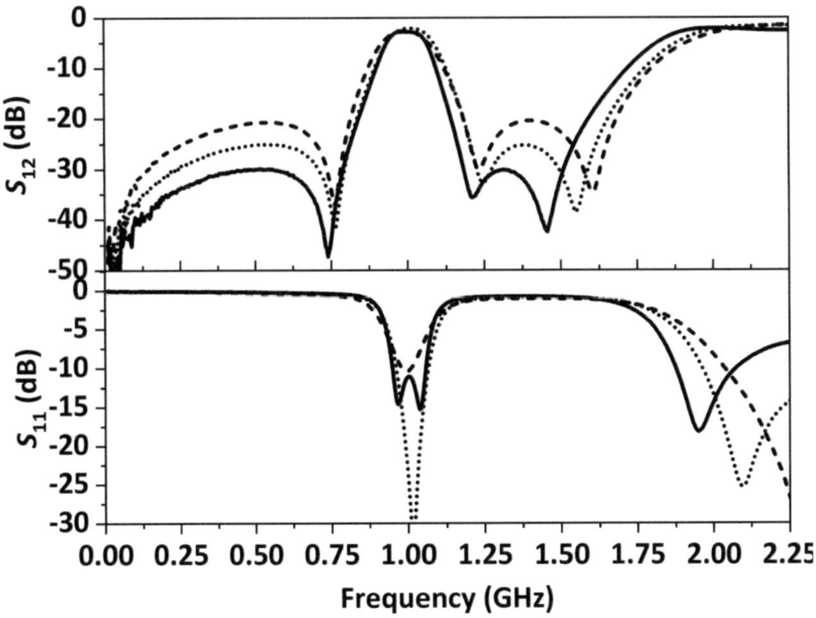

Figure 14.
Measurement curves for tuning the RL with constant CF and BW.

shown in **Figure 9**. **Figure 14** shows the RL reconfiguration results of the filter. As shown, the passband is fixed at 1 GHz with 130-MHz BW, while the RL is reconfigured as prescribed by the EVCM (**Figure 10**) and the rejection is reshaped. The agreement between the experimental circuit and EVCM is obtained. Note that the practical circuit generates the transmission zeros to enforce $Q_{exo} = Q_{exe}$, which causes the major disagreement in the stopband. These three transmission zeros are owing to the added circuit configuration, which introduces an out-band resonant mode that forms one more coupling path from source to load and generates transmission zeros there.

3. Synthesis of the constant-BW filter using EVCM

With the introduction of the new matrix, the tunable filter can correspond to an EVCM uniquely. Thus, the tunable filter with constant BW can be synthesized with a few steps of the mathematical manipulations. First, for a tunable resonator, the adjustment of the *CF* is realized by tuning the loading effect, which can be treated as adding or subtracting the susceptance from the resonator. Thus, the tunable parameter of a resonator can be modeled by the corresponding diagonal element of an EVCM as:

$$m_{tt} = \left(\frac{f_d}{f_t} - \frac{f_t}{f_d}\right) \tag{11}$$

For the tunable coupled-resonator filters, it is reasonable to define $m_{tt1} = m_{tt2} = m_{tt3} = \ldots = m_{tt}$, and thus, when the filter is adjusted, the frequencies of all the resonators will be tuned with the same tuning step. Additionally, to maintain the constant BW over a wide tuning range, when adjusting CF, the entries of the EVCM (including the external quality factor Q_e) are required to change in a specific manner synchronously.

Based on the earlier discussion, one may assume that the EVCMs are the linear functions of the variable m_{tt}, and thus the EVCM $[m_\Delta]_t$ is converted as (12, 13). $[m_\Delta]$ is the frequency-fixed part of the EVCM, which is unchanged when the frequency is tuned. $[c]$ is a factor matrix and determines the passband variation rate of the filter. When tuning CF, the total matrix varies as predefined by $[c]$, which results in a tunable passband that varies in a linear function according to the predefined response and BW variation:

$$[m_\Delta]_t = \begin{bmatrix} m_{11} + c_{11}m_{tt} & m_{12} + c_{12}m_{kk} & \cdots & m_{1n} + c_{1n}m_{tt} \\ m_{21} + c_{21}m_{tt} & m_{22} + c_{22}m_{tt} & \cdots & m_{2n} + c_{2n}m_{tt} \\ \vdots & \vdots & \vdots & \vdots \\ m_{n1} + c_{n1}m_{tt} & m_{n2} + c_{n2}m_{tt} & \cdots & m_{kk} + c_{kk}(m_{tt}) \end{bmatrix} = ([m_\Delta] + m_{tt}[c]),$$
$$Q_e = \Delta Q_e + m_{kk}c_Q \tag{12}$$

$$[c] = \begin{bmatrix} c_{11} & c_{12} & \cdots & c_{1n} \\ c_{21} & c_{22} & \cdots & c_{2n} \\ \vdots & \vdots & \vdots & \vdots \\ c_{n1} & c_{n2} & \cdots & c_{kk} \end{bmatrix} \tag{13}$$

The linear EVCM that represents the tunable filter with a specified tuning range and reflects the passband variation or the passband tuning behavior is defined. Now, extracting parameters of the EVCM from the general filtering function is of great importance to design the tunable passband with constant *BW*. The extraction method is derived as follows.

First, the tuning range from f_l to f_h is given, and then the mapping frequency can be defined as $f_d = (f_l + f_h)/2$. The RL and BW are, respectively, prescribed at *RL* and *ABW*. Then, the conventional coupling matrix $[M]_d$ is extracted at f_d. Note here that the diagonal elements are all zeros because the resonant frequencies of all resonators are the same (with symmetric distribution) for the simpler case demonstration:

$$\begin{cases} [M]_d = \begin{bmatrix} 0 & M_{12} & M_{13} & \ldots & M_{1N} \\ M_{21} & 0 & M_{23} & & M_{2N} \\ M_{31} & M_{32} & \ddots & & \vdots \\ \vdots & & & \ddots & M_{N-1,N} \\ M_{N1} & & & & 0 \end{bmatrix} \\ q_e \end{cases} \tag{14}$$

The coupling matrix response can be mapped to f_d, scaled by ABW/f_l and moved to f_l. Thus, the coupling matrix is expressed as:

$$\begin{cases} [m]|(m_{tt} = m_{ttl}) = [M]_d \cdot (ABW/f_l) + [W] \cdot m_{ttl} \\ Q_{el} = q_e/(ABW/f_l) \end{cases} \tag{15}$$

$[W]$ is an identity matrix. Similarly, the coupling matrix response can be mapped to f_d, scaled by ABW/f_h, and moved to f_h. Then the coupling matrix is written as:

$$\begin{cases} [m]|(m_{tt} = m_{tth}) = [M]_d \cdot (ABW/f_h) + [W] \cdot m_{tth} \\ Q_{eh} = q_e/(ABW/f_h) \end{cases} \tag{16}$$

Using linear interpolation, the EVCM can be extracted as:

$$\begin{cases} [m_\Delta]_t = ([m_\Delta] + m_{tt}[c]) \\ Q_e = \Delta Q_e + m_{tt} c_Q \end{cases} \tag{17}$$

where

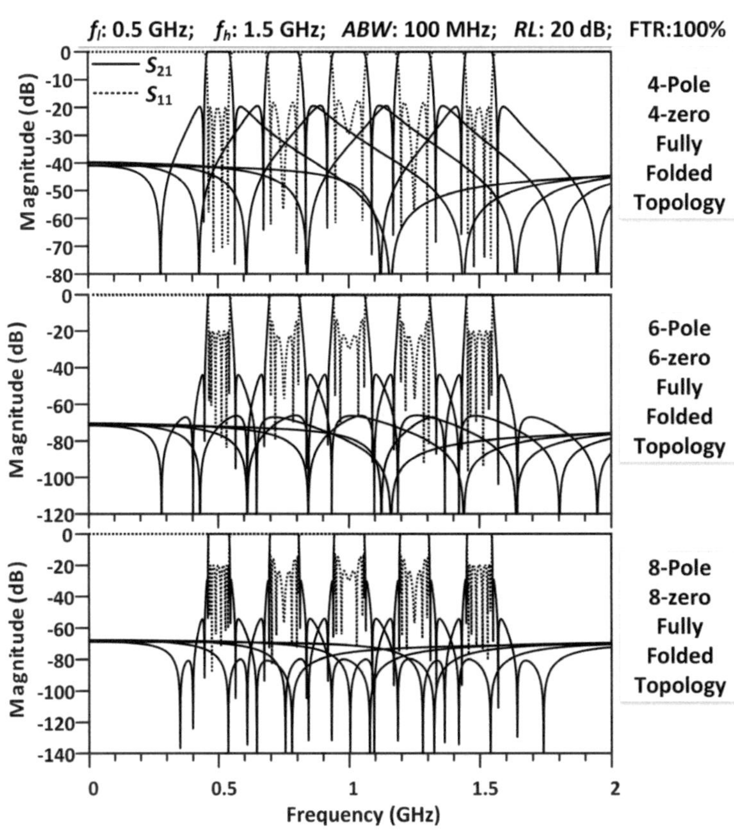

Figure 15.
Extracted 100% tunable 4th-/6th-/8th-order responses with constant 100-MHz ABW.

$$\begin{cases} [m_\Delta] = ABW \cdot [M]_d \cdot \left(\dfrac{(f_h + f_l)}{f_d^{\,2} + f_h f_l} \right) \\ [c] = [M]_d ABW \cdot \dfrac{f_d (f_l\text{-}f_h)}{(f_l f_d^{\,2} - f_h f_d^{\,2} - f_l f_h^{\,2} + f_l^{\,2} f_h)} + [W] \\ \Delta Q_e = \dfrac{q_e}{ABW} \left(\dfrac{f_d^{\,2}(f_h + f_l)}{f_d^{\,2} + f_h f_l} \right) \\ c_Q = \dfrac{q_e}{ABW} \left(- \dfrac{(f_d f_h f_l)}{f_d^{\,2} + f_h f_l} \right) \end{cases} \tag{18}$$

Now the extraction process is completed and the constant-BW tunable filter is synthesized by the EVCM.

For illustration, the 4th-/6th-/8th-order fully canonical folded filters are extracted with 20-dB RL, a 100% frequency tuning range (from 0.5 GHz to 1.5 GHz), and a constant 100 MHz *ABW*. The extracted EVCMs are given as follows:

Four-pole filter:

$$\begin{cases} [m]_{4-pole} = \begin{cases} m_{12} = 0.09259 + m_{kk} \cdot 0.0463 \\ m_{23} = 0.095338 + m_{kk} \cdot 0.04767 \\ m_{14} = -0.04064 - m_{kk} \cdot 0.0203 \end{cases} \\ Q_{e4-pole} = 11.114465 - m_{kk} \cdot 4.1679, m_{sl} = -5.15\text{e-}04 \end{cases} \tag{19}$$

Six-pole filter:

$$\begin{cases} [m]_{6-pole} = \begin{cases} m_{12} = 0.09487 + m_{kk} \cdot 0.04744 \\ m_{23} = 0.066177 + m_{kk} \cdot 0.03309 \\ m_{34} = 0.080465 + m_{kk} \cdot 0.04023 \\ m_{25} = -0.01828 - m_{kk} \cdot 0.0091 \\ m_{16} = 0.00137 + m_{kk} \cdot (6.867\text{e-}04) \end{cases} \\ Q_{e6-pole} = 11.5384 - m_{kk} \cdot 4.3269, m_{sl} = -1.5\text{e-}05 \end{cases} \tag{20}$$

Eight-pole filter:

$$\begin{cases} [m]_{8-pole} = \begin{cases} m_{12} = 0.09232 + m_{kk} \cdot 0.04616 \\ m_{23} = 0.0657 + m_{kk} \cdot 0.03285 \\ m_{34} = 0.0469 + m_{kk} \cdot 0.02345 \\ m_{45} = 0.09855 + m_{kk} \cdot 0.04927 \\ m_{36} = -0.048 - m_{kk} \cdot 0.02404 \\ m_{27} = 0.009 + m_{kk} \cdot 0.0045 \\ m_{18} = -7.28\text{e-}04 - m_{kk} \cdot 3.64\text{e} - 04 \end{cases} \\ Q_{e8-pole} = 11.84219 - m_{kk} \cdot 4.4408, m_{sl} = 2.057\text{e} - 05 \end{cases} \tag{21}$$

Figure 15 presents the calculation responses of three extracted EVCMs. It can be seen that the prescribed responses are tuned from 0.5GHz to 1.5 GHz, and their responses are kept nearly constant. Even though such a wide tuning range and 8th-order function are predefined, the maximum estimation error of this synthesis method only affects the RL. When the frequency of the passband is tuned closer to the middle point of the tuning range [$f_d = (f_h+f_l)/2$], the most significant RL deviation will be introduced, which will lead to deterioration of the high-order filter. However, it is still acceptable for the high-order tunable design with a wide FTR (100%).

4. Planar and 3D realizations of the constant-BW tunable filter based on the EVCM synthesis

The synthesis method can be used to extract the EVCM according to the prescribed filter specifications. This section will present the planar and 3D tunable filters to realize the extracted EVCMs practically.

4.1 Planar realization example of the constant-BW tunable filter

The resonator of the planar filter example is shown in **Figure 16**, and the frequency range design method is given in **Figure 17**. The resonant frequency is tuned by changing variable capacitor C_t, and the tuning range can be predefined by l_1. The capacitor loading position l_t does not noticeably affect the tuning frequency.

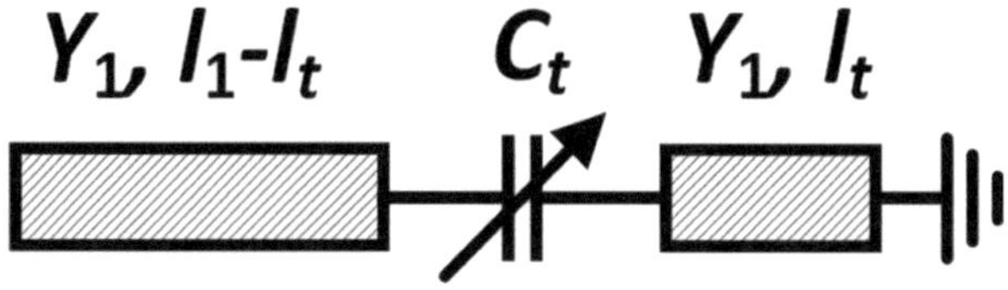

Figure 16.
Transmission line model of the synchronously tunable filter resonator.

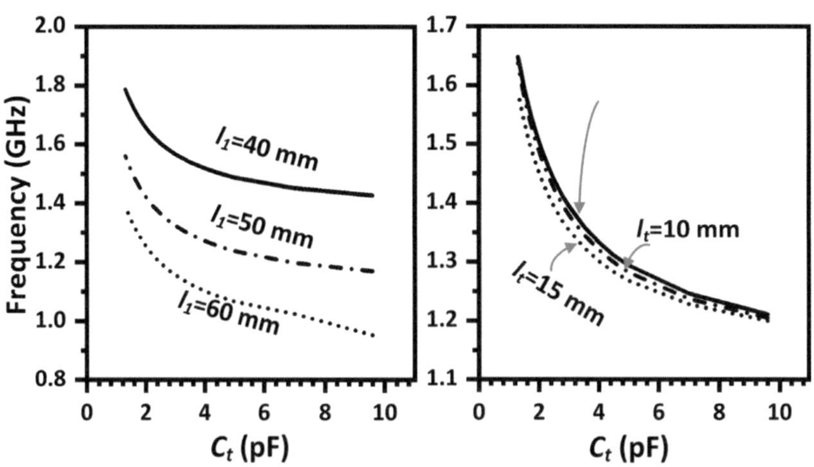

Figure 17.
Tunable resonant frequency vs. C_t, with different l_1 or l_t for (a) l_t = 11 mm (b) l_1= 48.6 mm.

Figures 18 and **19**, respectively, present the electrical coupling (EC) and magnetic coupling (MC) configurations between two planar resonators. Their corresponding coupling coefficient curves extracted from the configurations are shown in **Figures 20** and **21**. As can be seen, both the slope and position of coupling coefficient curves can be independently controlled with these two coupling configurations. Thus, the coupling elements in the EVCM can be physically realized, and both EC and MC are available.

Two feeding structures, i.e. short-end and open-end feeding structures, can be employed to feed this type of filter, as shown in **Figure 22a** and **b**. **Figures 23** and **24**, respectively, present the extracted external quality factor Q_e curves for these two configurations. It can be seen that both the slope and position of Q_e curves are approximately defined by adjusting their physical dimensions independently.

With all filter parts mentioned earlier, two tunable filters using EC and MC as the mainline coupling path to implement the constant BW are demonstrated. Two filters

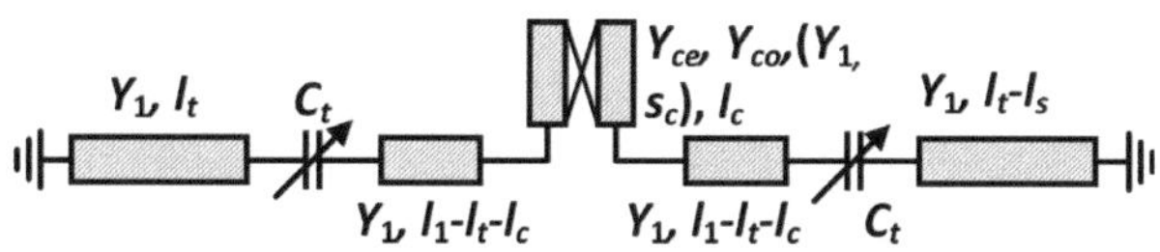

Figure 18.
EC configuration between resonators.

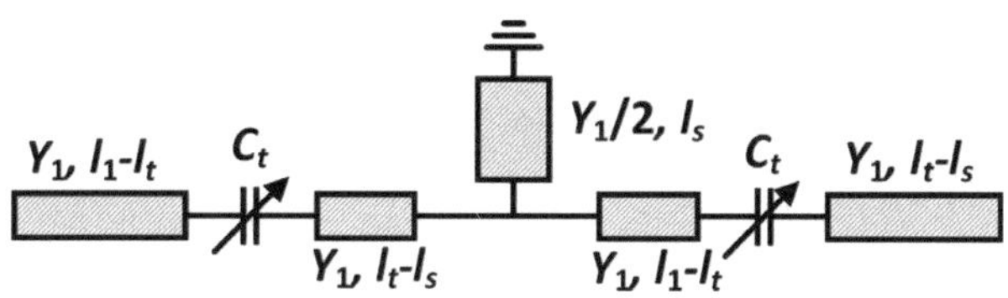

Figure 19.
MC configuration between resonators.

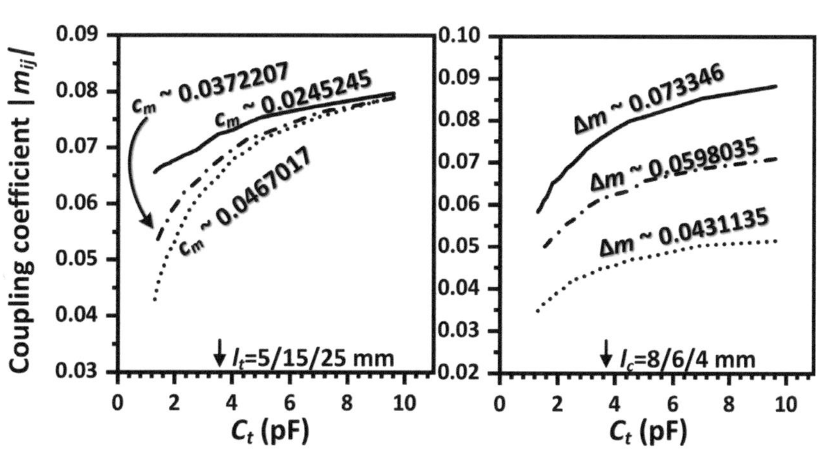

Figure 20.
*Coupling coefficient extracted from the EC configuration (**Figure 18**) with different l_t or l_c when s_c = 0.3 mm for (a) l_c = 6.8 mm (b) l_t = 15 mm.*

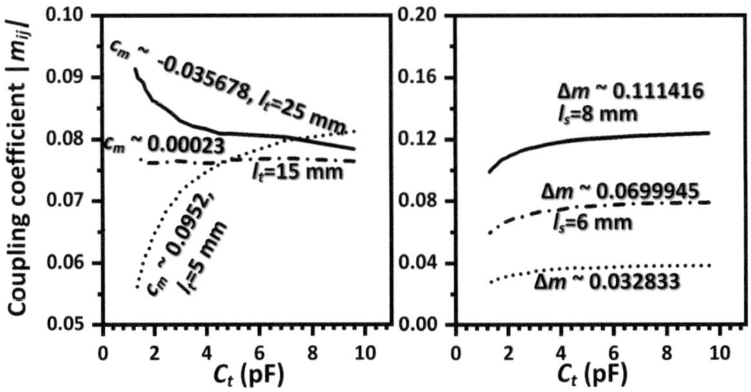

Figure 21.
*Coupling coefficient extracted from the MC configuration (**Figure 19**) with different l_t or l_s for (a) l_s = 5 mm (b) l_t = 8 mm.*

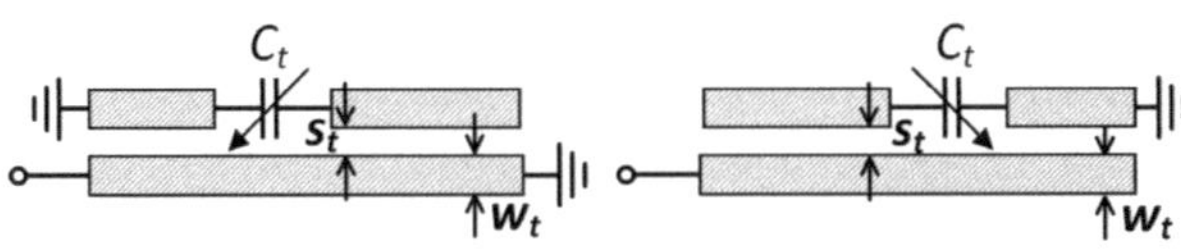

Figure 22.
(a) S-EF configuration. (b) O-EF configuration.

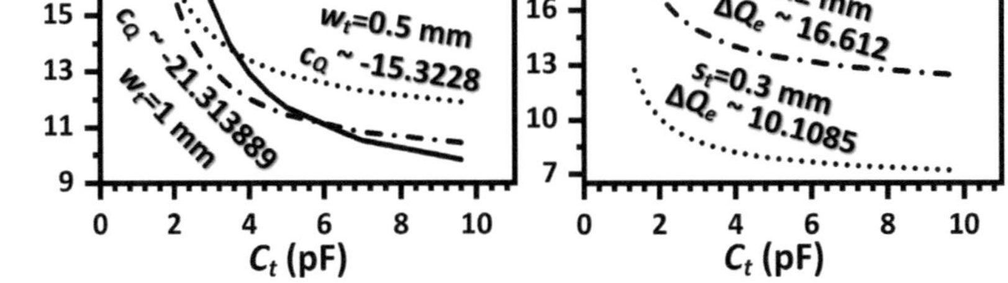

Figure 23.
*External quality factor Q_e extracted from S-EF configuration (**Figure 22a**)with different w_t or s_t for (a) s_t = 0.2 mm (b) w_t = 0.4 mm.*

are the four-pole cascade quartet topology, and their prototypes are the 4th-degree General Chebyshev polynomials with 20-RL and 130-MHz ABW. The transmission zeros of two filters are prescribed at [−10j, −2j, 2j, 10j]. The tuning range of the EC filter is predefined from 1.2 GHz to 1.6 GHz, and the EC filter is from 1.05 GHz to 1.45 GHz. According to the presented synthesis method, two EVCMs can be extracted, which are

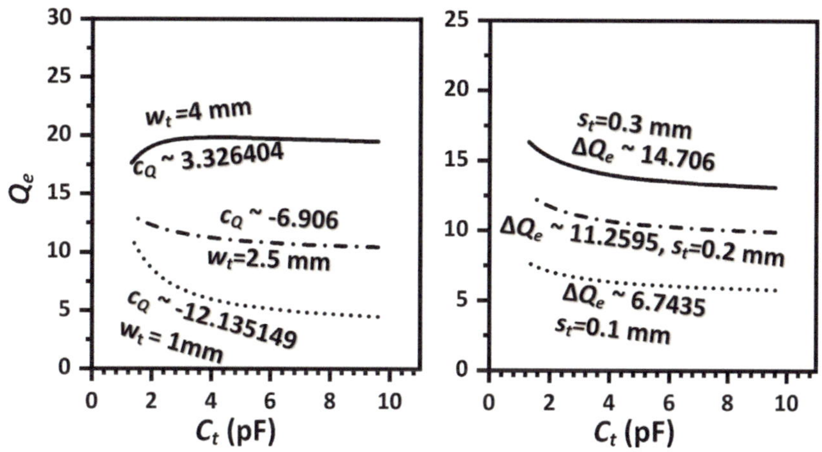

Figure 24.
*External quality factor Q_e extracted from O-EF configuration (**Figure 22b**) with different w_t or s_t for (a) s_t = 0.22 mm (b) w_t = 2.5 mm.*

$$\begin{cases} [m]_{4-pole} = \begin{cases} m_{12} = 0.091646 + m_{kk} \cdot 0.04582 \\ m_{23} = 0.080922 + m_{kk} \cdot 0.04046 \\ m_{14} = -0.0184 - m_{kk} \cdot 0.0092 \end{cases} \\ Q_{e4-pole} = 9.30255 - m_{kk} \cdot 4.5322, m_{sl} = 0.00016 \end{cases} \tag{22}$$

for EC filter and

$$\begin{cases} [m]_{4-pole} = \begin{cases} m_{12} = 0.081612 + m_{kk} \cdot 0.040806 \\ m_{23} = 0.072063 + m_{kk} \cdot 0.036031 \\ m_{14} = -0.01638 - m_{kk} \cdot 0.00819 \end{cases} \\ Q_{e4-pole} = 10.39153 - m_{kk} \cdot 5.0897, m_{sl} = 0.00019 \end{cases} \tag{23}$$

for MC filter. Besides, all filter examples are designed on the Rogers RT/duroid 5880 (h = 0.787 mm, ε_r = 2.2, tanδ = 0.0009) and the variable capacitors are SMV1234 (C_t = 1.3-9.6 pF, R_s = 0.8 Ω).

Figures 25 and **26** present the EC and MC filter examples where the cascade quartet configurations are used, thus yielding the prescribed tunable responses with constant BW. The measurement results of the two design examples are shown in **Figures 27** and **28**. The BW of the two filters is approximately 133 MHz and the tuning ranges are over 27%. Good agreement between simulations and measurements is

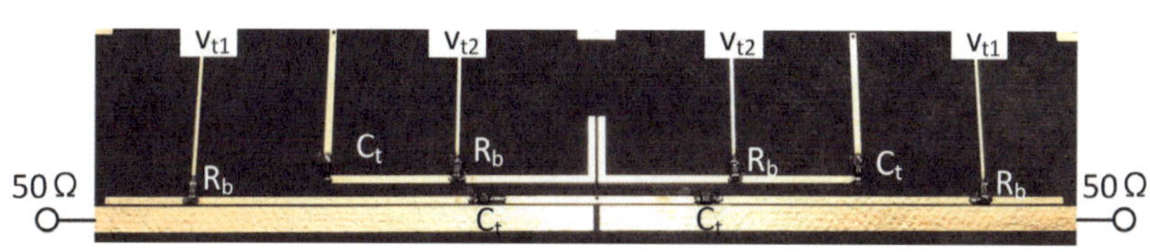

Figure 25.
EC filter example.

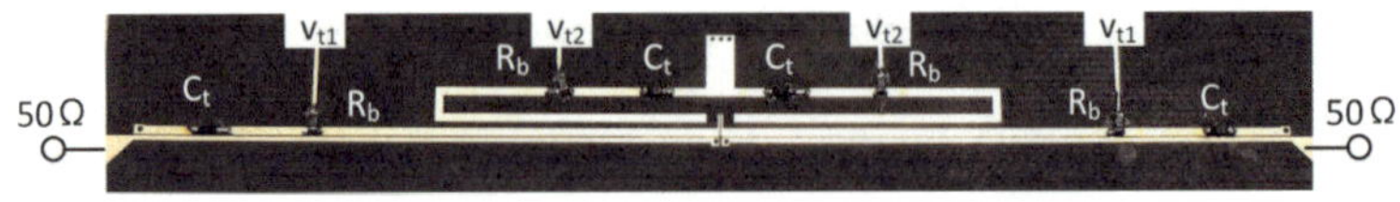

Figure 26.
MC filter example.

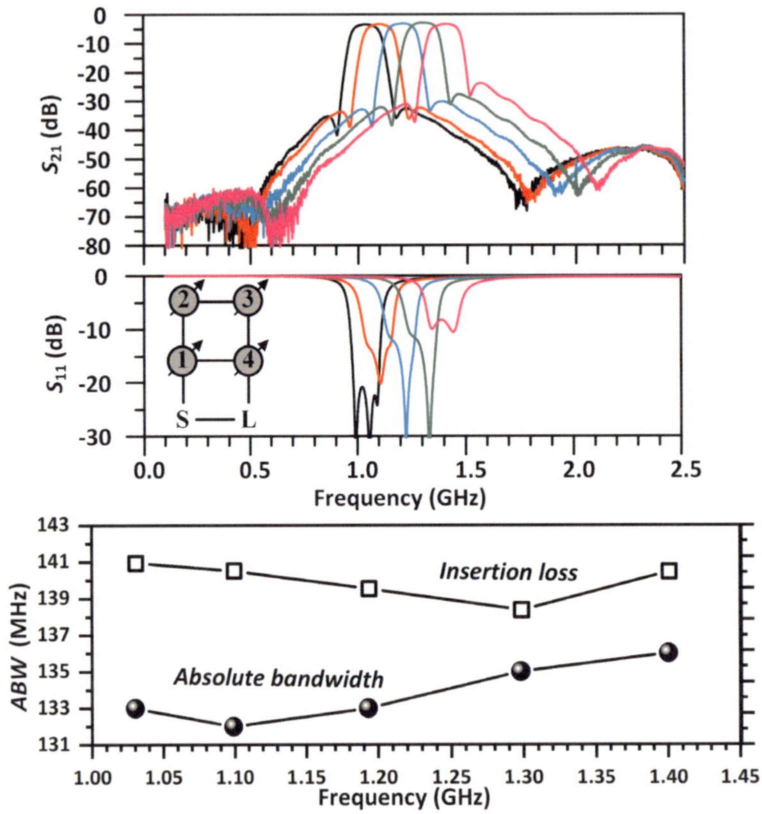

Figure 27.
Measurement results of the EC filter example ***(Figure 25)****.*

achieved, and the design objective is fully implemented. It is noted that the MC filter has a more stable BW because of the more flexible choosing range of coupling and Q_e curves with the wider adjusting ranges.

4.2 3D realization example of the constant-BW tunable filter

The ceramic monoblock waveguide filter featured a low-cost and competitive Q_u/size ratio that attracted a great deal of attention for 5G applications [39]. The 3D realization example of the constant-BW tunable filter presented in this section will be based on the ceramic monoblock waveguide structure and the high-Q mechanically tunable mechanism.

Figure 29 present the tunable ceramic monoblock waveguide resonator. It generally contains the fixed block and the movable cylinder. The low-dielectric (ε_r = 19.5)

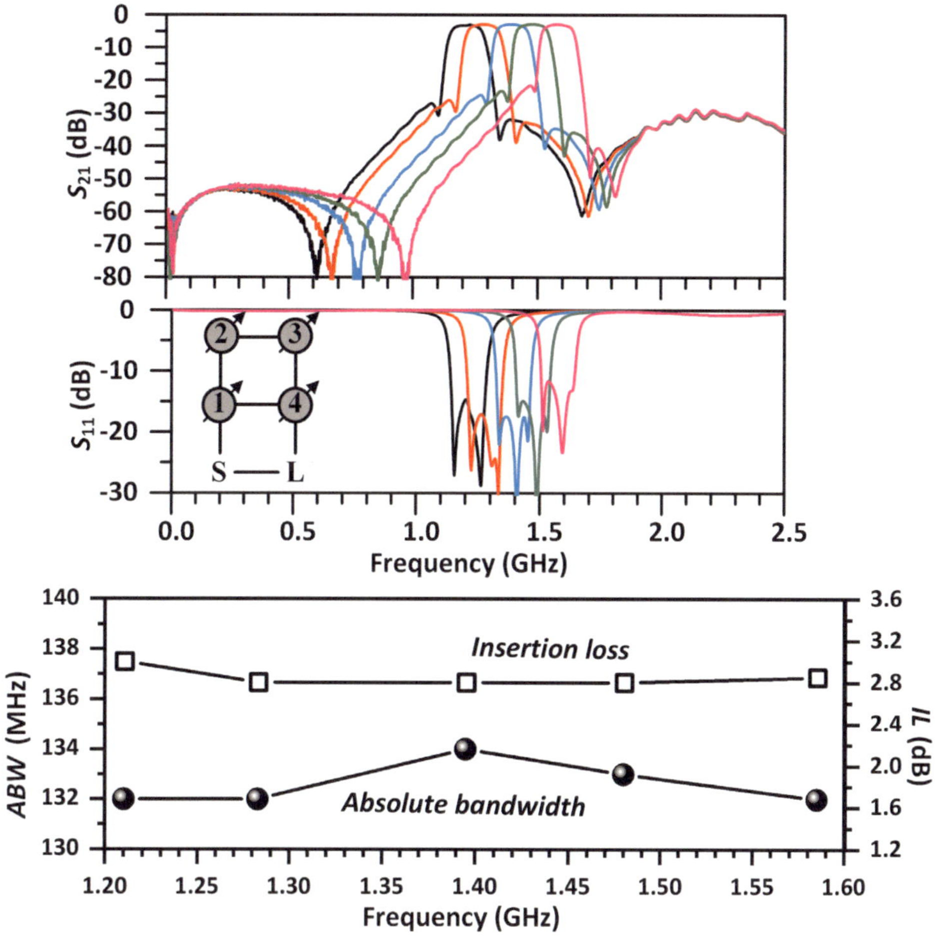

Figure 28.
*Measurement results of the MC filter example (**Figure 26**).*

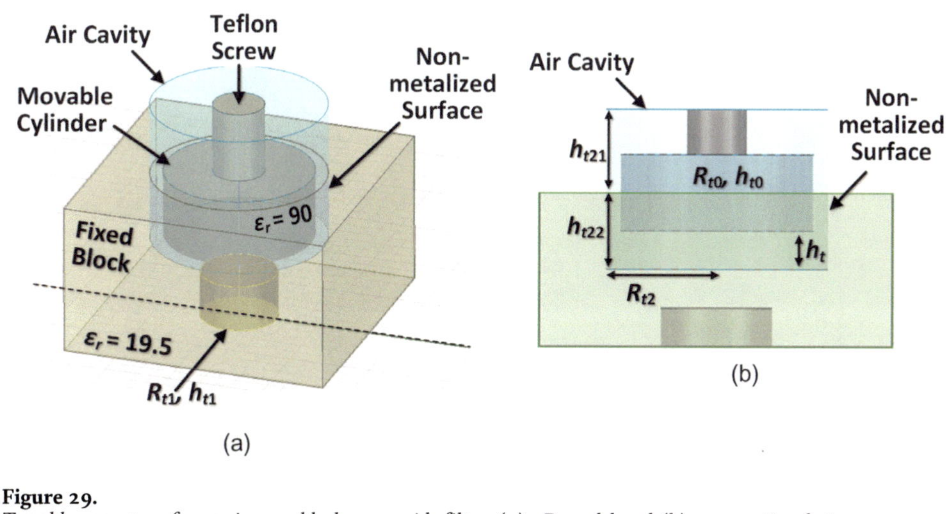

Figure 29.
Tunable resonator of ceramic monoblock waveguide filter. (a) 3D model and (b) cross-sectional view.

fixed block is coated with the silver layer except for the big center hole (h_{t21}, h_{t22}, R_{t2}). The hole on the bottom (h_{t1}, R_{t1}) is metalized. The movable cylinder (ε_r = 90) is not coated with the silver layer and can be dragged up and down in the non-metalized hole. The Teflon screw (M0.8) is employed to control the movable cylinder via a driving mechanism. The non-metalized hole with a lid constructs an air cavity to accommodate the movable cylinder. As the ceramic cylinder moves in the air cavity, the resonator's resonant frequency is tuned. **Figure 30** presents the tuning frequency ranges and the Q_u performance by changing h_{t1}. As expected, the frequency is tuned by changing the positions of the movable cylinder, and the Q_u is always kept on a high level. The frequency tuning range can be predefined by h_{t1}.

The coupling configuration between two resonators is shown in **Figure 31**. The extracted coupling coefficient curves with different key dimensions are presented in **Figure 32**. It is seen that the coupling coefficient curve is controlled by h_{m1} and R_{m1}. h_{m1} and R_{m1} have more influence in the coupling curve's high-frequency area and low-frequency area, respectively.

The feeding configuration for the tunable ceramic waveguide resonator is given in **Figure 33**, where the resonator is fed by a coplanar waveguide (CPW) line on the bottom of the fixed block. The external Q_e can be evaluated by the single-end group delay of the fed resonator structure using $Q_e = 2\pi f_0 \cdot \tau_{11}(f_0)/4$[20]. **Figure 34** shows the

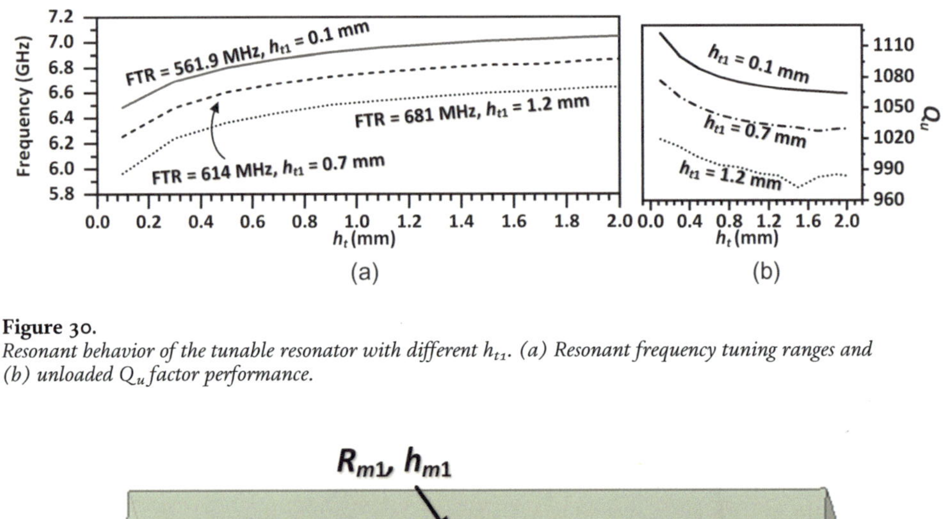

Figure 30.
Resonant behavior of the tunable resonator with different h_{t1}. (a) Resonant frequency tuning ranges and (b) unloaded Q_u factor performance.

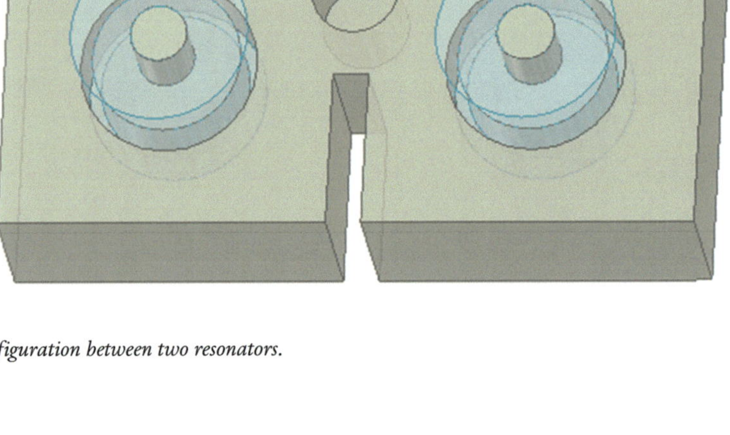

Figure 31.
Coupling configuration between two resonators.

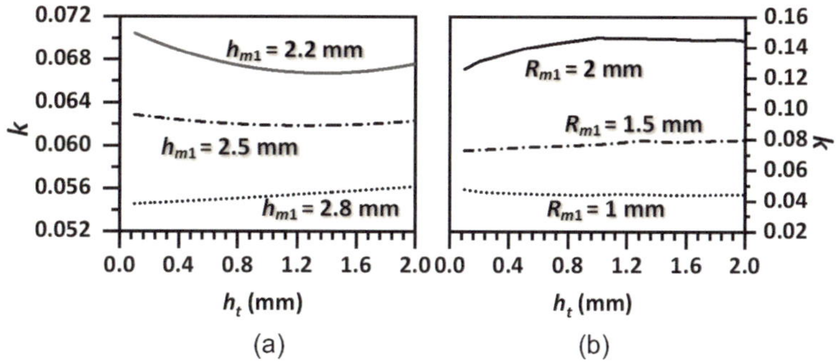

Figure 32.
Extracted coupling coefficient curves with different (a) h_{m1} and (b) R_{m1}.

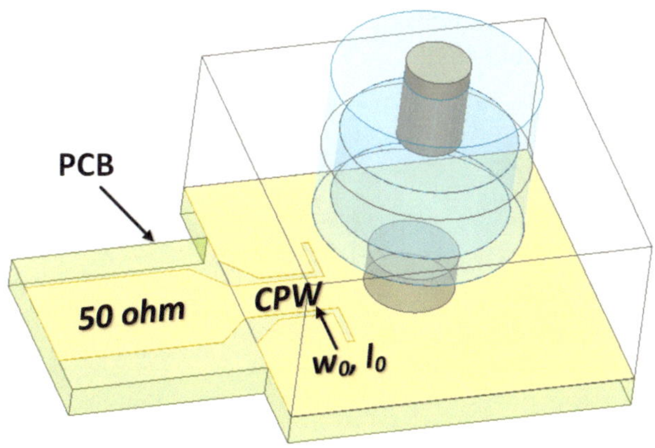

Figure 33.
Feeding structures excited using stripline.

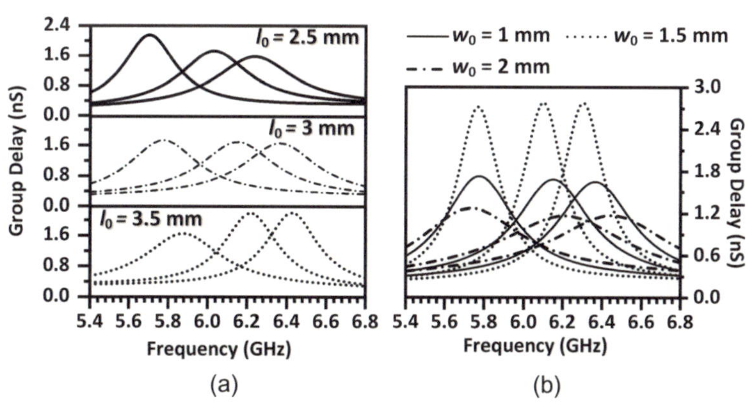

Figure 34.
Group delay responses of the feeding structures with different (a) l_o and (b) w_o.

extracted group delay. The observation implies that the variation tendency of the Q_e is controlled by l_0 and the magnitude is controlled by w_0 independently.

With the tunable resonator, coupling structure, and feeding configuration discussed earlier, the ceramic waveguide tunable filter can be constructed. For the demonstration, the six-degree Chebyshev low-pass prototype with 17-dB return loss and two TZs (±1.19511j) is employed to form the two-section cascade-triplet topology. The EVCM is extracted for the 5.8–6.2 GHz constant-400-MHz bandwidth filter as:

$$\begin{cases} m_{s1} = m_{6l} = 0.031m_{kkt} + 0.062 \\ m_{12} = -0.0194m_{kkt} - 0.0388 \\ m_{23} = -0.0134m_{kkt} - 0.0268 \\ m_{34} = -0.0189m_{kkt} - 0.0379 \\ m_{45} = -0.0134m_{kkt} - 0.0268 \\ m_{56} = -0.0194m_{kkt} - 0.0388 \end{cases}, \text{and} \begin{cases} m_{13} = -0.0171m_{kkt} - 0.0341 \\ m_{46} = 0.0171m_{kkt} + 0.0341 \\ m_{22} = 1.0247m_{kkt} + 0.0493 \\ m_{33} = 0.9986m_{kkt} - 0.0029 \\ m_{44} = 1.0014m_{kkt} + 0.0029 \\ m_{55} = 0.9753m_{kkt} - 0.0493 \end{cases} \quad (24)$$

EM design according to the extracted EVCM is carried out, and the optimized filter structure with the manufactured design sample is given in **Figure 35**. The feeding circuit is on Rogers RT/duroid 5880.

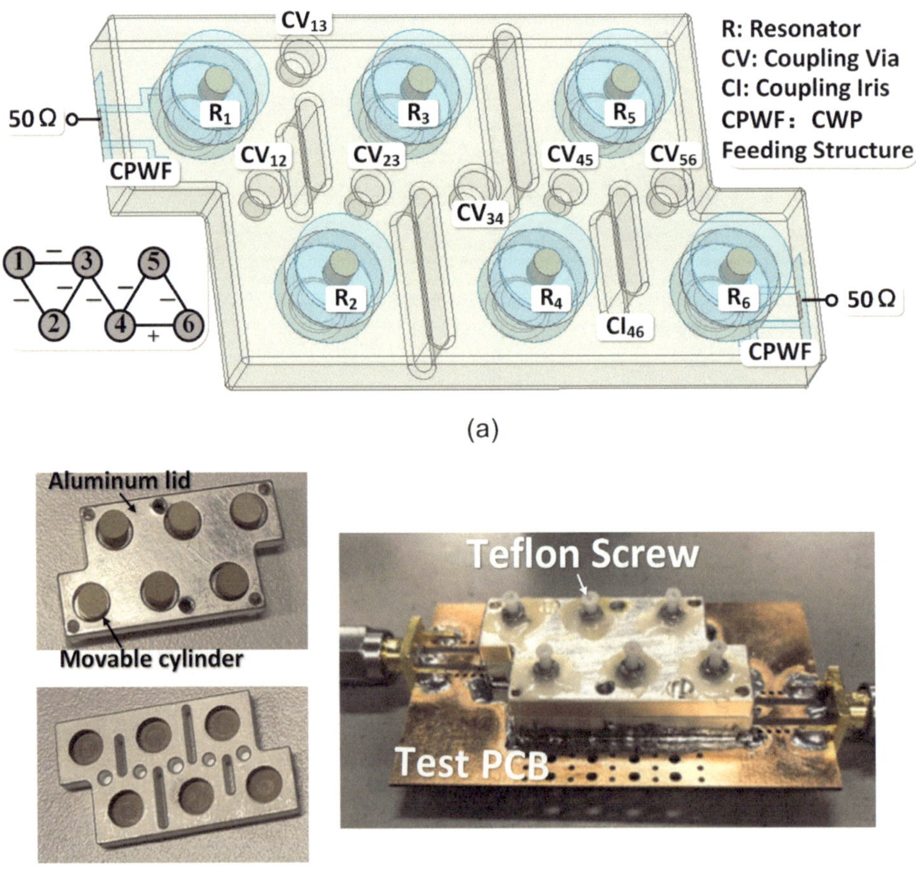

Figure 35.
Six-pole quasi-elliptic ceramic waveguide filter with constant absolute bandwidth. (a) Filter structure and (b) filter photograph.

DOI: http://dx.doi.org/10.5772/intechopen.104391

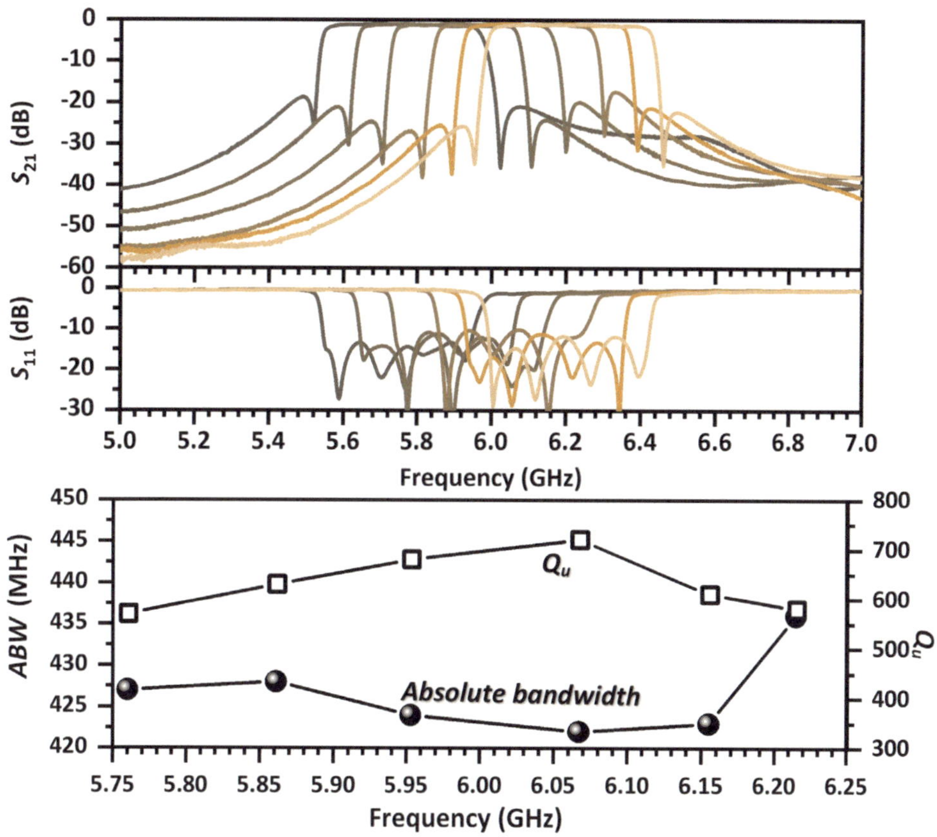

Figure 36.
Measurement results of the manufactured design sample.

Figure 36 presents the measured responses of the tunable ceramic waveguide filter. The measured passband moves from 5.76 GHz to 6.22 GHz, but the 3-dB bandwidth maintains 429 ± 7 MHz. The rectangular factor of two skirts is better than 12 dB/20 MHz because of two symmetric transmission zeros. The fitted Q_u of the tunable filter is kept from 570 to 720.

5. Conclusions

EVCM is introduced to represent the tunable filter with its tuning behavior. The relationship between the matrix and physical circuit is established, and this correspondence is investigated in detail. The synthesis approach based on EVCM extraction techniques is presented, and the planar as well as the ceramic waveguide tunable filters are designed according to the approach. The experiment is carried out to confirm the theory.

Author details

Di Lu
Southern University of Science and Technology, Shenzhen, China

*Address all correspondence to: ludi888abc@hotmail.com

References

[1] Liu X. High-Q RF-mems tunable resonators and filters for reconfigurable radio frequency front-ends [Doctor of Philosophy thesis]. Purdue University; 2010

[2] Aigle M, Hechtfischer G, Hohenester W, et al. A systematic way to YIG-filter-design. In: 2007 European Microwave Conference. Munich, Germany. 2007. pp. 668-671

[3] Adam JD, Davis LE, Dionne GF, et al. Ferrite devices and materials. IEEE Transactions on Microwave Theory and Techniques. 2002;**50**:721-737

[4] George LM, Leo Y, Jones EMT. Microwave Filters, Impedance-Matching Networks, and Coupling Structures. Dedham, MA: Artech House; 1980

[5] Cameron RJ, Kudsia CM, Mansour RR. Microwave Filters for Communication Systems: Fundamentals, Design, and Applications. New York: Wiley-Interscience; 2007

[6] Laplanche E, Delhote N, Périgaud A, et al. Tunable filtering devices in satellite payloads: A review of recent advanced fabrication technologies and designs of tunable cavity filters and multiplexers using mechanical actuation. IEEE Microwave Magazine. 2020;**21**: 69-83

[7] Fouladi S, Huang F, Yan WD, et al. High-Q narrowband tunable combline bandpass filters using MEMS capacitor banks and piezomotors. IEEE Transactions on Microwave Theory and Techniques. 2013;**61**:393-402t

[8] Iskander MA, Nasresfahani M, Mansour RR. A constant-Q tunable combline bandpass filter using angular tuning technique. 2014 44th European Microwave Conference. 2014. p. 1103–1106

[9] Mansour BG. A tunable quarter-wavelength coaxial filter with constant absolute bandwidth using a single tuning element. IEEE Microwave and Wireless Components Letters. 2021;**31**:658-661

[10] Basavarajappa G, Mansour RR. Design methodology of a high- Q tunable coaxial filter and diplexer. IEEE Transactions on Microwave Theory and Techniques. 2019;**67**:5005-5015

[11] Yassini B, Yu M, Keats B. A ka -band fully tunable cavity filter. IEEE Transactions on Microwave Theory and Techniques. 2012;**60**:4002-4012

[12] Arnold C, Parlebas J, Zwick T. Reconfigurable waveguide filter with variable bandwidth and center frequency. IEEE Transactions on Microwave Theory and Techniques. 2014;**62**:1663-1670

[13] Ossorio J, Vague J, Boria VE, et al. Exploring the tuning range of channel filters for satellite applications using electromagnetic-based computer aided design tools. IEEE Transactions on Microwave Theory and Techniques. 2018;**66**:717-725

[14] Gowrish B, Mansour RR. A dual-mode frequency reconfigurable waveguide filter with a constant frequency spacing between transmission zeros. In: 2020 IEEE/MTT-S International Microwave Symposium (IMS). 2020. pp. 811-814

[15] Basavarajappa G, Mansour RR. Design methodology of a tunable waveguide filter with a constant absolute bandwidth using a single tuning element. IEEE Transactions on Microwave Theory and Techniques. 2018;**66**:5632-5639

[16] Wang C, Blair WD. Tunable high-Q dielectric loaded resonator and filter. In: Proceedings RAWCON 2002 2002 IEEE Radio and Wireless Conference (Cat No02EX573). 2002. pp. 249-252

[17] Pance K, Rochford G. Multiple band and multiple frequency dielectric resonators tunable filters for base stations. In: 2008 38th European Microwave Conference. 2008. pp. 488-491

[18] Shen T, Zaki KA, Wang C. Tunable dielectric resonators with dielectric tuning disks. IEEE Transactions on Microwave Theory and Techniques. 2000;**48**:2439-2445

[19] Hunter IC, Rhodes JD. Electronically tunable microwave bandstop filters. IEEE Transactions on Microwave Theory and Techniques. 1982;**30**:1361-1367

[20] Hong J-SG, Lancaster MJ. Microstrip Filters for RF/Microwave Applications. NJ, USA: Wiley; 2004

[21] Guyette A. Controlled agility: Frequency-agile planar filters with advanced features. IEEE Microwave Magazine. 2014;**15**:32-42

[22] Islam H, Das S, Bose T, et al. Diode based reconfigurable microwave filters for cognitive radio applications: A review. IEEE Access. 2020;**8**: 185429-185444

[23] Lu D, Tang X, Barker NS, et al. Synthesis-applied highly selective tunable dual-mode BPF with element-variable coupling matrix. IEEE Transactions on Microwave Theory and Techniques. 2018;**66**: 1804-1816

[24] Rebeiz G, Entesari K, Reines I, et al. Tuning in to RF MEMS. IEEE Microwave Magazine. 2009;**10**:55-72

[25] Tombak A, Maria J-P, Ayguavives FT, et al. Voltage-controlled RF filters employing thin-film barium-strontium-titanate tunable capacitors. IEEE Transactions on Microwave Theory and Techniques. 2003;**51**:462-467

[26] Entesari K, Saghati AP, Sekar V, et al. Tunable SIW structures: Antennas, VCOs, and filters. IEEE Microwave Magazine. 2015;**16**:34-54

[27] Park S-J, Reines I, Patel C, et al. High-Q RF-MEMS 4–6-GHz tunable evanescent-mode cavity filter. IEEE Transactions on Microwave Theory and Techniques. 2010;**58**:381-389

[28] Kang Q, Shamsaifar K. Electronically Tunable Block Filter with Tunable Transmission Zeros. Google Patents; 2004

[29] Anand A, Small J, Peroulis D, et al. Theory and design of octave tunable filters with lumped tuning elements. IEEE Transactions on Microwave Theory and Techniques. 2013;**61**:4353-4364

[30] Xu J, Yang L, Yang Y, et al. High- $Q $ -factor tunable bandpass filter with constant absolute bandwidth and wide tuning range based on coaxial resonators. IEEE Transactions on Microwave Theory and Techniques. 2019;**67**:4186-4195

[31] Yang Z, Psychogiou D, Peroulis D. Design and optimization of tunable silicon-integrated evanescent-mode bandpass filters. IEEE Transactions on Microwave Theory and Techniques. 2018;**66**:1790-1803

[32] Mansour RR. High-Q tunable dielectric resonator filters. IEEE Microwave Magazine. 2009;**10**:84-98

[33] Xu J, Zhang XY, Li H, et al. Narrowband single-pole double-throw filtering switch based on dielectric

resonator. IEEE Microwave and Wireless Components Letters. 2018;**28**:594-596

[34] Pelliccia L, Cacciamani F, Farinelli P, et al. High-Q tunable waveguide filters using ohmic RF MEMS switches. IEEE Transactions on Microwave Theory and Techniques. 2015;**63**:3381-3390

[35] Park SJ, Rebeiz GM. Low-loss two-pole tunable filters with three different predefined bandwidth characteristics. IEEE Transactions on Microwave Theory and Techniques. 2008;**56**:1137-1148

[36] Lu D, Tang X, Barker NS, et al. Single-band and switchable dual-/single-band tunable BPFs with predefined tuning range, bandwidth, and selectivity. IEEE Transactions on Microwave Theory and Techniques. 2017;**66**:1215-1227

[37] Lu D, Yu M, Barker NS, et al. Advanced synthesis of wide-tuning-range frequency-adaptive bandpass filter with constant absolute bandwidth. IEEE Transactions on Microwave Theory and Techniques. 2019;**67**:4362-4375

[38] Amari S. Synthesis of cross-coupled resonator filters using an analytical gradient-based optimization technique. IEEE Transactions on Microwave Theory and Techniques. 2000;**48**:1559-1564

[39] Chen Y, Zhang Y, Wu K-L. A dual-mode monoblock dielectric bandpass filter using dissimilar fundamental modes. IEEE Transactions on Microwave Theory and Techniques. 2021;**69**: 3811-3819

Chapter 8

Analysis and Design of Miniaturized Substrate Integrated Waveguide CSRR Bandpass Filters for Wireless Communication

Nitin Muchhal, Shweta Srivastava and Mostafa Elkhouly

Abstract

Microwave planar bandpass filters are indispensable in wireless communication systems in most applications. With miniaturization and limited spectrum, there is a great need for a compact, selective bandpass filter with a wide stopband performance. Substrate integrated waveguide (SIW) has become a potential technology for designing and developing microwave and millimeter-wave components, circuits, and systems. This chapter presents novel, compact metamaterial-based bandpass filters with improved stopbands. Several filters' design methodology and performance are evaluated using broadside-coupled complementary split-ring resonators (BC-CSRR) and edge-coupled complementary split-ring resonators (EC-CSRR) techniques. A comprehensive method to evaluate negative permittivity and permeability for designing the proposed metamaterial structure is also described. These filters have not only compact size but also a wider upper stopband resulting from bandstop resonator characteristics.

Keywords: substrate integrated waveguide, miniaturization, BPF, EC-CSRR, BC-CSRR

1. Introduction

Filters play a vital role in numerous microwave applications. A microwave bandpass filter (BPF), in general, is a class of filter that is utilized to operate on the frequency response within the range of frequencies lying between 300 MHz and 300 GHz and allowing the best signal transmission at desired frequencies (passband), while eliminating signals at redundant frequencies (stopband) [1]. Among various techniques to design a bandpass filter, substrate integrated waveguides (SIWs) [2] are becoming more popular recently. SIW is a planar structure that is fabricated by using two periodic rows of conducting cylindrical vias implanted in a dielectric substrate, as shown in **Figure 1**. Hence, it acts as a bridge between planar and nonplanar technology.

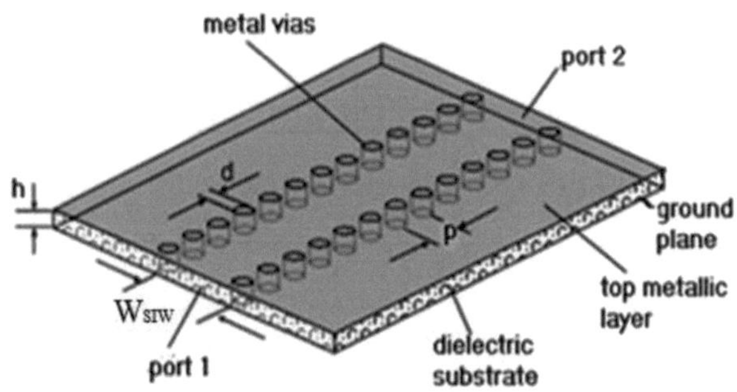

Figure 1.
Conventional substrate integrated waveguide.

To design efficient and well-performing wireless systems, there is a great need to design compact, lightweight microwave components. Over the past few years, various SIW miniaturization techniques have been proposed by researchers. Recently, [3] has reviewed the recent trends and various miniaturization techniques of SIW. Recently, folded SIW (FSIW) technique (C & T type FSIW) has been proposed by [4, 5]. Miniaturization was achieved using half mode SIW and Hilbert fractal for 5G applications [6]. Further, [7] proposed a ridge SIW to achieve miniaturization and suppress the harmonics.

From the design Equations [8] for a substrate integrated waveguide (SIW), d as the diameter of the vias and p as distance between the vias known as pitch, the equivalent width of dielectric-filled rectangular waveguide,

$$W_{EQ} = \frac{c}{2f_c\sqrt{\varepsilon_r}} \tag{1}$$

Width of SIW,

$$W_{SIW} = W_{EQ} + \frac{d^2}{0.95p} \tag{2}$$

Also, for choosing the value of p and d, the following inequalities should be satisfied.

$$p < 4d \text{ and } p < \frac{\lambda_O}{2}\sqrt{\varepsilon_r} \tag{3}$$

2. Metamaterial

A metamaterial is a word derived from the Greek word—it is a combination of the words “meta” and “material,” in which “meta” means something beyond normal, altered, changed, or something advanced. It is an artificial material designed to obtain the physical properties that do not exist in natural materials. A metamaterial [9] is an artificially engineered material with desirable properties not found in nature. A metamaterial affects electromagnetic waves by having structural features smaller than the wavelength of the medium of electromagnetic interaction. Metamaterials rely mainly on their physical structure to manipulate the electromagnetic waves to exhibit superior characteristics.

Figure 2.
The array of split-ring resonators plus wire assemblies.

- In 1999, John Pendry was the first to identify a practical way to make a left-handed metamaterial. Pendry's theoretical idea was that metallic wires aligned along the direction of a wave could provide negative permittivity ($\varepsilon < 0$), and a split ring with its axis placed along the direction of wave propagation could do so could provide negative permeability. In 2000, Smith et al. reported the experimental demonstration of functioning electromagnetic metamaterials by stacking, periodically, split-ring resonators and thin wire structures as shown in **Figure 2**.

2.1 Metamaterials' characteristics

Metamaterials with negative RI have numerous interesting properties. Several physical phenomena are reversed in LH media and at the intersection between LH and RH media due to the opposite sign of phase and group velocities. Some of the effects are:

- Reversal of Snell's law

- Reversal of Doppler effect

- Reversal of Vavilov-Cherenkov radiation

- Lensing effect (convex lenses produce diverging rays, which is opposite to RH lenses)

- The time-averaged Poynting vector (S) is antiparallel to phase velocity

2.2 Metamaterial's classification based on their properties

Russian scientist Veselago first proposed the metamaterial classification by considering the permittivity, ε, and the permeability, μ of a homogeneous material.

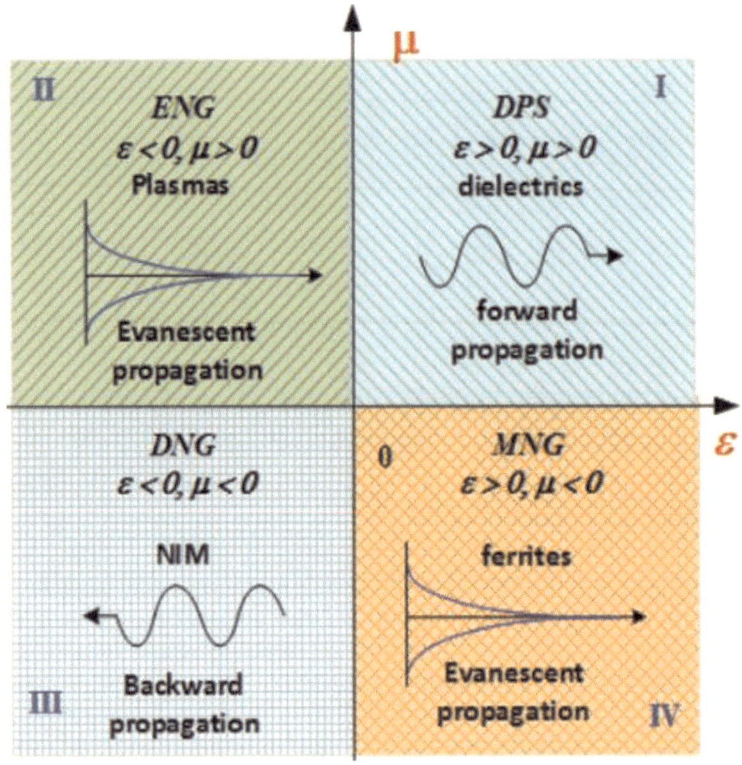

Figure 3.
Metamaterial classification.

The relationship between the refractive index and the constituent parameters ε and μ is given by the formula:

$$n = \pm\sqrt{\varepsilon_r \mu_r} \tag{4}$$

where ε_r and μ_r are the relative permittivity and permeability of the material. From Eq. (4), sign ± of n can get 1 in the four cases, which depends on the pairs of the sign of ε_r and μ_r. The electromagnetic metamaterials are classified based on each case of the pair sign ε and μ; they are shown in **Figure 3**.

In quadrant I, both parameters ε and μ are positive and are called double positive (DPS) or right-handed medium (RHM). In quadrant II, the parameters are $\varepsilon < 0$—negative, and $\mu > 0$—positive, and such material is called epsilon negative (ENG) medium and is represented by plasma. In quadrant III, parameters $\varepsilon < 0$—negative, and $\mu < 0$—negative, this region is called double-negative (DNG) or left-handed medium (LHM), and such material could not be found in nature. The quadrant IV $\varepsilon > 0$—positive, and $\mu < 0$—negative, such material is called μ—negative (MNG), represented by ferrite materials.

2.3 Types of metamaterial

A split-ring resonator (SRR) is a type of metamaterial, which is artificially created. SRR cell is made up of a pair of enclosed loops of nonmagnetic metals that split at

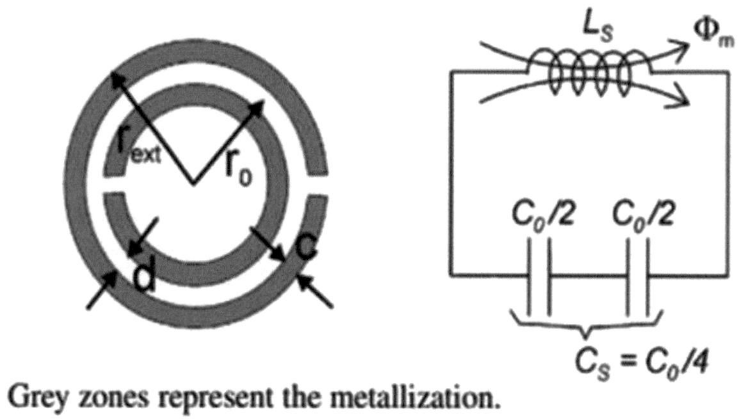

Figure 4.
Split-ring resonator with its equivalent circuit.

opposite ends, as shown in **Figure 4**. When these materials are exposed to the magnetic field of electromagnetic waves, they give strong magnetic coupling unavailable in conventional materials. When SRRs are arranged periodically (array), they provides negative permeability.

The above structure of SRR is known as edge-coupled split-ring resonator (EC-SRR) structure, which comprises concentric metal split rings printed on the same side of the dielectric substrate. EC SRR benefits of strong magnetic polarizability near resonance and easy fabrication. However, it has certain drawbacks: (i) Its electric size cannot be reduced below one-tenth of wavelength; (ii) it suffers from cross-polarizability/ bianisotropic effect. Another type of SRR overcomes these limitations, called broadside-coupled SRR (BC-SRR) [10]. In the BC-SRR configuration, the rings are etched on both faces of the substrate, as shown in **Figure 5a**. Similar to EC-SRR, charges formed in the lower half of the BC-SRR are the replica of charges formed in the upper half, as shown in **Figure 5b**. Though this formation of charge does not create an electric dipole, BC SRR is non-bianisotropic. Since both rings are of identical dimension and keep inverse symmetry, for this reason, cross-polarizability tensor vanishes.

The application of the Babinet principle leads to the origin of its counterpart known as a complementary split-ring resonator (CSRR) in which the rings are

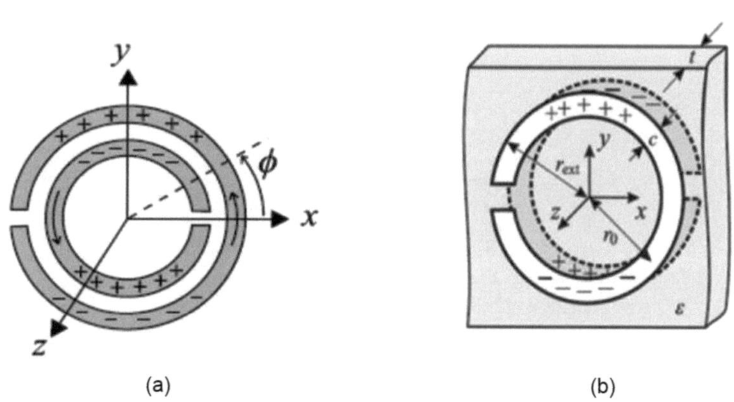

Figure 5.
Charge distributions in (a) EC-SRR (b) BC-SRR. r_{ext} is the outer radius of ring and r_o is the inner radius of the ring.

engraved on the conductive surface, and its magnetic and electric characteristics are changed when compared with SRR.

3. SIW loaded with basic EC-CSRR structure

A SIW bandpass filter based on edge-coupled CSRRs was proposed for the first time in 2007 by [11]. The SIW filter consisted of the tapered transition line with the CSRR. As SIW possesses high-pass characteristic, whereas a CSSR manifests band-stop characteristic, therefore by integrating CSRR with SIW, a bandpass SIW filter is designed. **Figure 6** depicts the structure of SIW with CSSR etched in the top side of the substrate.

3.1 Equivalent circuit model

Figure 7 depicts the equivalent lumped equivalent circuit for **Figure 6**. CSRRs etched in the center are excited by the electric field induced by the SIW. Therefore, this coupling can be labeled by connecting the SIW capacitance to the CSRR. In these models, L is the inductance of SIW vias, and C is the coupling capacitance between the CSRR and SIW. The resonator is represented by a parallel LC tank, where Lc and Cc represent the reactive elements, and R accounts for losses.

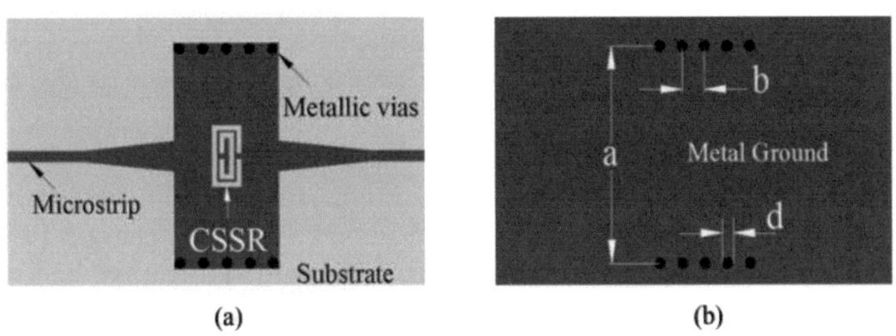

Figure 6.
(a) Top and (b) bottom view of basic unit cell [11] [reproduced courtesy of the electromagnetics academy].

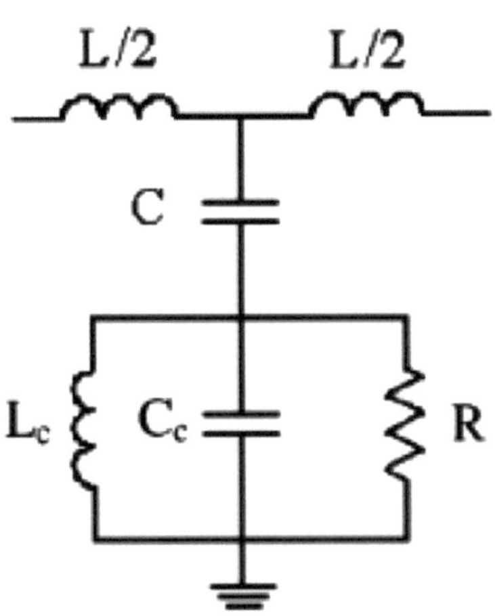

Figure 7.
The equivalent circuit model.

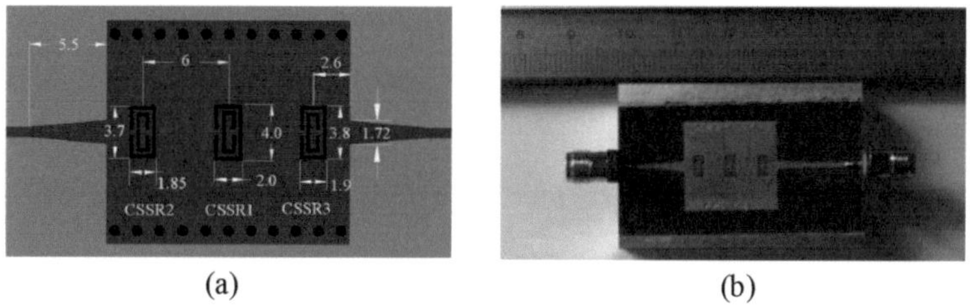

Figure 8.
(a) Dimensional layout of BPF and (b) fabricated BPF [11] [reproduced courtesy of the electromagnetics academy].

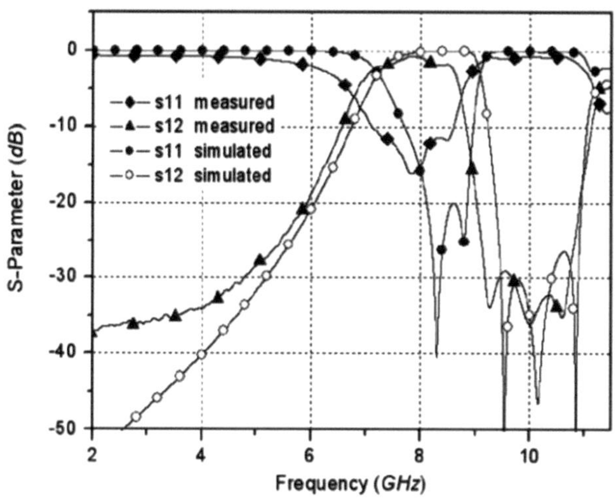

Figure 9.
Comparison of simulated and measured results [11] [reproduced courtesy of the electromagnetics academy].

Figure 8a shows the dimensional geometry of the proposed SIW-CSSRs bandpass filter [11], and **Figure 8b** shows the photograph of the fabricated design. The substrate used in the filter is RT/Duroid 5880, with a permittivity of 2.2 and a height of 0.254 mm.

Figure 9 compares the simulated and measured results of the filter. The measured insertion and return losses are about 2.16 dB and 11.6 dB, respectively. The filter shows a wide bandwidth ranging from 6.2 to 8.6 GHz (FBW of 32.4%).

4. Study on effect of orientation of CSRR ring on frequency response

The effect of changing the orientations of the CSRR ring was exhaustively studied by [12], which was verified by simulations and experiments that modify CSRR's orientations, different passband characteristics can be obtained. The orientation was specified with respect to the direction of the outer ring's split, as shown in **Figure 10**. Hence, they are aligned face to face, back to back, and side by side. The side-by-side type has also been divided into two cases with the CSRRs reversely or equally oriented.

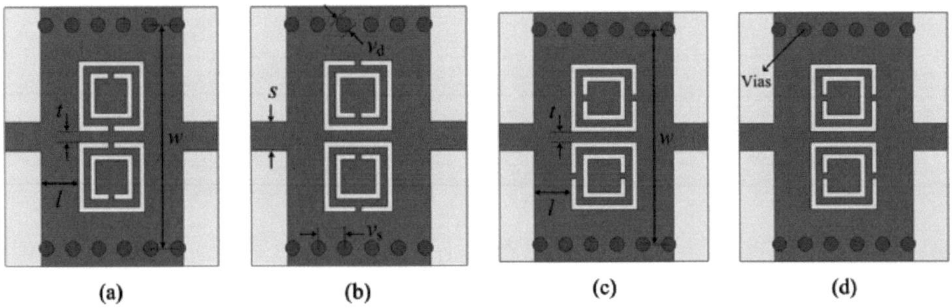

Figure 10.
Configurations of various SIW-CSRR unit cells in which the CSRRs are: (a) face to face, (b) back to back, (c) side-by-side reversely oriented, and (d) side-by-side equally oriented.

After simulation of various orientations, it was found that by altering the configuration of the CSRRs in a particular position (face-to-face orientation), the propagation of TE10 mode can be suppressed, resulting in enhanced selectivity and stopband rejection of the filter. The waveguide width was chosen as w = 12.3 mm to keep the cutoff frequency of the initial SIW at about 8.7 GHz. The Rogers substrate RT/Duroid 5880 with a thickness of 0.508 mm and a relative permittivity of 2.2 is used in the design. The metalized vias have a diameter of 0.8 mm and a center-to-center spacing of 1.48 mm.

After the simulation of various configurations, it was found that the unit cells with face-to-face and back-to-back oriented CSRR exhibit a similar kind of passband with one transmission zero and one pole located above the passband. Nonetheless, for the second case, the transmission zero is close to the pole leading to a steep upper side transition but with large insertion loss due to the weak coupling. For the third case, two rings are arranged side by side in opposite directions, and two transmission poles with two transmission zeros in the upper band are achieved. The propagation is quite weak for the fourth case due to weak magnetic coupling.

Eventually, a two-stage filter using the unit cell aligned face to face is simulated and fabricated using Rogers RT/Duroid 5880. A distance of 8.8 mm separates the two cells. The proposed bandpass filter achieves one transmission zeros at 6.4 GHz in the upper band, resulting in high selectivity and a wide upper stopband. The two-pole filter has a measured center frequency of 5.0 GHz and a 3-dB bandwidth of 0.33 GHz (3.2% FBW).

5. Bandpass filter with diamond-shaped edge-coupled CSRR (EC-CSRR)

Recently, a novel bandpass filter using diamond-shaped edge-coupled CSRR was proposed [13]. This section discusses the design methodology of single-stage and two-stage bandpass filters with diamond-shaped EC-CSRR structures.

5.1 Single-stage BPF with diamond-shaped CSRR

The physical construction of CSRR is shown in **Figure 11**, where the upper orange part is conducting layer, and the light gray part is the substrate. The CSRR structure consists of two diamond-shaped split resonant rings with their openings opposite (face to face) to each other for tight coupling between them. As CSRRs are integrated

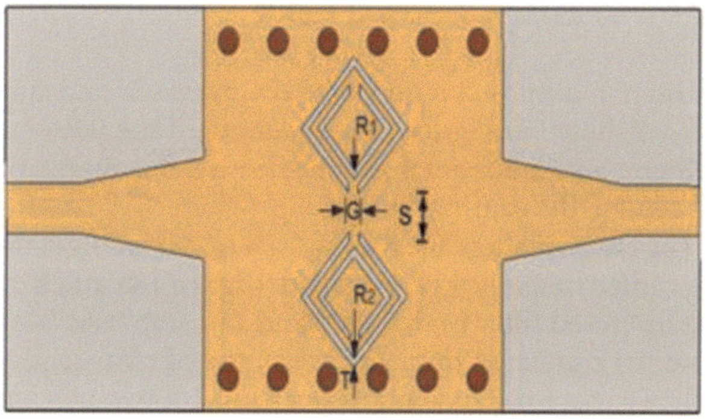

Figure 11.
Schematics of single-stage SIW BPF.

with SIW, a passband with an evanescent resonant mode lower than the SIW's cutoff frequency is created, miniaturizing the size of the conventional SIW [13]. **Figure 11** shows the dimensional view of a single-stage SIW filter loaded with diamond-shaped CSRR. The optimized dimensions of filter are: length of single-stage SIW L_{SIW} = 10 mm, width of SIW W_{SIW} = 8.5 mm, the inner radius of ring R1 = 1.0 mm, the outer radius of ring R2 = 1.6 mm, the thickness of ring T = 0.25 mm, the gap between open ends of outer ring G = 0.40 mm, the perpendicular distance between outer rings S = 1.25 mm. **Figure 12** shows the frequency response of single-stage CSRR incorporating SIW filter. The figure shows that in a single-stage SIW filter, one passband is formed with a center frequency of 8.75 GHz, below the waveguide cutoff frequency causing miniaturization by approximately 33%. The passband has 3-dB bandwidth of 0.42 GHz with an in-band insertion loss of 0.62 dB. The maximum value of return loss is −24.4 dB. Also, the stopband created has a high rejection level at the upper stopband.

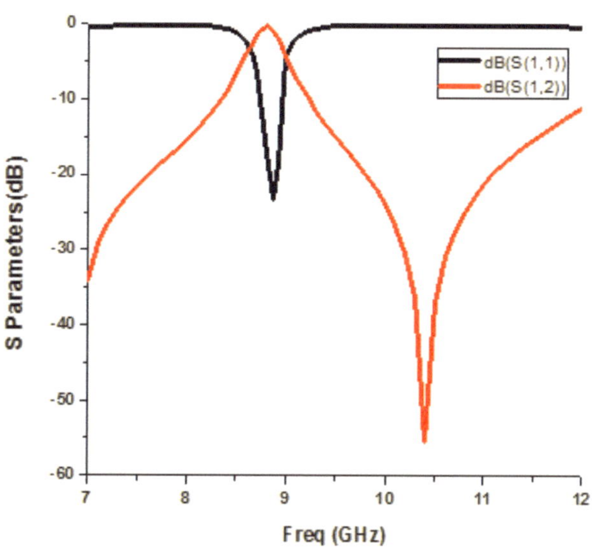

Figure 12.
Frequency response of single-stage SIW BPF.

5.2 Two-stage BPF with diamond-shaped CSRR

The two-stage filter is proposed to improve the passband and stopband performance of the filter, as shown in **Figure 13**. The length of the two-stage SIW filter is taken as $L_{SIW'}$ = 18 mm with horizontal distance between centers of rings, L = 8 mm. Other dimensions remain the same as in the single-stage SIW bandpass filter. Hence, the total filter size of the two-stage is 1.2 $\lambda_0 \times 0.28\ \lambda_0$, where λ_0 is the free space wavelength of the center frequency of passband. **Figure 14a** and **b** show the current distribution of the proposed filter in the passband and stopband, respectively.

Figure 15 shows the simulated frequency response of two-stage SIW BPF. The response clearly shows that one passband is formed with two poles and transmission zero. The passband has a center frequency of 8.86 GHz with 3-dB bandwidth of 0.74 GHz and an in-band insertion loss of 0.48 dB. The maximum return loss is −29.4 dB. Further, the stopband rejection is more than 60 dB, which is relatively better than a single-stage filter. In the second stage of transmission, zero is in proximity to poles leading to a high roll-off rate of 72.5 dB/GHz and 40.5 dB/GHz at the upper and lower edge, respectively.

6. Novel substrate integrated waveguide bandpass filter with broadside-coupled complementary split ring resonators

A novel SIW BPF using broadside-coupled complementary split-ring resonator (BC-CSRR) pairs was implemented for the first time by [14]. **Figure 16** (left) shows

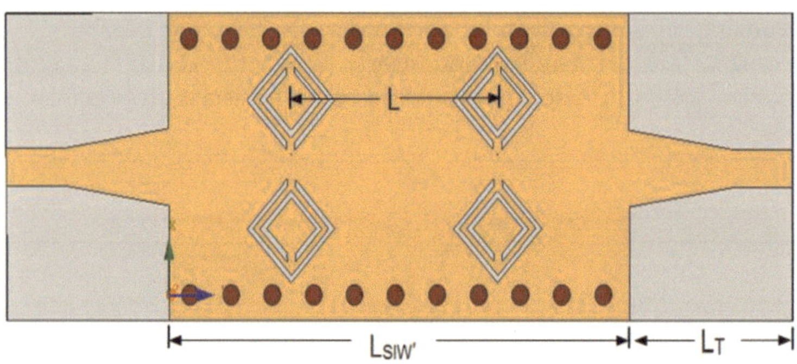

Figure 13.
Schematics of two-stage SIW BPF.

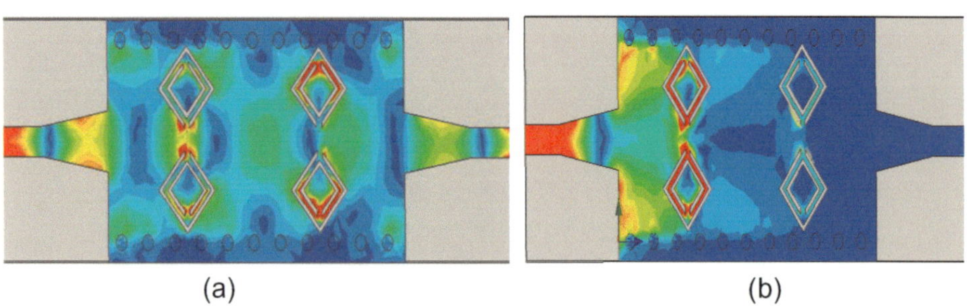

Figure 14.
(a) Current distribution in the passband. (b) Current distribution in the stopband.

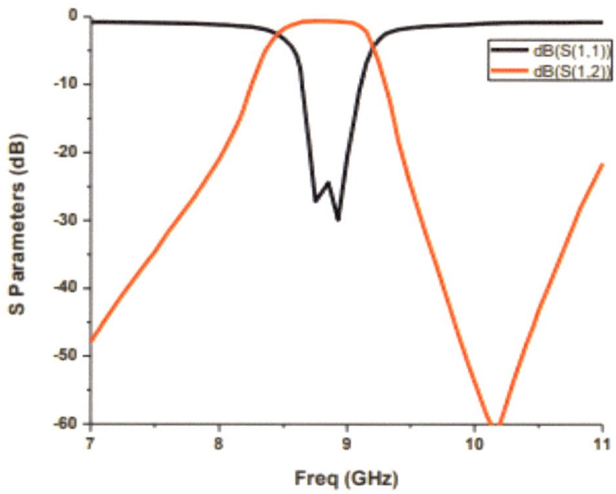

Figure 15.
Frequency response of two-stage SIW BPF.

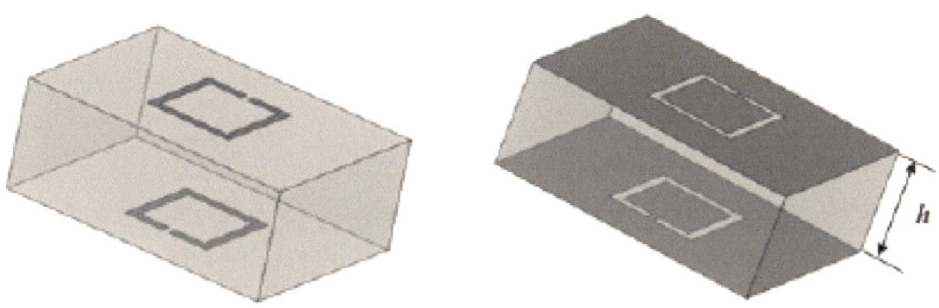

Figure 16.
Broadside-coupled SRR (BC-SRR), left and broadside-coupled CSRR (BC-CSRR), right.

the structure of the BC-SRR (broadside-coupled split-ring resonator. It can be derived from EC-SRR by substituting one of the rings with another ring situated precisely at the opposite side of the substrate. From the duality principle, the negative image of the BC-SRR is termed as the broadside-coupled complementary split-ring resonator (BC-CSRR), as shown in **Figure 16** (right).

6.1 The SIW BC-CSRR pair

Figure 17 depicts the layout of the proposed SIW BC-CSRR. It is evident that two BC-CSRRs are aligned side by side with opposite orientations to each other. A microstrip feed line is used to excite the SIW cavity. For the selected dielectric substrate with ε_r = 2.65 and waveguide cutoff frequency of 8.15 GHz, the width of the SIW (w) is calculated to be 12.5 mm. **Figure 18** shows the simulated transmission response for the SIW integrated with the unit cell. It is evident from the response that it creates a passband with a center frequency of 5.6 GHz, which is below waveguide cutoff frequency.

6.2 Two-stage BPF with BC-CSRR unit cell

Figure 19 depicts the proposed two-stage BC-CSRR BPF with separation between rings (l_d = 7 mm).

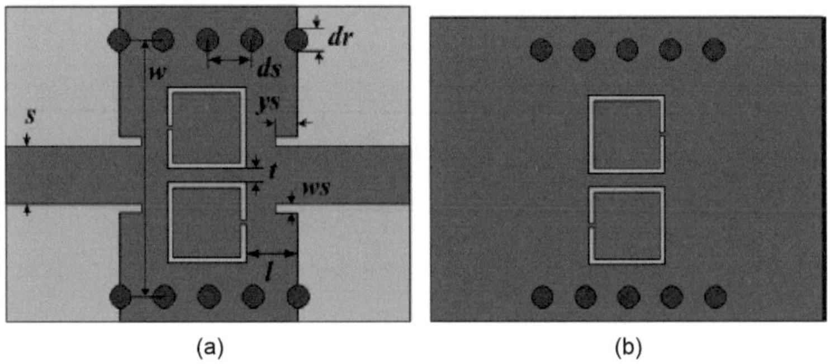

Figure 17.
Structure of the proposed SIW BC-CSRR unit cell [14]. (a) Top view; and (b) bottom view. w: Width of SIW, ds: distance between consecutive vias, dr: diameter of via, t: spacing between rings, s: thickness of microstrip line, l: distance of ring edge from SIW edge.

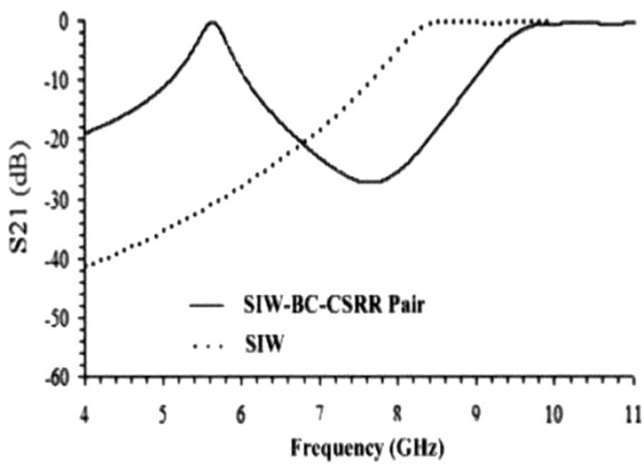

Figure 18.
Simulated frequency response of the original SIW and SIW BC-CSRR pair [14] (with parameters $h = 1$ mm, $g = 0.3$ mm, $b = 3.9$ mm, $d = 0.34$ mm, $w = 12$ mm, $t = 0.5$ mm, $s = 2.65$ mm, $y_s = 1.5$ mm, $w_s = 0.5$ mm, $l = 2.5$ mm, $dr = 1.2$ mm, $ds = 2.2$ mm).

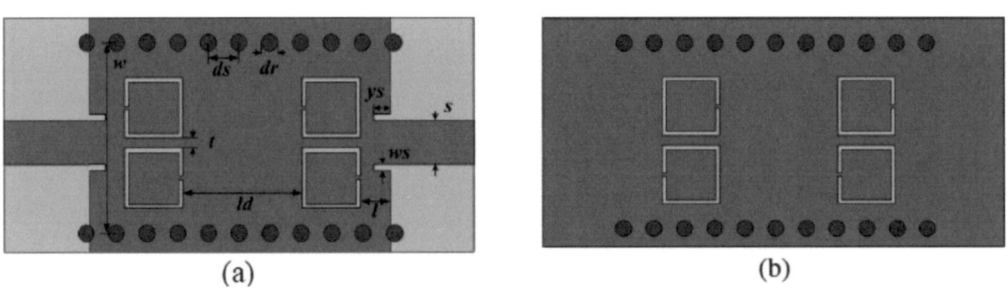

Figure 19.
Structure of the proposed SIW BC-CSRR unit cell [14]. (a) Top view; and (b) bottom view.

Figure 20 shows the photograph of the fabricated filter using a substrate with εr = 2.65 and a thickness of 1 mm. **Figure 21** compares the simulated and measured frequency response of the BPF. The measured center frequency and 3-dB bandwidths are 5.75 GHz and 0.32 GHz, respectively. The measured in-band return loss is below 12 dB. The dimension of the filter is 20 mm x 13 mm (0.38 x 0.25 λo2).

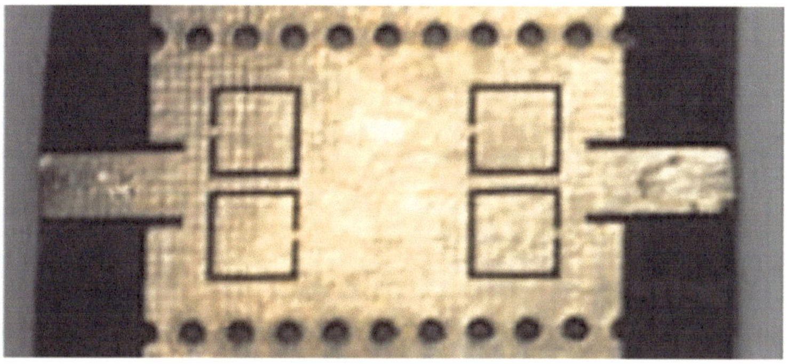

Figure 20.
Snapshot of the SIW BPF with BC-CSRR pairs [14].

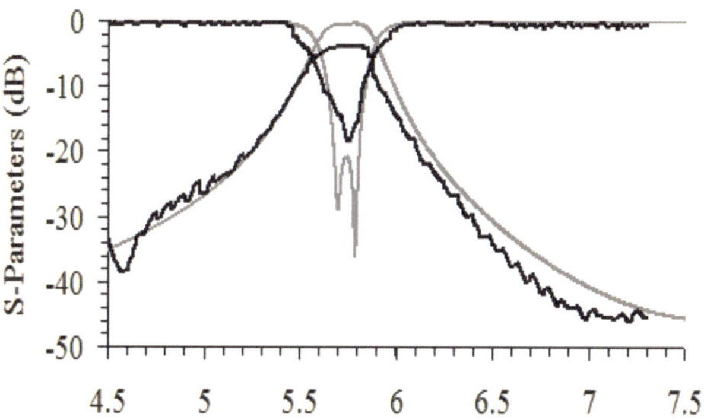

Figure 21.
Comparison of simulated and measured result [14].

7. Miniaturized and selective SIW bandpass filter with S-shaped broadside-coupled complementary split-ring resonators (BC-CSRR)

This work [15] proposes the design of a substrate integrated waveguide (SIW) bandpass filter (BPF) incorporated with a novel broadside-coupled complementary split-ring resonator (BC-CSRR). The complementary double S shape as metamaterial is carved on the top and broad bottom walls of SIW with orientation 180° to each other. The proposed filter is designed for X band using substrate alumina with a relative permittivity of 9.8 and height of 0.508 mm. Further, the width of the SIW, W_{SIW} is set to 5.4 mm to keep the nominal cutoff frequency of the waveguide to 10 GHz using SIW design equations.

7.1 Analysis and design of metamaterial

For designing the proposed S-shaped metamaterial, a double S-shaped structure was placed one above the another in an antisymmetrical manner over a dielectric layer forming a shape of 8 [16]. S on both sides of the dielectric forms metamaterial that simultaneously provides negative permeability and permittivity. The side length of the S shape is kept equal to λg/4 (A = 2.25 mm), and thickness T is kept equal to 0.35 mm, as shown in **Figure 22**.

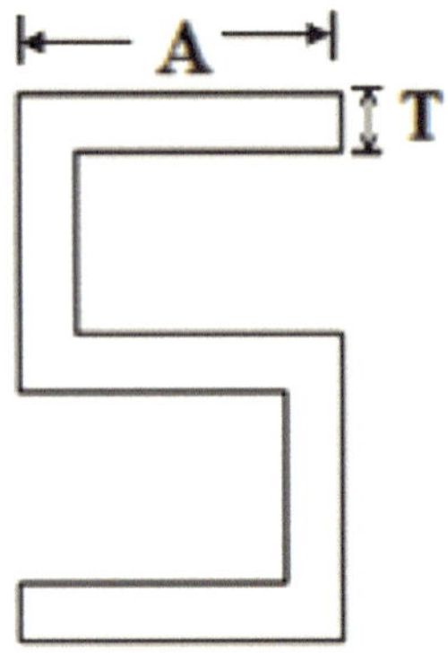

Figure 22.
Geometry of S structure.

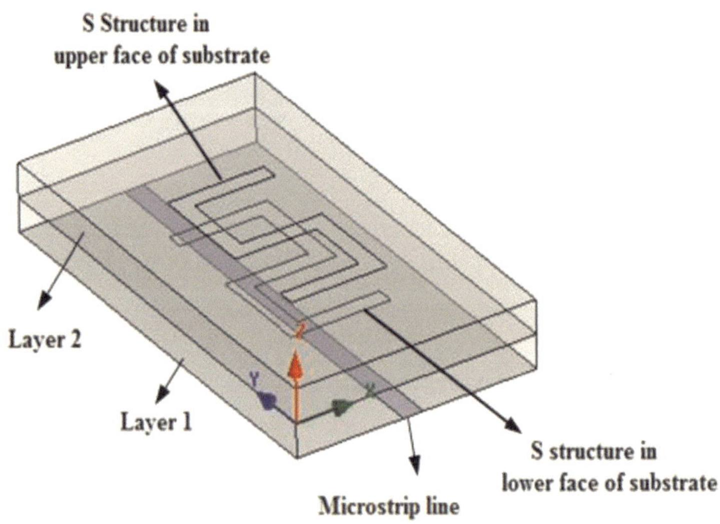

Figure 23.
Geometry of double "S"-shaped structure with microstrip line at the bottom.

Figure 23 depicts the setup to get S parameters of complementary S-shaped metamaterial using HFSS. For this, two-layered dielectric substrates (alumina) having relative permittivity 9.8 of thickness 0.508 mm are stacked over each other. The S-shaped structure is placed on the opposite side of the top dielectric substrate one above the other (in a complementary manner) to form **Figure 8**. A 50 Ω microstrip line is provided at the bottom of the lower substrate.

In HFSS, first, simulate the metamaterial structure by providing the solution frequency. Then get S-parameters (S11, S21) in tabular form as follows:

Result- > Create Modal Simulation Data Report - > Data Table.

Create a data table for S(1,1) containing magnitude and angle in rad (phase).

Similarly, create a data table for S (2,1). These files have extension .csv (comma-separated values).

Export these .csv files to the same folder where MATLAB code is kept. Now, call these files S(1,1).csv and S(2,1).csv in parameter extraction MATLAB code [17] in function referred as DATA_READ specifying the path locations of files.

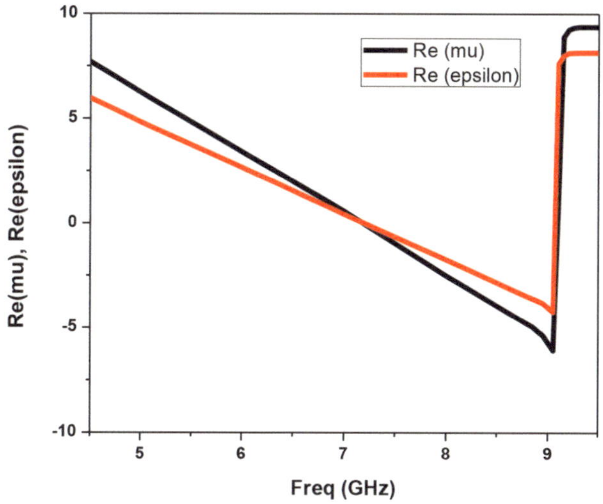

Figure 24.
Graph of real values of μ and ε.

Successful execution of MATLAB code [17] for the parameter extraction results led to permittivity and permeability, as shown in **Figure 24**. The graph indicates that the permeability and permittivity are negative simultaneously for the frequency range between 7.25 GHz and 9.15 GHz. It illustrates that the structure has metamaterial characteristics for the frequency range between 7.25 GHz and 9.15 GHz.

7.2 Design of single-stage SIW BC-CSRR bandpass filter

Figure 25 shows single-stage BC-CSRR BPF, which has a pair of identical "S"-shaped etched on the SIW top and broad bottom walls but at 180° to each other. A tapered microstrip feed line has been used for exciting the SIW. The design parameters are taken as: W_{SIW} = 5.4 mm, L_{SIW} = 4.2 mm, P = 1.6 mm, D = 0.8 mm, L_T = 4 mm, W_T = 2 mm, L_M = 2 mm, W_M = 0.50 mm, A = 2.25 mm, and T = 0.35 mm.

Figure 26 shows the equivalent circuit of the single-stage BC-CSRR BPF. The equivalent circuit of the S-shaped SRR structure is given by [18], in which S-SRR is modeled by a series L-C circuit in each half ring of the eight-shaped structure through a common capacitor. Since CSRR is complementary to the SRR structure, the

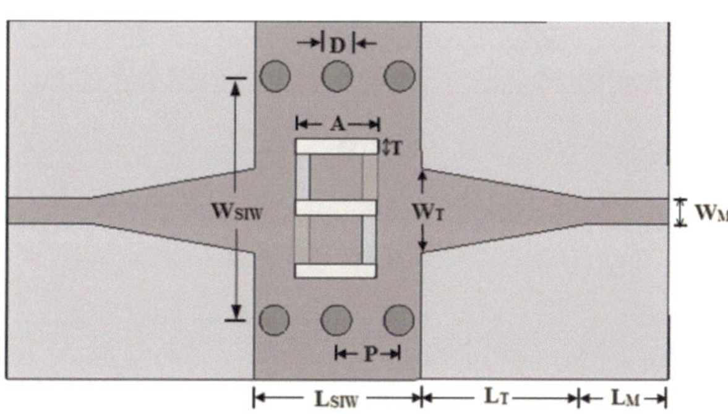

Figure 25.
Schematics of single-stage SIW BC-CSRR BPF.

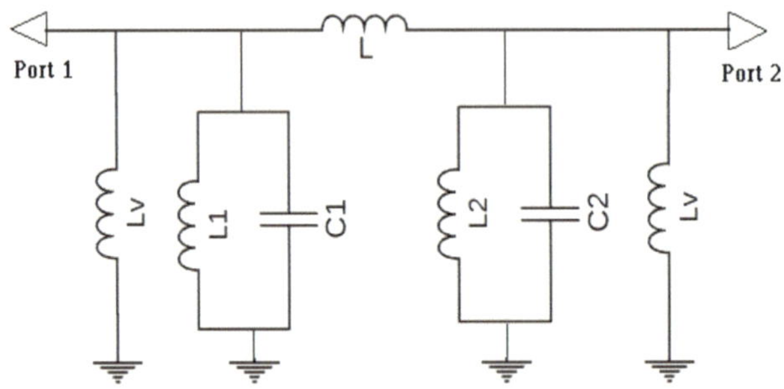

Figure 26.
Equivalent circuit of BC-CSRR BPF.

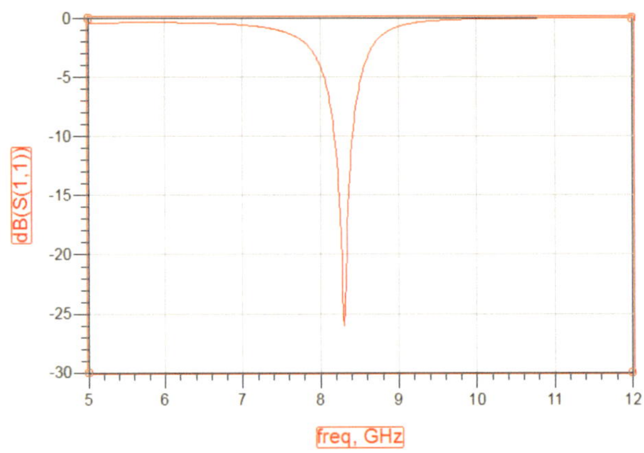

Figure 27.
Frequency response (S_{11}) of an equivalent lumped circuit of single-stage BC-CSRR SIW filter.

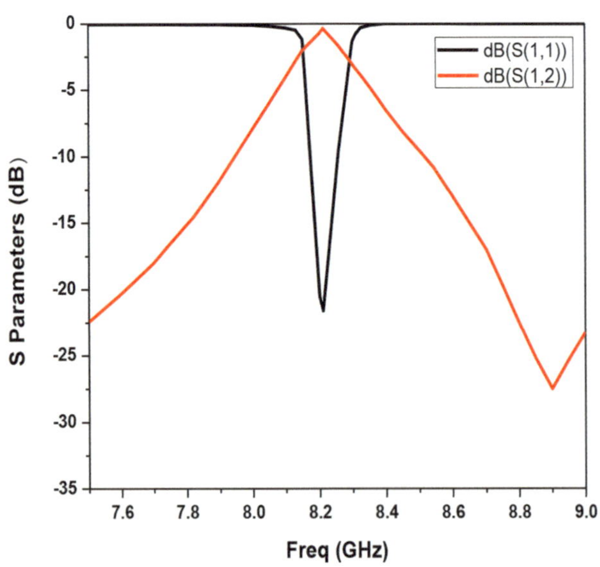

Figure 28.
Frequency response of single-stage BC-CSRR SIW filter.

equivalent circuit of single unit BC-CSRR will be dual of S-SRR. The metallic vias of the SIW are modeled as L_V.

Figure 27 shows the simulated result of the equivalent lumped circuit using ADS.

Figure 28 shows the frequency response of single-stage BC-CSRR incorporated SIW filter. The figure shows that by etching the S structure in SIW, a passband is obtained with a center frequency of 8.2 GHz and 3-dB bandwidth of 0.15 GHz. The maximum return loss is 21.55 dB, and insertion loss is 0.32 dB at the center frequency. It can be seen that the resonant frequency of the SIW BC-CSRR element is well below the cutoff frequency of the original SIW, causing its miniaturization.

7.3 Design of two-stage SIW BC-CSRR bandpass filter

In order to improve roll-off factor and order of filter, cascaded connection [19] of two identical BC-CSRR structures is used to form two-stage BPF. **Figure 29** shows the

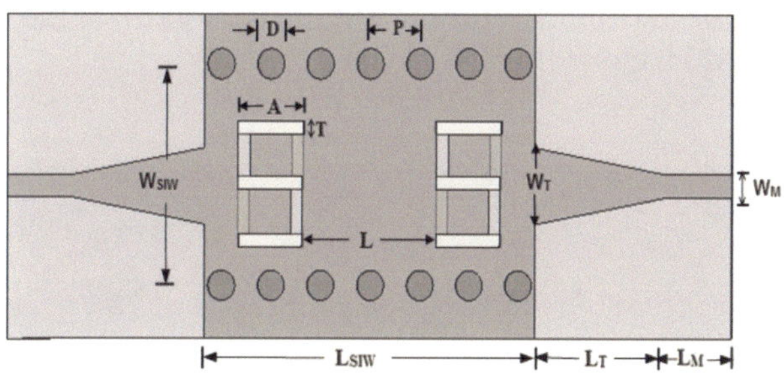

Figure 29.
Schematics of two-stage SIW BC-CSRR BPF.

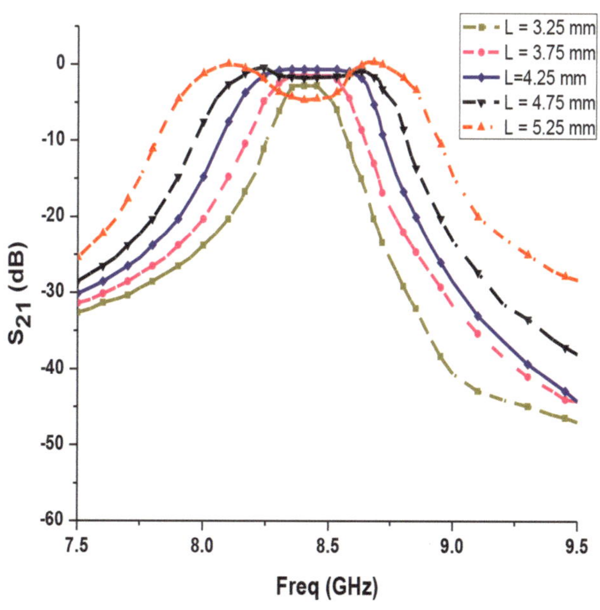

Figure 30.
Parametric analysis of return loss for varying side length "a."

structure of two-stage BC-CSRR BPF with design parameters taken as: W_{SIW} = 5.4 mm, L_{SIW} = 4.2 mm P = 1.6 mm, diameter of via D = 0.8 mm, L_T = 4 mm, W_T = 2 mm, L_M = 2 mm, W_M = 0.50 mm, A = 2.25 mm, T = 0.35 mm, and L = 4.25 mm.

The distance (L) between two BC-CSRRs has a vital influence on the performance of the proposed two-stage filter. **Figure 30** shows the parametric analysis of return loss with varying values of L (for L = 3.25, 3.75, 4.25, 4.75, 5.25 mm). It is clear from **Figure 30** that the filter shows optimum performance for L = 4.25 mm. For other small or big values of L, its response becomes undesirable.

Figure 31a and **b** depicts the current distribution in passband and stopband, respectively. As seen from the current distribution, it is clear that when the filter is passing the signal, the center resonator is resonant and has a large current that couples the signal through to the output.

Figure 32 shows the frequency response of two-stage BS-CSRR. From the response, it can be observed that a passband with 3-dB bandwidth of 0.385 GHz is obtained. The simulated insertion loss is 0.32 dB, and the simulated roll-off rate at the lower and upper edge of the passband is calculated to be 78.26 dB/GHz and 65.5 dB/GHz, respectively. The maximum return loss value is 24.85 dB at the center frequency of 8.4 GHz with a 3-dB bandwidth of 0.38 GHz.

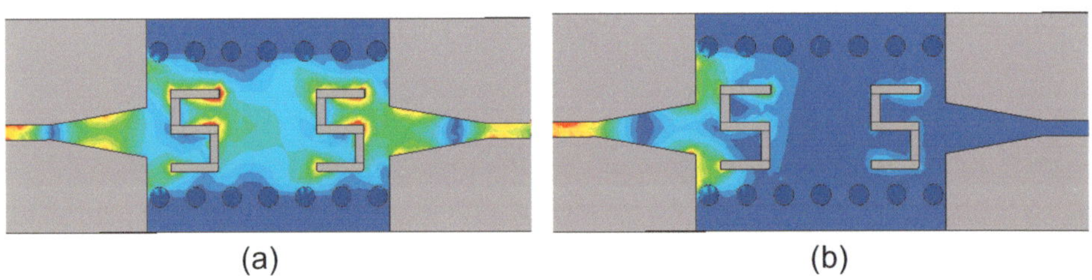

Figure 31.
Current distribution in (a) passband and (b) stopband.

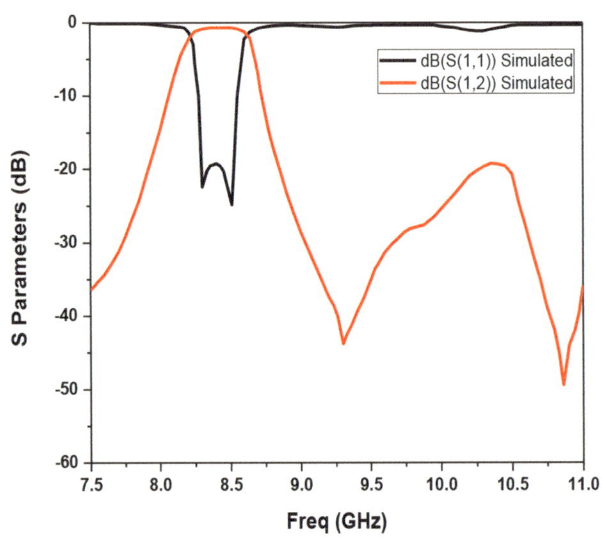

Figure 32.
Frequency response of single-stage BC-CSRR SIW filter.

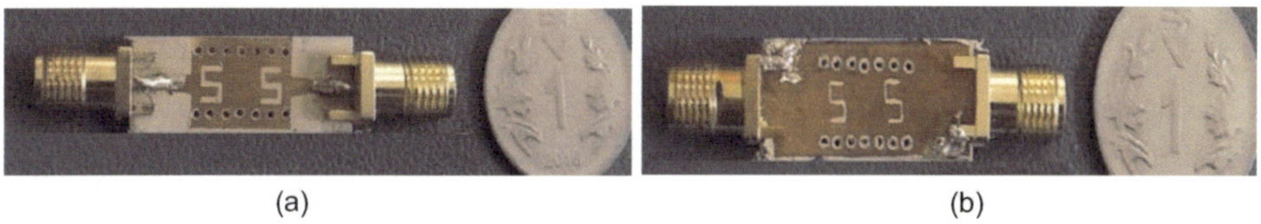

Figure 33.
(a) Top and (b) bottom view of the fabricated bandpass filter.

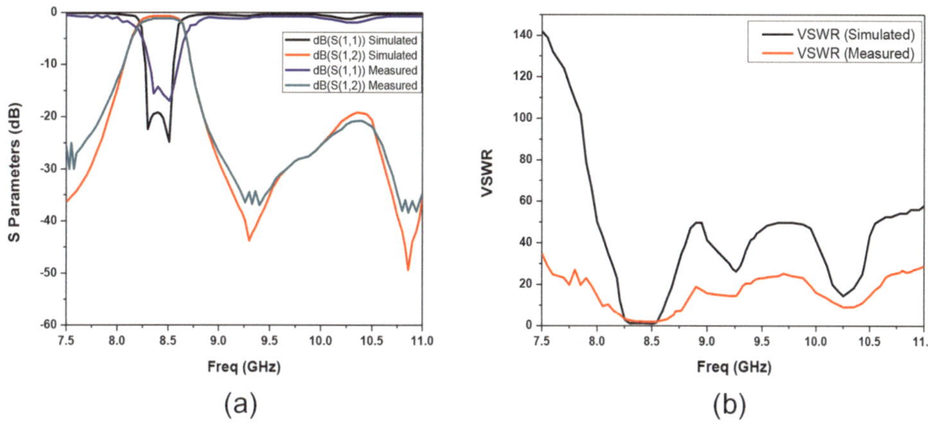

Figure 34.
Comparison of (a) the simulated and measured result S parameters and (b) VSWR.

7.4 Fabrication and results

The proposed filter is fabricated using substrate material alumina with a relative dielectric constant of 9.8, tan δ = 0.001, and thickness of 0.508 mm to validate the result. **Figure 33a** and **b** shows the photograph of the top and bottom layer of the assembled filter with overall dimensions as 10 mm (length excluding transition) $\times$ 8.5 mm (width).

The scattering parameters of the fabricated filter are measured by a vector network analyzer Anritsu S 820E. A two-port SOLT (short- open- load and thru) calibration has been done to consider cable losses between the VNA and the DUT. The measured and HFSS simulated results are compared and depicted in **Figure 34a**. It can be seen that the measured passband of the filter is from 8.20 GHz to 8.74 GHz with 3-dB bandwidth of 0.54 GHz. The maximum return loss value is 17.2 dB with an insertion loss of 0.92 dB in almost the entire passband. It achieves good attenuation (>20 dB) in the upper stopband. The measured roll-off rate is 58.5 dB/GHz and 60.2 dB/GHz at the lower and upper edge of the passband, respectively. **Figure 34b** depicts the simulated and measured VSWR plot for the entire range.

8. Conclusions

The metamaterials can be applied to enhance bandwidth, create a compact structure or multifrequency bands, etc. In this chapter, various compact and selective CSRR integrated SIW bandpass filters have been analyzed, demonstrating their performance. To apply metamaterials, the first step is to design their unit cells, creating special metamaterial properties at the desired frequency. The size of the unit cells is calculated, simulated, and optimized using the HFSS software. First, three edge-coupled CSRR (EC-CSRR) BPFs have been analyzed for the design and performance. Then two broadside-coupled CSRR (BC-CSRR) BPFs have been analyzed elaborately and evaluated for performance.

Acknowledgements

This work was carried out during the tenure of 'The European Research Consortium for Informatics and Mathematics (ERCIM) Alain Bensoussan' Fellowship' programme.

Author details

Nitin Muchhal[1*], Shweta Srivastava[2] and Mostafa Elkhouly[1]

1 Fraunhofer Institute of Integrated Circuits, Ilmenau, Germany

2 Jaypee Institute of Information Technology, Noida, India

*Address all correspondence to: nmuchhal@gmail.com

References

[1] Hong JS, Lancaster MJ. Microstrip Filters for RF/Microwave Applications. NewYork: John Wiley & Sons; 2001

[2] Deslandes D, Wu K. Single-substrate integration technique of planar circuits and waveguide filters. IEEE Transactions on Microwave Theory and Techniques. 2003;**51**(2):593-596

[3] Muchhal N, Srivastava S. Review of recent trends on miniaturization of substrate integrated waveguide (SIW) components. In: 2017 3rd International Conference on Computational Intelligence & Communication Technology (CICT), Ghaziabad, India: IEEE; 2017. pp. 1-6. DOI: 10.1109/CIACT.2017.7977389

[4] Muchhal N, Chakraborty A, Vishwakarma M, Srivastava S. Slotted folded substrate integrated waveguide band pass filter with enhanced bandwidth for Ku/K band applications. Progress in Electromagnetics Research M. 2018;**70**:51-60

[5] Muchhal N, Srivastava S. Design of miniaturized high selectivity folded substrate integrated waveguide band pass filter with Koch fractal, Electromagnetics. Oct 2019;**39**(8):571-581

[6] Muchhal N, Chakraborty A, Agrawal T, Srivastava S. Miniaturized and selective half-mode substrate integrated waveguide bandpass filter using Hilbert Fractal for sub-6 GHz 5G applications. IETE Journal of Research. 2022;**68**:1-8. DOI: 10.1080/03772063.2021.2021816

[7] Kumar R, Singh SN. Design and analysis of ridge substrate integrated waveguide bandpass filter with octagonal complementary split ring resonator for suppression of higher order harmonics. Progress In Electromagnetics Research C. 2019;**89**:87-99

[8] Kordiboroujeni Z, Bornemann J. Designing the width of substrate integrated waveguide structures. IEEE Microwave and Wireless Components Letters. 2003;**23**(10):518-522. DOI: 10.1109/LMWC.2013.2279098

[9] Kumari V, Bhowmik W, Srivastava S. Design of high-gain SIW and HMSIW H plane horn antenna using metamaterial. International Journal of Microwave and Wireless Technologies. 2014;**7**(6):713-720

[10] Ricardo M, Francisco M, Jesús M, Francisco M. Comparative analysis of edge- and broadside-coupled split ring resonators for metamaterial design—Theory and experiments. IEEE Transactions on Antennas and Propagation. 2003;**51**(10):2572-2582

[11] Zhang X-C, Yu Z-Y, Xu J. Novel band-pass substrate integrated waveguide (SIW) filter based on complementary split ring resonators (CSRRs). Progress In Electromagnetics Research. 2007;**72**:39-46. DOI: 10.2528/pier07030201

[12] Yuan DD, Tao Y, Itoh T. Substrate integrated waveguide loaded by complementary split-ring resonators and its applications to miniaturized waveguide filters. IEEE Transactions on Microwave Theory and Techniques. 2009;**57**(9):2211-2223. DOI: 10.1109/tmtt.2009.2027156

[13] Muchhal N, Srivastava S. Design of miniaturized diamond shaped substrate integrated waveguide CSRR band pass filter for X band applications. In: IEEE

2019 International Conference on Signal Processing and Communication (ICSC), NOIDA, India: IEEE; 2019. pp. 113-116

[14] Huang L. Novel Substrate Integrated Waveguide Filters and Circuits [PhD Thesis]. Woodhouse, Leeds, United Kingdom: The University of Leeds; 2013

[15] Muchhal N, Srivastava S. Compact selective substrate integrated waveguide bandpass filter with S-shaped broadside-coupled complementary split ring resonators (BC-CSRR). International Journal of Microwave and Optical Technology. 2020;**15**(5):440-448

[16] Shatabdi C, Srivastava S. Design of compact annular ring antenna on metamaterials for improved radiation pattern and gain. International Journal of Microwave and Optical Technology. 2013;**8**(3):155-163

[17] Ahmad BN, Mohammad SS. Extraction of material parameters for metamaterials using a full-wave simulator. IEEE Antennas and Propagation Magazine. 2013;**55**(5): 202-211

[18] Chen H, Ran LX, Huang-Fu J, Kong JA. Magnetic properties of S-shaped split-ring resonators. Progress In Electromagnetics Research. 2005;**51**: 231-247

[19] Mondal P, Parui SK. Design of higher order miniaturized bandpass filter using two cascaded new multimode resonators. In: 2017 IEEE Asia Pacific Microwave Conference (APMC), Kuala Lumpur, Malaysia: IEEE; 2017. pp. 813-816

Chapter 9

Power Divider/Combiner

Tadashi Kawai, Ayumu Tsuchiya and Akira Enokihara

Abstract

With the remarkable progress in the use of Internet of Things (IoT) and 5G, there is a demand for higher performance such as miniaturization, broadband/multiband, low loss, and high integration for several microwave circuits. This chapter treats microwave power dividers/combiners used in amplifiers, mixers, phase shifters, antenna feeding networks, and so on. Here, the treated circuits are composed of LC-ladder circuits and an absorption resistor. It shows that multiband (dual-band and tri-band) and broadband can be achieved by changing the number of stages of the LC-ladder circuit. In addition, the effectiveness of this design method is demonstrated by electromagnetic simulations and prototype experiments.

Keywords: Wilkinson power divider, lumped-element, LC-ladder circuit, compact, broadband/multiband

1. Introduction

In recent years, wireless communication equipment has been rapidly researched and developed such as Internet of Things (IoT), wireless local area network (LAN), and 5G, and it is expected that the demand for wireless communication equipment will become more widespread in the future. Along with this, various microwave circuit elements mounted on wireless communication devices are also required to have higher performance, such as miniaturization, low loss, high integration, and wideband/multiband. The authors are paying attention to the power divider/combiner that divides/combines microwave signals among various microwave circuit elements. The reason is that the power divider/combiner is considered to be an important circuit element that is directly linked to its performance in microwave circuits.

As a power divider/combiner for a three-port network, the Wilkinson power divider (hereinafter referred to as a conventional circuit) composed of two quarter wavelength transmission lines at a design frequency and an absorption resistor connected between two output ports is widely used at several microwave/millimeter-wave circuit system such as a balanced amplifier, a mixer, a phase shifter, an antenna feeding network, and so on [1]. However, since the circuit size depends on the wavelength due to the distributed circuit configuration, there arises a problem that the area occupied by the circuit system becomes especially large in a low-frequency band. As a method for reducing the size of a microwave circuit, a method of replacing a transmission line with a Π-type/T-type circuit equivalent to that at the design frequency is often used, but the equivalence between the two circuits is guaranteed only at the design frequency [2]. Therefore, such a circuit generally has a narrow band characteristic. In addition to that, as a method of shortening the transmission line, methods of loading parallel capacitances or parallel open-circuited stubs at both ends or the center of the transmission line have been reported [3–6]. In addition, some miniaturization design methods using composite right-/left-handed transmission lines and lumped elements have also been proposed [7–10]. However, their operation bands are still narrower than that of the conventional circuit. Therefore, it is considered difficult to achieve both miniaturization and wide bandwidth of the circuit at the same time. On the other hand, our research group proposes a configuration using an LC-ladder circuit as a lumped-element circuit type Wilkinson power divider. It has been analytically and experimentally clarified that a configuration using a two-stage LC-ladder circuit on the input side can realize frequency characteristics equal to or higher than those of the conventional circuit. Furthermore, ultra-wideband power dividers, unequal power dividers, and *N*-way power dividers, etc., in a circuit configuration, using LC-ladder circuits have been also reported [11–15].

This chapter shows how to design a power divider that can be matched at arbitrary two frequencies with a simple circuit configuration with 9 lumped elements. The circuit is designed for application in IoT (920 MHz) and 5G (sub6 band: 3.7 GHz). The influence of the self-resonant frequency of the chip element used in the circuit configuration is considered in the SHF-band, so the inductance is realized using a meander line or a bent line. Electromagnetic field simulations and prototype experiments confirm the effectiveness of the two-frequency matching circuit with a quasi-lumped-element circuit configuration. It should be noted that this circuit also has a feature that high-pass or low-pass characteristics can be selected by replacing the inductance L and the capacitance C of the components.

In the circuit configuration described above, the frequency characteristic of either the high-frequency or the low-frequency band becomes a narrow band. Therefore, the number of stages of the LC-ladder circuit was increased, and a circuit with 15 elements in which an LC-ladder circuit and an LR/CR circuit were connected in parallel between the output ports enabled three-frequency matching. It was shown that by moving the matching frequency in the middle of the three matching frequencies closer to the low-frequency side or the high-frequency side, a divider having an absolute constant bandwidth in the low-frequency and high-frequency bands becomes possible.

Furthermore, ultra-wideband characteristics are possible by increasing the number of stages in the LC-ladder circuit. As a method for widening the bandwidth of impedance transformers, quarter wavelength multistage transformers are also described in Pozer's book and are often used. By using the concept of this multistage impedance transformer [16] and L-type matching circuit [17], a circuit with a relative bandwidth exceeding 100% in the UHF band was realized. Specifically, we have experimentally confirmed an ultra-wideband divider with a relative bandwidth of 100% or more, which covers the 80 MHz–370 MHz band used for public radio in Japan, with a lumped-element circuit configuration.

2. Two-stage LC-ladder divider

This section shows a lumped-element power divider that realizes the same frequency characteristics as the conventional Wilkinson power divider. Furthermore, a two-frequency matching divider operating in the UHF and the SHF band will be described.

2.1 Circuit configuration

Figure 1a shows the dual-band power divider with arbitrary two matching frequencies treated in this section [18, 19]. This circuit consists of an LC-ladder circuit ($C_{1–2}$, $L_{1–2}$) connected between Port1 and Port2 (3), an L (L_3), and an LRC circuit (C_3, L_4, R) connected between the output ports in parallel. Each parameter in the figure is normalized by the center angle frequency ω_0 and the characteristic impedance Z_0 of the input/output port, and each design value is obtained by $R_{nor}\,Z_0$, $L_{nor}\,Z_0/\omega_0$, and $C_{nor}/(Z_0\,\omega_0)$. In order to apply the even/odd mode excitation method to the proposed circuit, **Figure 1b** shows an equivalent circuit symmetric with respect to the plane AA'. At that time, each input/output port is represented by terminal resistors R_1, R_2, and R_3, and since Port1 is parallelized, it is twice as large as the output port.

2.2 Design method and scattering matrix

2.2.1 Even mode

When a signal of the same phase and amplitude is applied to each output port Port2/3 of the equivalent circuit shown in **Figure 1b**, the plane AA' becomes a magnetic wall. It is not necessary to consider the inflow of current to the L and RLC parallel circuits, and the signal applied to the output port propagates to the input port side while maintaining its potential. Therefore, the equivalent circuit can be simplified as shown in **Figure 2a**. The following equation expresses the signal non-reflection condition at the input end for conjugate matching of the terminal resistance of Port1

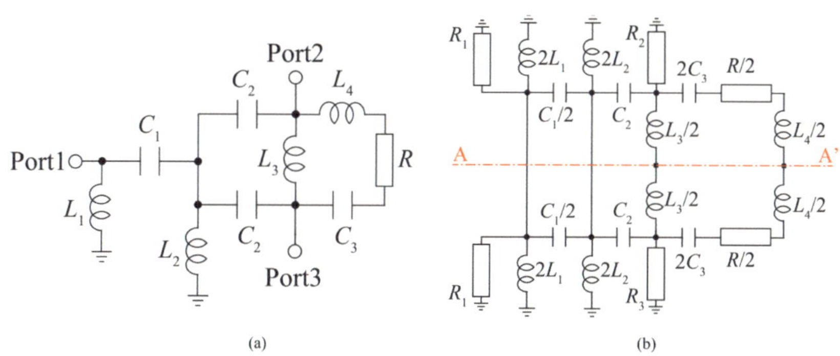

Figure 1.
Circuit configuration. (a) Schematic of two-section LC-ladder divider and (b) its equivalent circuit with onefold symmetry.

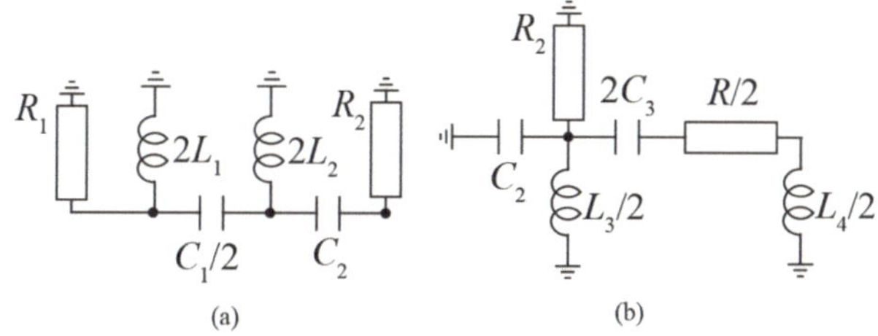

Figure 2.
Equivalent circuits at (a) even- and (b) odd-mode excitations.

and the input impedance seen from Port1 according to the theorem of maximum power supply.

$$\frac{1}{R_1} = \frac{1}{j2L_1} + \frac{1}{\frac{2}{jC_1} + \frac{1}{\frac{1}{j2L_2} + \frac{1}{\frac{1}{jC_2} + R_{(2,3)}}}} \tag{1}$$

For each of the real and imaginary parts, by determining the parameters, so that Eq. (1) is satisfied with two matching frequencies, the circuit parameters on the input side $L_{1,2}$, $C_{1,2}$ can be derived.

2.2.2 Odd mode

In the odd-mode excitation in which a signal of opposite phase and the same amplitude is applied to the output port of the circuit shown in **Figure 1b**, the plane AA' becomes an electric wall, and when the potential becomes 0 on the plane AA'. Therefore, the inflow of current to the input side can be ignored. Therefore, in this case, the equivalent circuit can be simplified as shown in **Figure 2b**.

$$\frac{1}{R_{(2,3)}} = jC_2 + \frac{2}{jL_3} + \frac{1}{\frac{1}{j2C_3} + \frac{jL_4}{2} + \frac{R}{2}} \tag{2}$$

By satisfying Eq. (2) for the real and imaginary parts and designing it to operate as an impedance transformer at the design frequency, the circuit parameters on the output port side $L_{3,4}$, C_3, and R can be calculated.

The above operation can derive all parameters, and it is possible to design an equal power divider that matches at arbitrary two frequencies. **Table 1** shows the normalized circuit parameters for some design frequency ratios.

2.2.3 Frequency characteristics of scattering parameters

The normalized circuit parameters obtained by the above procedure are C_1 = 1.23, C_2 = 1.20, C_3 = 0.54, L_1 = 1.20, L_2 = 0.61, L_3 = 1.74, L_4 = 1.93, R = 1.62 when the matching frequency ratio (f_1/f_2) is 0.8/1.2. **Figure 3a–c** shows the frequency

Frequency ratio	0.8/1.2	0.6/1.4	0.4/1.6
C_1	1.23	1.52	2.46
C_2	1.20	1.11	0.90
L_1	1.20	1.11	0.90
L_2	0.61	0.76	1.23
C_3	0.54	0.96	1.85
L_3	1.74	2.14	3.46
L_4	1.93	1.24	0.83
R	1.62	1.12	0.92

Table 1.
Normalized circuit parameters for each design frequency ratio.

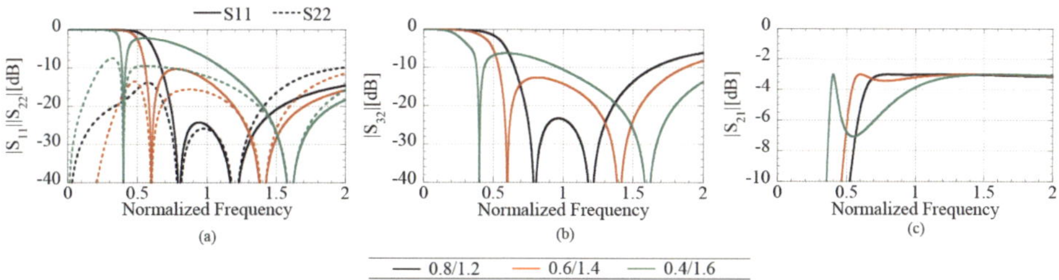

Figure 3.
Frequency characteristics of scattering parameters for two-section LC-ladder dividers with several matching frequency ratios. (a) Input/output port reflection, (b) isolation, and (c) power division characteristics.

characteristics of the scattering matrix of a dual-band power divider with several matching frequency ratios. Here, the arbitrary matching frequency ratios f_1/f_2 are set to 0.8/1.2, 0.6/1.4, and 0.4/1.6, respectively. Due to the symmetry of the circuit, the scattering matrix elements shown in the figure show the reflection characteristics (S_{11}, S_{22}) in (a), the isolation characteristics (S_{32}) in (b), and the power division characteristics (S_{21}) in (c). The relative bandwidth is defined as the value obtained by dividing the band where the reflection characteristics and isolation characteristics are −20 dB or less by the matching frequency. The relative bandwidths of the two matching frequency bands are 51.8%/77.7%, 6.6%/23.5%, and 1.9%/25.0% for each matching frequency ratio described above, and good power division characteristics can be confirmed within each bandwidth. It can also be seen that the relative bandwidth is wide on the high-frequency side of two matching frequencies and narrow on the low-frequency side. Therefore, in order to realize a wideband divider, a prototype experiment is conducted with a matching frequency of 0.8/1.2, which has a common operating frequency band.

2.3 Simulation and experimental results

In order to confirm the validity of the circuit design method, we designed a broadband power divider in the 920 MHz band using a commercial electromagnetic simulator (Sonnet em). The design conditions are a dielectric substrate with a relative permittivity of 2.2, a thickness of 0.787 mm, and each port has a microstrip line configuration with a characteristic impedance of 50 Ω. Since the commercially available 1005 size chip inductor has a self-resonant frequency in the UHF/SHF bands, it is difficult to use it in circuit design above the UHF band. Therefore, as a lumped element model, the circuit pattern is designed using a spiral inductor that directly reproduces the metal pattern on the dielectric substrate and a commercially available chip capacitor. In addition, the circuit pattern was determined by trial and error to reduce the influence of the land pattern on the characteristics while securing the land pattern for soldering required for the chip element. In addition, S-parameter data related to GRM series capacitors are used for the simulation. **Figure 4a** shows a circuit pattern with a board area of 8.8×14.8 mm^2. From the figure, the short-circuited part of the inductor is connected to the ground conductor by a via hole. **Figure 4b** shows the frequency characteristics of the scattering matrix obtained from the circuit pattern using the electromagnetic simulator. The figure shows the input/output reflection characteristics (S_{11}, S_{22}, S_{33}), isolation characteristics (S_{32}), power division characteristics (S_{21}, S_{31}), and output phase difference characteristics ($\arg(S_{21}/S_{31})$). The relative bandwidth for -20 dB reflection/isolation characteristics was 45.8%. This value is

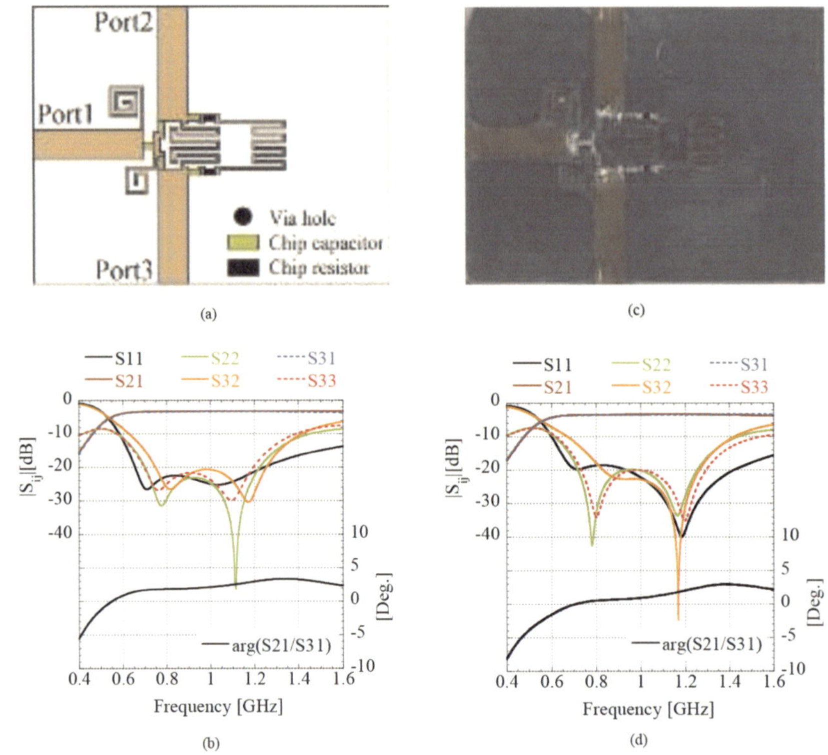

Figure 4.
Experimental results for broadband divider. (a) Simulation pattern, (b) its analysis result, (c) photograph of fabricated circuit, and (d) measured S-parameters.

larger than that of the conventional circuit based on the distributed circuit theory. Furthermore, the maximum output phase difference in the band was 2.9°.

Considering the practical application of the proposed circuit, a prototype experiment was conducted under the same conditions using the circuit pattern shown in **Figure 4a**. A conductor pattern was formed on the dielectric substrate Rogers/Duroid 5880 using a substrate processing machine (ProtoMat S63) made by LPKF. The chip elements used are the same 1005 size commercially available chip capacitors (GRM series) and thick film chip resistors (MCR series) used in the simulation and soldered to the conductor pattern. In addition, the via hole part of the simulation pattern is short-circuited with the ground conductor by making a hole with a diameter of 0.3 mm at the desired position, inserting silver paste, and sintering it. **Figure 4c** shows a prototype circuit photograph. **Figure 4d** shows the frequency characteristics of the scattering matrix of the prototype circuit measured using a vector network analyzer. From the figure, the measured results are almost the same as the electromagnetic simulation results, but the relative bandwidth with reflection characteristics and isolation characteristics of −18 dB or less is about 45.9%, and some deterioration can be seen. This is thought to be due to the tolerance of each chip element and manufacturing error of the spiral inductor. However, in the actual measurement, it was confirmed that the two frequencies were matched, and the power division characteristics were flat around the matching frequency band, and the maximum output phase difference was 2.6°.

On the other hand, by separating the matching two frequencies, it is possible to realize a divider that operates in two bands. Here, the results of electromagnetic field

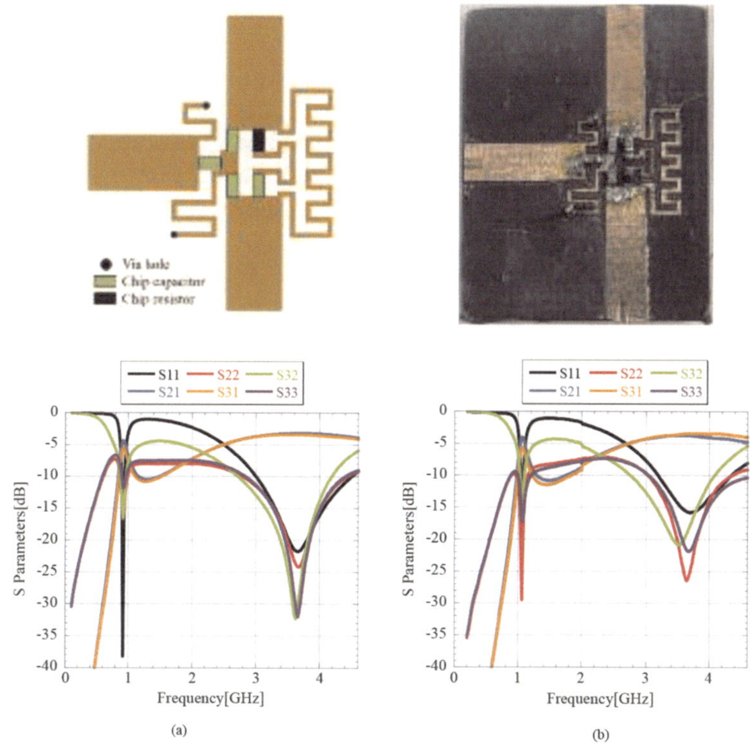

Figure 5.
Experimental results for dual-band divider. (a) Simulation pattern and its analysis result and (b)photograph of fabricated circuit and its measured S-parameters.

simulations and prototype experiments for the divider shown by the green line in **Figure 3** are introduced. The two matching frequencies are selected as 920 MHz and 3.68GHz used in IoT and 5G (sub6 band). **Figure 5a** shows the simulation pattern and its analysis results, and **Figure 5b** shows the prototype circuit photograph and measurement results. It can be confirmed that the matching frequency on the low-frequency side is slightly shifted to the higher side, and the matching frequency on the high-frequency side is slightly different. However, the measurement results and the simulation results are in good agreement.

2.4 Summary

This section has proposed a design method for a Wilkinson-type dual-band power divider with a new configuration using an LC-ladder circuit. It is known that a power divider using lumped elements in order to reduce the circuit area in a relatively low-frequency band generally has a narrow band frequency characteristic. We conducted a trial experiment of a power divider with lumped elements design in the 920 MHz band and showed that a wide operating frequency band with a relative bandwidth of about 45.9% could be obtained. Furthermore, it was shown that a divider operating in two separate bands (920 MHz/3.68GHz) could be realized. The proposed circuit is useful in reducing the circuit area in the UHF/SHF band. Next, we will conduct a prototype experiment in the 5G (Sub6) band to confirm its usefulness further.

3. Multiband divider

Since research on circuits and devices that support multiband systems is also actively conducted [20, 21], this section describes power dividers that can be matched at arbitrary three frequencies. Further, as an application thereof, it is shown that a dual-band power divider having an absolute constant bandwidth can be realized by moving the middle frequency closer to the low-frequency side among any three matching frequencies.

3.1 Circuit construction and design method

Figure 6a shows the circuit configuration of a power divider with three sections of LC-ladder circuits at the input port side [22]. The circuit parameters are normalized as described above. In designing a circuit using the even-/odd-mode excitation methods, consider an equivalent circuit having a onefold symmetry with respect to the plane AA’ in **Figure 6b**. Each input/output port is represented as a terminal resistor. When the circuit structure is vertically symmetrical, $L_{1,2,3,6}$, $C_{1,2,4,5}$, and R are considered as two elements.

3.1.1 Even-/odd-mode analysis

Since the circuit in **Figure 6b** is also symmetric with respect to the plane AA’, the even-/odd-mode analysis can be applied as in 2.2.1 and 2.2.2. **Figure 7** shows equivalent circuits at even-/odd-mode excitations. The conditional equations for obtaining the circuit parameters are as follows.

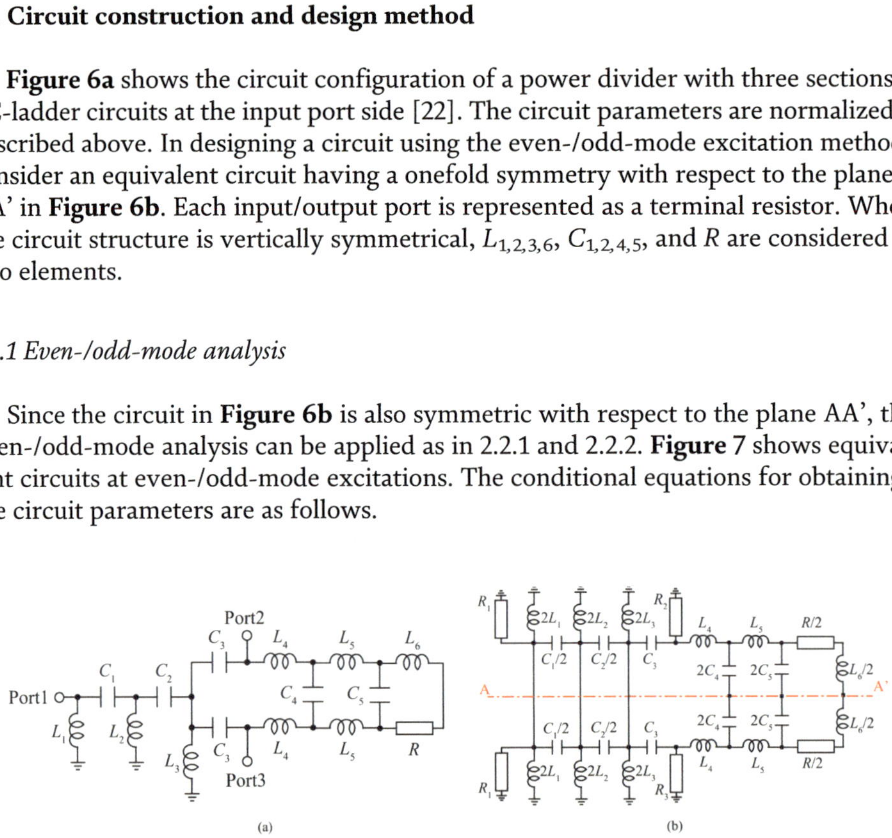

Figure 6.
(a) Schematic of LC-ladder divider with three matching frequencies and (b) its equivalent circuit with onefold symmetry.

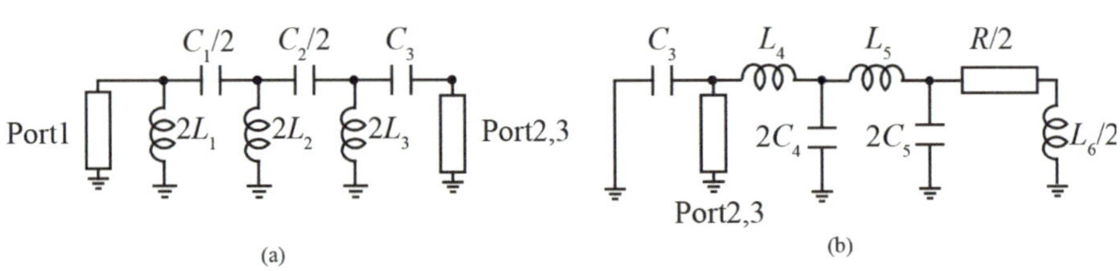

Figure 7.
*Equivalent circuit of **Figure 6** at (a) even- and (b) odd-mode excitations.*

$$\frac{1}{2R_1} = \frac{1}{j2L_1} + \cfrac{1}{\frac{2}{jC_1} + \cfrac{1}{\frac{1}{j2L_2} + \cfrac{1}{\frac{2}{jC_2 +} + \cfrac{1}{\frac{1}{j2L_3} + \cfrac{1}{\frac{1}{jC_3} + R_{(2,3)}}}}}} \tag{3}$$

$$\frac{1}{R_{(2,3)}} = jC_3 + \cfrac{1}{jL_4 + \cfrac{1}{\frac{jC_4}{2} + \cfrac{1}{jL_5 + \cfrac{1}{j2C_5 + \cfrac{1}{\frac{jL_6}{2} + \frac{R}{2}}}}}} \tag{4}$$

All parameters can be derived by the above operation, and an equal power divider with arbitrary three matching frequencies can be designed.

3.1.2 Scattering parameters

If the three normalized matching frequencies $f_1/f_2/f_3$ are set to 0.4/0.6/1.6, 0.4/0.8/1.6, and 0.4/1.0/1.6 according to the above design procedure, the circuit parameters of the equal power divider are shown in **Table 2**. The frequency characteristics of the scattering matrix for each power divider are shown in **Figure 8a-c**. The scattering matrix elements shown in this figure refer to reflection characteristics (S_{11}, S_{22}), division characteristics (S_{21}), and isolation characteristics (S_{32}). Good power division characteristics and reflection/isolation characteristics at the desired frequency can be confirmed for each circuit. Next, we are studying a dual-band power divider with absolute constant bandwidth in the UHF/SHF band. Here, the bandwidth of the UHF band is expanded by bringing two of the abovementioned three matching frequencies closer to each other. **Figure 9** shows the frequency characteristics when the design frequency ratio is 0.4/0.45/1.6. As shown in this graph, it can be seen that the band of the 920 MHz band is expanded. In the next section, when designing a dual-band power divider in the UHF/SHF band, the circuit pattern is examined by referring to each element value when the matching frequency ratio is 0.4/0.45/1.6 in **Table 2**.

Frequency ratio	0.4/0.6/1.6	0.4/0.8/1.6	0.4/1.0/1.6	0.4/0.45/1.6
L_1	2.07	1.75	1.42	1.51
L_2	1.11	4.62	2.41	0.93
L_3	0.97	1.40	1.83	1.24
C_1	1.92	2.74	2.55	2.49
C_2	2.12	7.61	8.13	1.87
C_3	2.02	1.64	1.43	1.52
L_4	1.26	1.02	0.81	0.64
L_5	0.65	0.63	0.75	0.09
L_6	1.49	1.43	1.38	1.49
C_4	0.57	0.70	0.77	0.46
C_5	1.49	1.15	0.84	0.21
R	1.05	0.89	1.01	1.04

Table 2.
Normalized circuit parameters for LC-ladder divider with various three matching frequencies.

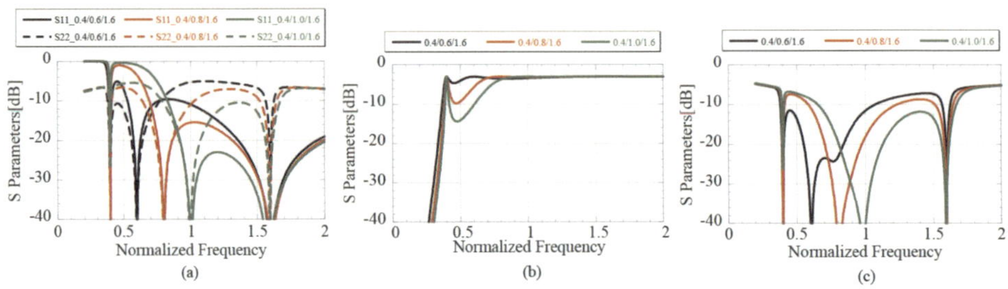

Figure 8.
Scattering matrix of LC-ladder divider with three matching frequencies. (a) Reflection, (b) power division, and (c) isolation characteristics.

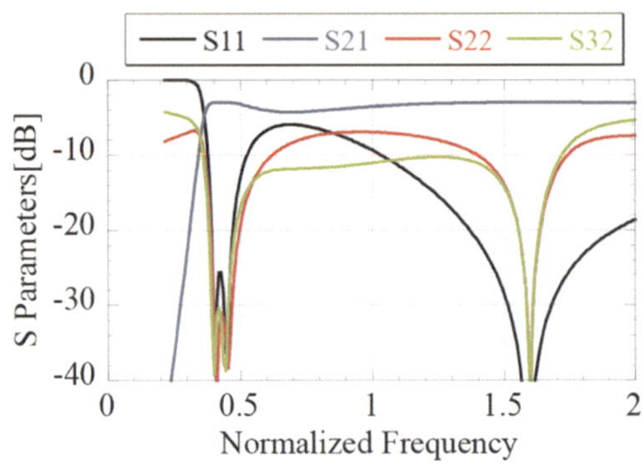

Figure 9.
Scattering matrix of LC-ladder divider with constant absolute bandwidth.

3.2 Simulation and experiment

Based on the circuit analysis using the even-/odd-mode excitation method mentioned above, we are studying the circuit pattern on the dielectric substrate by an electromagnetic analysis as a preliminary step to the trial production. The circuit pattern is shown in **Figure 10a**. The design frequency is 0.4/0.45/1.6 with a frequency ratio to the center frequency of 2.3 GHz, that is, 920 MHz/1.03 GHz/3.68 GHz, and the conditions for the electromagnetic simulation are the same as in Section 2.3. Assume the use of GRM Series and MCR series for capacitors and resistors, respectively. Assuming the influence of the self-resonant frequency of the element on the circuit characteristics, the inductor is arranged by the bending pattern of the line instead of the chip element. In **Figure 10a**, the circuit size is $7.0 \times 9.3\ mm^2$ excluding the input/output ports. The frequency characteristics of the scattering matrix obtained by the electromagnetic simulation of the circuit pattern in **Figure 10a** are shown in **Figure 10b**. In addition to good reflection/isolation characteristics and division characteristics at the desired design frequency, the absolute constant bandwidth based on the center frequency in the UHF/SHF band is about 8.6%/7.4%. A prototype experiment is being conducted under the same conditions for the circuit pattern examined by the electromagnetic simulation above. A circuit is realized by forming a conductor pattern on a dielectric substrate using a commercially available substrate processing machine (ProtoMat S63/LPKF) and soldering each element to a predetermined position. A conductor pin is inserted into the via hole with a diameter of 0.3 mm and sintered to connect to the

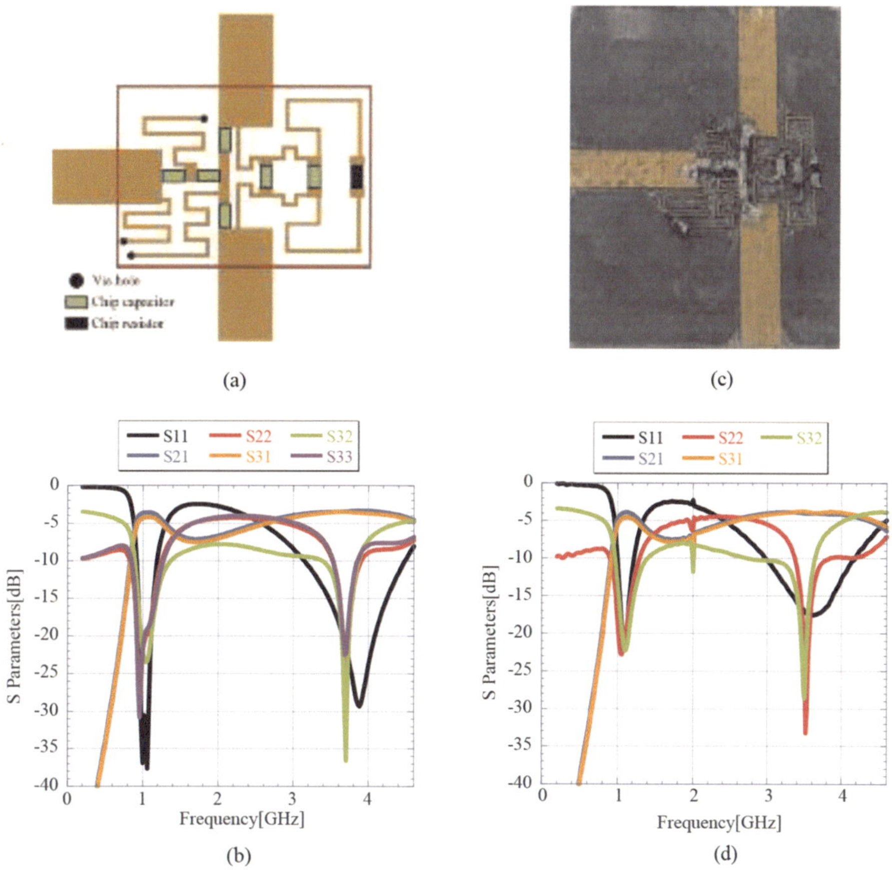

Figure 10.
Experiment of LC-ladder divider with constant absolute bandwidth. (a) Simulation pattern, (b) its analysis results, (c) photograph of fabricated divider, and (d) experimental results.

ground conductor. **Figure 10c** shows a photograph of the prototype circuit. The frequency characteristics of this circuit were measured using a vector network analyzer. The results are shown in **Figure 10d**. Good characteristics are almost identical to the analysis results, and the absolute bandwidth of 5.6%/4.8% can be confirmed in both operating bands.

3.3 Summary

As mentioned above, the usefulness of the circuit that can be designed by arbitrarily determining the three matching frequencies using the LC-ladder type configuration was examined. It was analytically and experimentally shown that a dual-band power divider with an absolute constant bandwidth in the UHF/SHF (sub6) band can be realized by closing the two matching frequencies to each other on the low-frequency side. The design method in this study is considered to be very useful in the situation where the use of the SHF band becomes more active due to social factors such as the spread of 5G. In the future, we plan to conduct an experimental study on the design of power dividers in the higher frequency range, such as the X-Band.

4. Multistage LC-ladder divider

4.1 Circuit construction and design results

By increasing the number of stages of the LC-ladder circuit, the operating band can be expanded, corresponding to the number of stages. In this section, the number of stages of the LC-ladder circuit on the input side is set to 8, and the ultra-wideband (relative bandwidth 100% over), an equal power division circuit with characteristics, will be described.

As a design method that uses lumped elements and realizes wideband characteristics while avoiding an increase in circuit size, there is a method that uses an LC-ladder impedance transformer. Here, we show the design method of the multiband LC-ladder divider proposed by Okada et al. and focus on the viewpoint of wideband and multiband. **Figure 11a** shows a multiband power divider consisting of multiple LC-ladder circuits and an RL series circuit on the output side. The figure shows an N-stage LC-ladder circuit (L_{1-N}, C_{1-N}) connected between Port1 and Port 2/3 and a ladder circuit with L and C interchanged between Port2/3 ($C_{(N+1)-(2N-1)}$, $L_{(N+1)-(2N-1)}$), RL_{2N} series circuit. Since the number of matching frequencies in this circuit corresponds to the number of stages N of the LC-ladder circuit, it can operate as a power divider with arbitrary N frequency matching according to the number of stages. In applying the even-/odd-mode excitation method to the circuit in **Figure 11a**, consider an equivalent circuit having a mirror image symmetry structure in the vertical direction as described above. **Figure 11b** shows the equivalent circuit of a multiband circuit with the plane AA' as the plane of symmetry. Considering this figure, each input/output port has terminal resistors R_1, R_2, and R_3, the input side element is doubled in parallel, and the output side

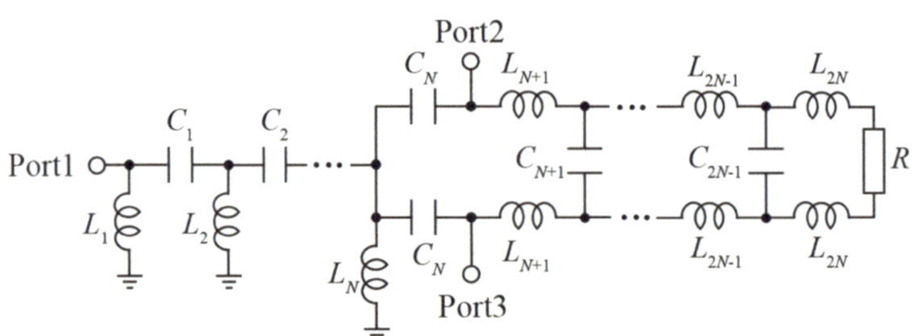

(a)

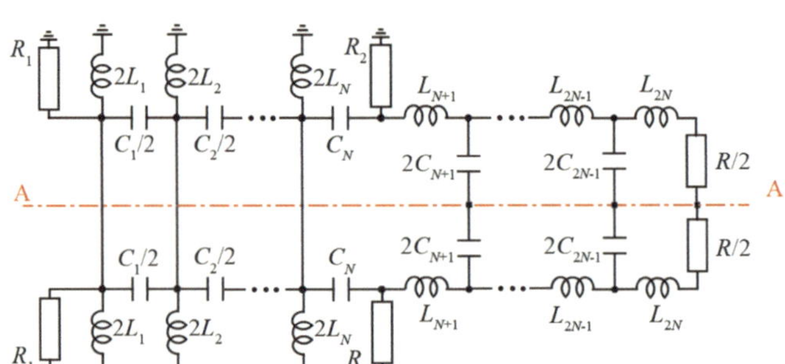

(b)

Figure 11.
(a) Schematic of N-section LC-ladder divider and (b) its equivalent circuit with onefold symmetry.

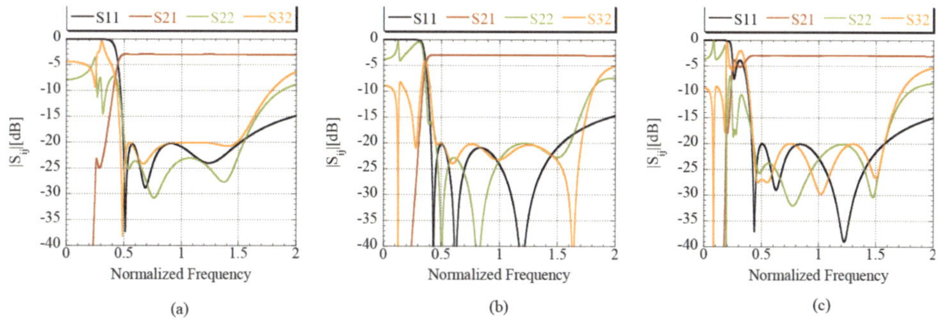

Figure 12.
Frequency characteristics of scattering parameters for N-section LC-ladder dividers: (a) four-section, (b) six-section, and (c) eight-section LC-ladder dividers.

element is divided, and the value is halved. The circuit design is possible using even-/odd-mode excitation methods as in Sections 2 and 3. As a result of designing the circuit, it can be seen that the relative bandwidth is expanded by increasing the number of stages of the LC-ladder divider. In **Figure 12a-c** are the cases where the number of stages is 4, 6, and 8, and the specific bandwidths are 98%, 108%, and 115%, respectively. By increasing the number of stages, the band becomes a wide band, but it becomes saturated to some extent when the number of stages is about 8.

4.2 Experiments

Figure 13 is a photograph of the prototype circuit of the circuit shown in **Figure 12** and the measurement result of its *S*-parameter. The chip elements used in each circuit are 19, 31, and 37, respectively, and the circuit areas excluding the input/output ports

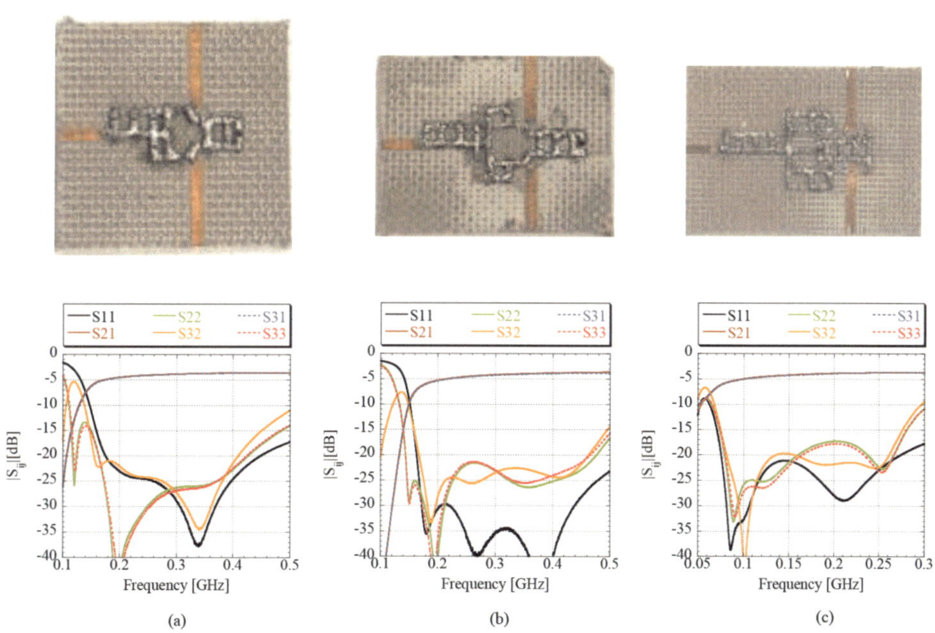

Figure 13.
Photographs of fabricated circuits and their measurement results for N-section LC-ladder dividers: (a) four-section, (b) six-section, and (c) eight-section LC-ladder dividers.

are 51.5mm^2, 138.4mm^2, and 182.4mm^2. Comparing the measurement results with the abovementioned analysis results, it can be seen that they are in good agreement. In manufacturing the prototype circuit, the circuit pattern is determined by electromagnetic simulations, as in the case of the two-section LC-ladder divider.

4.3 Summary

As mentioned above, it has been shown that the operating band of the power divider is expanded by increasing the number of stages of the LC-ladder circuit. Setting eight stages, an ultra-wideband circuit with a relative bandwidth exceeding 100% becomes possible. The effectiveness was also shown experimentally in a prototype experiment in the VHF band. With a wideband characteristic of over 100%, it is available for public radio in Japan.

5. Conclusions

By using a multistage impedance transformer and an L-type matching circuit design method to downsize the microwave power divider, the characteristics are equal to or higher than those of the conventional Wilkinson power divider designed based on the distributed circuit theory. First, it was shown that in a circuit capable of dual-band operation, a wideband circuit or two-frequency operation is possible by moving the operating two frequencies closer to or further away from each other. Next, a circuit configuration that enables three-frequency matching is shown, and by utilizing the feature that the frequency at the center of the three matching frequencies can be arbitrarily selected, a divider having an absolute constant bandwidth in the UHF/SHF band was made possible. Finally, it was shown that the operating band can be expanded by increasing the number of stages of the LC-ladder circuit and that an ultra-wideband circuit exceeding 100% can be realized if the number of stages is 8. These results have been clarified analytically and experimentally.

Author details

Tadashi Kawai*, Ayumu Tsuchiya and Akira Enokihara
Graduate School of Engineering, University of Hyogo, Himeji, Japan

*Address all correspondence to: kawai@eng.u-hyogo.ac.jp

References

[1] Wilkinson EJ. An N-way hybrid power divider. IEEE Transactions on Microwave Theory and Techniques. 1960;**8**(1):116-118

[2] Noriega F, Gonzales PJ. Designing LC Wilkinson Power Splitters. RF Interconnects/Interfaces. 2002;**48**:18-24

[3] Hirota T, Minakawa A, Muraguchi M. Reduced-size branch-line and rat-race hybrids for uniplanar MMIC's. IEEE Transactions on Microwave Theory and Techniques. 1990;**38**(3):270-275

[4] Tang C-W, Chen M-G. Realization of ultra-compact planar microstrip branch-line couplers with high-impedance open stubs. In: IEEE MTT-S International Microwave Symposium Digest. London: IEEE; 2007. pp. 995-998

[5] Pu T, Cui J, Chang H, Lu J. "A Novel ultra compact Luzzatto power divider using miniaturized stubs and folded structure," 2020 IEEE MTT-S International Wireless Symposium, 2020. IEEE London

[6] Scarddelletti MC, Ponchak GE, Weller TM. Miniaturized Wilkinson power dividers utilizing capacitive loading. IEEE Microwave and Wireless Components Letters. 2002;**12**(1):6-8

[7] Deutschmann B, Jacob A. Compact ultra-broadband power dividers with integrated resistors. In: Proceedings of the 48th European Microwave Conference. London: IEEE; 2018. pp. 620-623

[8] Lu LH, Bhattacharya P, Katehi LPB. X-Band and K-Band Lumped Wilkinson power dividers with a micromachined technology. In: IEEE MTT-S International Microwave Symposium Digest. London: IEEE; 2000. pp. 287-290

[9] Song A, Wang X, Ma Z, Ohira M. Design theory of dual-band Wilkinson power divider with different frequency ratio ranges. In: Proceedings of 2018 Asia-Pacific European Microwave Conference. London: IEEE; 2018. pp. 1486-1491

[10] I. Chen, W. Lee, and Y. Pang, "A Compact Power divider using dual composite right−/left-handed resonators (D-CRLH) with filtering response," Proceedings of 2018 Asia-Pacific European Microwave Conference, pp. 1366–1368, 2018. IEEE London

[11] Kawai T, Nakashima Y, Kokubo Y. Okada, and Ohta Lumped-element wilkinson power dividers using LC-ladder circuits. In: Proceedings of International Symposium RAD System Space Place, C1-C1407. London: IEEE; 2007. pp. 35-38

[12] Kawai T, Nagano K, Enokihara A. A 920MHz lumped-element Wilkinson power divider utilizing LC-ladder circuits. IEICE Transactions on Electronics. 2018;**E101-C**(10):801-804

[13] Okada Y, Kawai T, Enokihara A. Wideband lumped-element wilkinson power dividers using LC-ladder circuits. In: Proceedings of the 45th European Microwave Conference. London: IEEE; 2015. pp. 115-118

[14] Okada Y, Kawai T, Enokihara A. Design method for multiband WPDs using multisection LC-ladder circuits. IEEE Microwave and Wireless Components Letters. 2017;**27**(10): 894-896

[15] Okada Y, Kawai T, Enokihara A. Design of multi-way LC-ladder dividers with multi-band operation. IEICE Transmission Electron. 2017;**E100-C** (10):893-900

[16] Pozar DM. Microwave Engineering. 4th ed. Hoboken, New Jersey, United States: John Wiley and Sons, Inc.; 2011. pp. 229-234

[17] Pozar DM. Microwave Engineering. 4th ed. Hoboken, New Jersey, United States: John Wiley and Sons, Inc.; 2011. pp. 246-254

[18] Nagano K, Tsuchiya A, Kawai T, Enokihara A. "Experimental study of quasi-lumped-element power divider utilizing LC-ladder circuits at SHF-Band," IEEE Transaction of Electronic Information and Systems, vol. 141, no. 2, pp. 94–98, Feb. 2021 (Japanese ed.)

[19] Kawai T, Nagano K, Enokihara A. Dual-band semi-lumped-element power dividers at UHF/SHF Bands. In: Proceedings of the 50th European Microwave Conference. London: IEEE; 2021. pp. 844-847

[20] Yang X, Wang Z, Xie J, Zhang X, Liao Z, Yang M, et al. A tri-band gysel power divider based on tri-band phase inverter. In: 2020 Cross Strait Radio Science & Wireless Technology Conference (CSRSWTC). London: IEEE; 2020

[21] Abdelrahman BM, Ahmed HN, Nashed AI. A novel tri- band Wilkinson power divider for multiband wireless applications. IEEE Microwave and Wireless Components Letters. 2017; **27**(10):891-893

[22] Tsuchiya A, Okuda T, Kawai T, Enokihara A. Experimental study of UHF/SHF dual-band semi-lumped-element power dividers. In: Proceedings of 2021 IEEE International Symposium on Radio-Frequency Integration Technology (RFIT)|IEEE Conference, 210773. London: IEEE; 2021

Section 4

Applications in Biomedicine

Chapter 10

Applications of Microwaves in Medicine and Biology

David Vrba, Jan Vrba, Ondrej Fiser, Jesus Cumana, Milan Babak and Jan Vrba Senior

Abstract

This chapter deals with the description of recent research activities oriented on the perspective of microwave technologies in medicine and biology. It brings new ideas about the possibilities of using microwaves in thermotherapy—above all toward hyperthermia in cancer treatment. Development of new types of hyperthermia applicators (based, e.g., on technologies such as metamaterials, evanescent modes in waveguides, and other types of transmission structures) will be discussed here. Furthermore, we would like to underline in this chapter perspectives of microwaves in medical diagnostics. It is possible to expect that, e.g., microwave differential tomography, UWB radar, and microwave radiometers (all three can be used both for medical diagnostic and for noninvasive temperature measurement) will soon play an important role in it. Finally, experimental equipment necessary for research on the biological effects of EM fields is presented.

Keywords: microwave thermotherapy, hyperthermia in cancer treatment, microwave medical diagnostics, noninvasive temperature measurement, microwave differential tomography, UWB radar in medical diagnostics

1. Introduction

Wide utilization of microwave thermotherapy can be observed in the countries of the European Union, the United States, Russia, China, Japan, and many others, including the Czech Republic. Interactions of the electromagnetic (EM) field with the human body have been utilized in medicine (e.g., cardiology, oncology, physiotherapy, and urology) since the late seventieth of the twentieth century. A very important role in this process plays scientific societies, e.g., the European Society for Hyperthermia Oncology (ESHO), which cooperates with STM (Society for Thermal Medicine), and ASHO (Asian Society of Hyperthermia Oncology).

Currently, EM fields are frequently used in a few well-established medical procedures already. Good examples in the area of medical diagnostics are, e.g., computer tomography (CT) and magnetic resonance imaging (MRI). In the area of therapy, we can mention, e.g., electrosurgery and radiofrequency (RF) heating in physiotherapy. Then microwave (MW) hyperthermia and RF + MW ablation in clinical therapy are being used for the treatment of cancer and other diseases.

According to the purpose of how microwaves are used, we can divide the medical applications of microwaves into the following three main groups [1–17]:

- Therapy (mostly so-called thermal effects are being used; so-called nonthermal effects are still under discussion).
- Diagnostics (based on permittivity measurements, very prospective can be, e.g., an MW differential tomography, UWB radar technology, and MW radiometers).
- MW is only a part of the medical system (e.g., microwave technology as the basis of the linear accelerator).

As told above, treatment applications of MW are represented mainly by those based on thermal effects. Thus, we can speak about the MW thermotherapy, which can be generally divided into several basic modalities with respect to the goal temperature level or interval:

- Diathermia: which means mild heating up to 41°C at maximum (clinically can be used, e.g., in physiotherapy).
- Hyperthermia: this means increasing the temperature in the tumor area up to the interval of 41–45°C (clinically mostly used in oncology).
- Thermoablation: this means to increase the temperature over 45°C (clinically can be used, e.g., in urology for BPH treatment and in cardiology for treatments of fibrillation and/or arrhythmia).

For the abovementioned methods of thermotherapy treatments, frequency interval from 1 up to 5600 MHz is mostly used.

As for diagnostics based on the EM field—significant importance for the near future can be identified for the following methods mainly:

- MRI and CT (both these diagnostics methods have been largely used since the 80th of the last century already; they both represent the highest possible level of medical diagnostics).
- Microwave differential tomography (very prospective as a diagnostics method for the near future, to be used mainly as diagnostics of cancer or for noninvasive temperature measurements).
- Microwave UWB diagnostic radar (can be used for the same purposes as microwave differential tomography, but as well for other purposes, such as monitoring of breathing).
- Microwave radiometry (can be used for the same purposes as microwave differential tomography, but it works on different physical principles).

As for the frequency spectrum of the EM field (**Figure 1**), then it is possible to see that MRI is working in the frequency band from 64 to 299 MHz (i.e., the upper part of the RF band); instead, CT then is working in hard X-ray band. The MW frequency

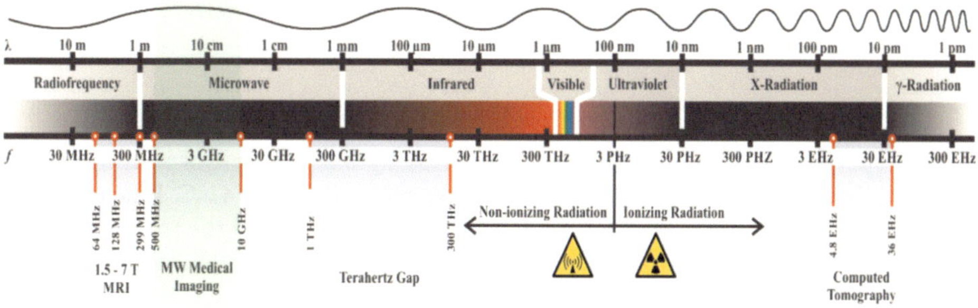

Figure 1.
Frequency spectrum of EM fields.

band is frequencies from 300 MHz to 300 GHz. The lower part of this frequency band, from 300 MHz to 6 GHz, is very prospective for MW medical imaging. Frequency band above approx. 100 GHz is very prospective for imaging with Terahertz waves. In **Figure 1**, there is a picture of the frequency spectrum of the EM field.

In this chapter, we will not describe the MRI and CT technology, as it is a well-known and broadly used application of the EM field in medical diagnostics. We will describe and discuss here other methods based on microwave technology mentioned above. The idea of MWs for medical diagnostics is a relatively new area but rapidly developing. The main advantages of MW technology with respect to CT and/or MRI are as follows:

- MWs belong to EM nonionizing radiation (such as MRI); instead, CT works in the ionizing part of the EM frequency spectrum.

- The system for MW diagnostics is very small and lightweight in comparison with MRI and CT (its dimension and weight are comparable to a notebook, so it will be possible to have it in an ambulance).

- MW diagnostic systems have the potential to be at least one order less expensive than either MRI or CT (since the MW technology is being massively used in mobile telecommunications).

And it is important to underline that for MW diagnostics, low power levels (1–20 mW) are used only.

2. Microwave thermotherapy

The first four of the following list of thermotherapeutical applications are just largely used in many countries around the world; the last three instead are at this moment in the phase of very promising projects:

- Oncology (cancer treatment)

Biological principle utilizes the fact that certain tumor cells are very sensitive to a temperature higher than 41°C, while normal cells generally survive elevated temperatures up to 45°C. And so heating of the tumor region at temperatures of 41–45°C can selectively destroy tumor cells.

- Physiotherapy (treatment of rheumatic and skeletal diseases)

Like in hyperthermia, therapeutic effect is caused by the principle of heating biological tissue, but to lower temperatures—usually only up to 41°C. It is used for treating pain in certain rheumatic and degenerative diseases and the treatment of chronic inflammations resistant to antibiotics, often used in rehabilitation and physical therapy.

- Urology (BPH treatment)

Microwave thermocoagulation—heating up to temperatures much higher than 45°C, usually around 70°C. An example can be given the microwave treatment of Benign Prostate Hyperplasia, which can replace a complicated surgery.

- Cardiology (arrhythmia and fibrillation treatment, microwave angioplasties)

Cardiac catheter thermal ablation is now the standard of care for various cardiac arrhythmia types (irregular heartbeat rhythm). The method uses a catheter terminated with a microwave antenna, which is introduced into either heart of the patient or into partially or totally blocked blood vessels. Heat gained by microwave energy either safely removes the cells inside the human heart causing arrhythmia and/or fibrillation or removes sclerotic plaques deposited on the walls of blood vessels.

- Surgery (microwave scalpel, growing implants)

- Ophthalmology (retina corrections)

- Neurology (stroke identification, brain stimulation)

For here mentioned thermotherapy treatments, it is important to mention that frequencies from the frequency band started approximately at 1 MHz and going up to 10 GHz are mostly used. This frequency range is given by the optimal depth of penetration of EM waves into biological tissue. Thus, this frequency band can achieve the needed depth of effective treatment.

2.1 Clinical applications of microwave hyperthermia

In Prague, the clinical applications of microwave hyperthermia for cancer treatment started in 1981, in cooperation with the Medical Faculty (the Charles University in Prague), the Radiotherapy Institute in Prague, and the Dept. of EM Field (the Czech Technical University in Prague). Since then, microwave hyperthermia has been clinically applied to more than 1000 cancer patients. Mostly added to radiotherapy (RT), a clinical study has been approved as a significantly positive contribution to RT treatment. Recently, a combination of hyperthermia added to proton therapy has been clinically applied in Prague.

Treatment of malignant tumors comprises several techniques usually. In some cases, tumors can be resected by surgery. Radiotherapy and/or chemotherapy can be applied when surgery is not possible or as part of a multidisciplinary approach. A less widely known treatment modality is hyperthermia. It is a therapeutic application of heat in which tumor temperatures are elevated in the range of 41–45°C. The heating of tumor tissue has a cell killing (cytotoxic) effect. However, the cytotoxic effect is small at temperatures below 45°C. Therefore, hyperthermia is always clinically combined with either radiotherapy or chemotherapy. The application of hyperthermia has been proven to increase the therapeutic effect of both radiotherapy and chemotherapy.

The effect of hyperthermia is strongly dependent on the achieved tumor temperatures and heating time. Preclinical research has shown that the cell-killing effect doubles every centigrade, e.g., 1 hour at 42°C is equivalent to half an hour at 43°C. Hypoxic tumors, i.e., tumors with a low level of oxygen, are more resistant to ionizing radiation than well-oxygenated tumors, while hyperthermia is particularly effective in hypoxic tumors.

Large solid tumors often contain hypoxic areas due to heterogeneous vascularization, making hyperthermia a useful addition to radiotherapy. The complementary effect of hyperthermia and radiotherapy is also because cells in the S-phase of the cell cycle are more sensitive to hyperthermia than the G1-phase, whereas cells are more resistant to radiotherapy in the S-phase.

Repair of DNA damage caused by radiotherapy is inhibited by hyperthermia. Hyperthermia also induces radiosensitization and chemosensitization. Furthermore, blood flow increases during hyperthermia improving tumor oxygenation and probably enhancing radiosensitivity. The increased blood flow also improves the uptake of cytostatics in tumor cells. Thus, the increased blood flow during hyperthermia is favorable for improving radiotherapy and chemotherapy effectiveness.

In clinical practice, we need to increase the temperature in a more or less circumscribed body region with tumor load. Treated volume ranges from a few cubic centimeters in case of thermoablation in lesions up to heating the whole body. Because of this, we need different types of applicators for each of the below-mentioned special cases. Thus, we can speak about different clinical modes of microwave hyperthermia.

First of all, we would like to offer an overview of the technical equipment needed for clinical applications of microwave thermotherapy in this chapter. Further, the main basic principles of EM field behavior inside the living biological system, selected from the point of view of physics related to microwave thermotherapy, will be mentioned. Moreover, we will provide the reader with references in literature, where detailed information on both physical and technical aspects of microwave thermotherapy (especially microwave hyperthermia) can be found.

2.2 Classification of clinical modes of microwave thermotherapy

According to ESHO guidelines following classification of different clinical modes of microwave hyperthermia (or thermotherapy in general) can be made:

- Local hyperthermia – Medical indications for local hyperthermia include chest wall recurrences, superficial malignant melanoma lesions, and lymph node metastases of head and neck tumors—all of which are validated in prospective randomized studies. The basic physical and technical descriptions will be given in the following text.

- Regional hyperthermia – Medical indications for regional hyperthermia include locally advanced and/or recurrent pelvis tumors, i.e., rectal carcinoma, cervical carcinoma, bladder carcinoma, prostate carcinoma, or soft tissue sarcoma. Some of these indications were validated in prospective studies. Basic physical and technical descriptions will be given in the following text.

- Part-body hyperthermia – Heated volume of a body region such as the whole pelvis, the whole abdomen, or (if clinically desirable) the upper abdomen or lower thorax or others. Basic physical and technical descriptions will be given in the following text.

- Whole-body hyperthermia (WBHT) –means to heat the whole body either up to 42°C for 60 minutes (so-called "Extreme WBHT") or only 39.5–41°C for a longer time, e.g., 3 hours (so-called "Moderate WBHT" or "Fever-like WBHT").

- Thermoablation – is performed with thin laser applicators or radiofrequency electrodes of a few millimeters. It is a minimally invasive procedure in every case, i.e., the applicators must be implanted in the lesions under CT or MR guidance. Achieved temperatures are high (up to 90°C), but the thermal gradients are pretty steep, and the effective range is 1–2 cm (i.e., lesions with diameters of 3–4 cm are the limit using standard techniques). Liver metastases (numbers up to 4) are the most frequent indication. The procedures are typically performed under MR control.

- Interstitial hyperthermia – an array of interstitial antennas or electrodes is implanted in inaccessible tumors, which might be located in deep or superficial tissues. The distance between the antennas must not exceed 1–2 cm, and therefore, lesions with diameters below 5 cm are suitable (in order to limit the number of puncturing tracks). From a physical point of view, we mostly want to create the best possible approximation through outward propagating spherical EM waves irradiated from each applicator. Thus, we can get the best approximation of the tumor dimensions and shape by dimensions and shape of the SAR distribution and thus, the best approximation of temperature distribution. Interstitial hyperthermia is an invasive procedure. Temperature measurements must be performed at the antennas and between them. In most systems, every single antenna is controlled by its generator. Dedicated systems have two or more segments per antenna or electrode controlled in phase and/or amplitude. Clinically interstitial hyperthermia has been applied for prostate carcinoma, recurrent breast cancer, and malignant brain tumors.

- Endoluminal hyperthermia – uses natural orifices to position various kinds of endocavitary applicators (microwave, radiowave, ultrasound, etc.) in direct contact with a tumor. From the physical point of view, we mostly want to create the best possible approximation with the aid of an outward propagating cylindrical EM wave irradiated from the applicator. For physical reasons, the penetration depth around those endoluminal applicators is limited and of the order of the applicator's diameter (in the centimeter range). Accessible tumors include esophageal carcinoma, prostate carcinoma, rectal and cervical carcinoma.

2.3 Local hyperthermia

The term "Local hyperthermia" means superficial treatment, and typically the clinical range is up to 3–4 cm. It can be performed with so-called superficial

DOI: http://dx.doi.org/10.5772/intechopen.105492

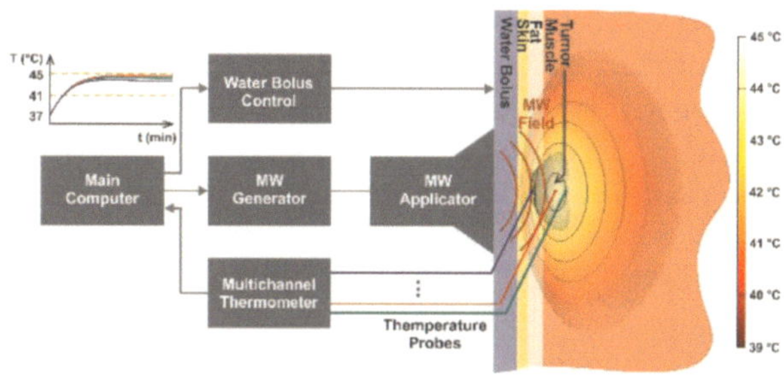

Figure 2.
MW hyperthermia system schematics.

applicators, e.g., based on EM waves in the lower part of the microwave frequency band (usually 434, 915, and 2450 MHz), ultrasound, and IR power. The technological base of EM wave applicators can be of different kinds: waveguides (water filled or with evanescent mode), microwave planar technology (e.g., patches and spirals), and according to results of this habilitation thesis, MTM applicators are very perspective as well.

From a physical point of view, we mostly want the superficial applicators to create the best possible approximation of a plane EM wave, which is the case of the deepest penetration of EM power into the area to be treated (at a given working frequency) and the best homogeneity of SAR distribution and thus, the best homogeneity of temperature distribution as well.

A system for local hyperthermia consists of a microwave (MW) power generator, an MW applicator for transfer of EM power into the treated area, see eq. (7) (tumor), a multichannel thermometer with several probes for temperature measurements in the tumor and its surroundings, and the main computer. See the schematics in **Figure 2**.

Invasive sensors then measure temperature, and according to it, MW power is being controlled in order to keep the temperature on a predetermined level.

The applicator is positioned upon the area to be treated and coupled to the tissue by a water bolus. The temperature and pressure of the water in the water bolus are possible to control, so it is possible to modify the temperature profile in the area to be treated.

In our discussion, it is essential to distinguish the following two important terms: "Depth of EM wave penetration" and "Depth of efficient treatment." The second one can have different definitions for different clinical applications of thermotherapy treatments (i.e., hyperthermia, physiotherapy, and ablations). Here, we will work with the definition for hyperthermia only.

2.4 Depth of EM wave penetration and depth of efficient treatment

For the initial estimation of EM wave penetration into biological tissue, we can take a model for the behavior of amplitude of the plane wave in a lossy media.

$$E(z) = E_0 e^{-\alpha z} \tag{1}$$

where E is electrical field intensity, E_0 is its value at the surface of biological tissue, z is the depth under the surface, and α means the attenuation constant of EM wave in lossy media.

The depth of the EM wave penetration d then has its definition in EM field theory based on the decrease of the amplitude of electrical field intensity to value E_o/e (where e is a basis of natural logarithm) when a plane EM wave enters into a lossy media; see the following equation.

$$d = \frac{1}{\sqrt{\pi \sigma \mu_0 f}}. \tag{2}$$

It can be seen that d is inversely proportional to the square root of σ and μ_0 of a given tissue and to the frequency f of the EM field. For the usual operating frequencies of the applicators 27, 70, 434, 915, and 2450 MHz (all of them except 70 MHz belong to ISM frequency bands reserved for industrial, scientific, and medical applications), it can be concluded that the lower the operating frequency, the deeper will be the penetration depth. The same conclusion can be made for the value of conductivity.

In the case of microwave hyperthermia, the depth of the efficient treatment is given by the distribution of temperature in the treated area—it is formulated as a 25% decrease of the SAR value with respect to the maximum value of SAR inside the treated area. That guarantees quick and quality heating of the treated area from 37°C to at least 41°C. Suppose the maximum temperature in the tumor will be at the level of 45°C, then the depth of the efficient treatment depends on the following factors:

1. Depth of EM wave penetration – At a given working frequency and a given type of biological tissue.

2. EM field distribution in the aperture of the applicator – Usually, we are trying to create in the aperture of the applicator the distribution of EM field very similar to plane wave—thus, it is possible to accomplish the deepest penetration depth for the particular frequency and aperture dimensions.

3. Aperture size of the applicator – Bigger aperture size helps to approach better the EM field distribution inside the applicator aperture to the case of a plane wave. Very good results are expected if the aperture dimension size is comparable or bigger than a half wavelength at the operating frequency.

4. 3D configuration of biological tissues in front of applicator aperture – Biological tissues can roughly be sorted into two categories: high or low water content. Tissues with high water content have higher attenuation than tissues with low water content. So, e.g., in the fat layer, there is EM wave less attenuated than in muscle tissue, and thus it penetrates deeper in fat than into muscle tissue.

5. Temperature of water in water bolus – This water can cool the surface of the area to be treated and thus improve the temperature profile inside this area.

There is a general rule for hyperthermia applicators optimization if we need to reach the maximal depth of efficient treatment and the best possible homogeneity of the temperature distribution inside the treated area. At least in the central part of their aperture, the distribution of the EM field should be very similar to plane wave; thus, it is possible to accomplish the deepest penetration depth for the particular frequency and aperture dimensions.

2.5 Regional hyperthermia

The term "Regional hyperthermia" means treating deep-seated tumors of the pelvis or lower extremities, etc. The so-called regional applicators can perform treatment, i.e., usually, an array of phase-controlled radiating elements typically working in the frequency range of 50–150 MHz. As radiating elements again, waveguides or dipoles are mostly being used. They surround the whole circumference; all possible directions are employed to deliver EM energy into the treated volume. The higher number of antennas and higher the frequency have the potential to control the heating 3D pattern. Several rings of antennas directed to the patient axis can be used to enable flexibility with respect to the anatomical structures for optimization.

A system for regional hyperthermia consists of a microwave (MW) or radio-frequency (RF) multichannel power generator (multiple power generators), an array of MW applicators for focusing EM power into the area to be treated (tumor), a multichannel thermometer with several probes for measurements of temperature in the tumor and its surroundings, and the main computer; see the schematics in **Figure 3**.

Applicators are positioned upon the area to be treated and coupled to the tissue by a water bolus. The water temperature in the water bolus is possible to control, so it is possible to modify the temperature profile in the area to be treated, like in the case of local hyperthermia. In the case of regional hyperthermia, the water temperature in the water bolus is usually below 10°C.

From a physical point of view, we mostly want to create the best possible approximation of a cylindrical or spherical EM wave irradiated from several (typically from 4

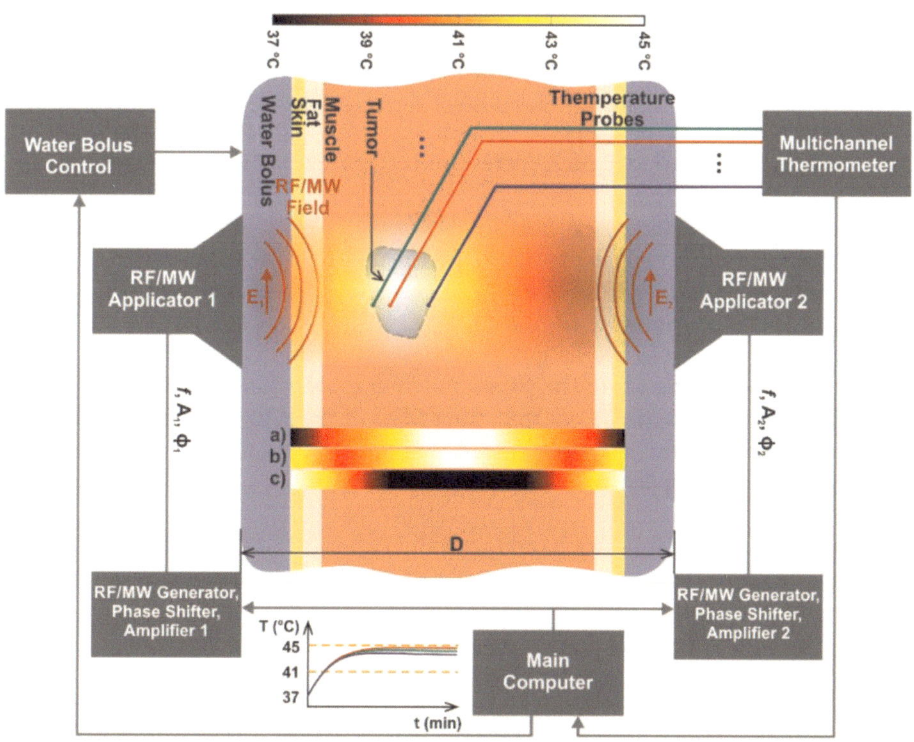

Figure 3.
Schematics of microwave system for regional or part-body hyperthermia.

up to 12) single applicators situated around the patient. Superposition of the waves from these single applicators then creates inward propagating cylindrical or spherical waves, enabling the focus EM power in the area to be treated. Thus we can get the best approximation of the tumor dimensions and shape by dimensions and shape of the SAR distribution and thus, the best approximation of temperature distribution.

In discussed case, when we have a cylindrical phantom surrounded by several above mentioned applicators, then for the thermotherapy, the most important component of the EM field will be longitudinal component Ez, which in the discussed case can be expressed by equation.

$$\mathbf{E}_z(r) = K\mathrm{H}_0^{(2)}(\gamma r), \tag{3}$$

where $\mathrm{H}_0^{(2)}$ is the Hankel function of zero order and second kind, K is a constant, r is a radius vector, γ is a complex propagation constant. Hankel function $\mathrm{H}_0^{(2)}$ can be calculated as a linear combination of the Bessel function J_0 and the Neumann function N_0:

$$\mathrm{H}_0^{(2)}(\gamma r) = \mathrm{J}_0(\gamma r) - \mathrm{j}\mathrm{N}_0(\gamma r), \tag{4}$$

As cylindrical agar phantom is mimicking muscle tissue, i.e., lossy medium, then the argument of Hankel function is a complex number. In **Figure 3**, there are three basic cases of temperature distribution inside an area treated by regional hyperthermia, which follows from the behavior of the Hankel function (see the narrow colored strips signed by letters *a*, *b*, and *c*):

- If the frequency is very low (from 27 to 80 MHz) and thus the depth of the penetration at this frequency is comparable to or bigger than the radius of the phantom *D*/2, then the shape of the temperature distribution may be approaching the case *a*, which corresponds to part-body treatment.

- If the frequency is increased to 100 MHz or higher (i.e., attenuation constant is increasing), then we expect to approach the shape of the temperature distribution given by case *b* of a very significant focus of temperature on the tumor. This case corresponds to regional treatment.

- Case *c* then corresponds to the situation when the depth of penetration of the studied EM wave is much less than the radius of the area to be treated. This typically happens for frequencies of, e.g., 434 MHz and higher. This case corresponds to deep local or superficial local treatment.

Given examples of frequency bands are valid either for the human body of average dimensions or for agar phantom with similar dimensions and dielectric parameters with values near to values valid for muscle tissue.

2.6 Part-body hyperthermia

Part-body hyperthermia is a technique derived from the regional approach and developed to heat a selected anatomical region in an extended manner up to 41–42°C under careful MR monitoring. From a physical point of view, we want mostly to create the best possible approximation of a spherical EM wave irradiated from several (typically from 12 up to 24) single applicators situated around the patient.

Superposition of the waves from these single applicators then creates inward propagating spherical waves enabling the focus EM power in the area to be treated. Thus, we can get the best approximation of the tumor dimensions and shape by dimensions and shape of the SAR distribution and thus, the best approximation of temperature distribution.

Due to safety reasons, the use of MR monitoring (to measure online temperature and perfusion) and a planning system is required at these higher power levels. Systems for part-body hyperthermia are called "hybrid systems" because they are based on the MR-compatible integration of a multiantenna applicator into an MR tomograph.

2.7 EM waves in biological tissue

The most important effect from the point of view of microwave thermotherapy is the propagation of EM waves through the biological tissue to be treated. It can be classified as lossy dielectrics. So the power (energy) of propagating EM wave will be changed into thermal power (energy). For more details, see [1–20].

EM energy turns to heat, particularly due to the following mechanism: When the alternate field takes effect, vibrating electric particles lag behind the exciting intensity of the electric field; the current is not entirely in phase with electric field intensity. It is possible to describe this phase mathematically in a way that we virtually split up the movement of electrons into:

- Component that follows electric field intensity.
- Component that is in phase with the difference of potentials on electrodes.

The first component mentioned above determines the real part of permittivity ε'. The other one is the cause of loss of current heating up the dielectric and determining the imaginary (conductive) part of permittivity ε''. From this, relative permittivity depends on the polarization charge value of the dipole, and alternate losses depend on the weight and volume of moving particles. The dielectric quality is thus given by the ratio of particle charge and particle mass.

When the electromagnetic energy goes through the biological tissue, it is absorbed and turned into heat, resulting in a temperature increase of biological tissue within the irradiated area. Spatial distribution of temperature induced the way mentioned above (with respect to depth of EM wave penetration and depth of efficient treatment) depends on various factors.

The interaction of the EM field with biological tissue studied from a physical point of view is well described in several references, e.g. [1–20], so we do not need to go into the details here. When studying these interactions to be used in clinical applications of thermotherapy, then usually it is necessary to determine by calculations or measurements following 3D or 4D distributions:

- The 3D spatial distribution of the values of the EM field main quantities (e.g., vector of electric field strength $\mathbf{E}(x,y,z)$, vector of magnetic field strength $\mathbf{H}(x,y,z)$, etc.) in a certain area of the biological tissue—area to be treated.

- The 3D spatial distribution of power P_a absorbed in a given biological object can be for a single point or elementary volume (voxel) calculated as

$$P_a(x,y,z) = \frac{\sigma}{2}|\mathbf{E}(x,y,z)|^2. \tag{5}$$

The 3D spatial distribution of specific absorption rate—the SAR [W/kg] indicates the EM energy absorbed in the biological tissue and, as shown by the unit, it is the power absorbed per 1 kg of tissue

$$SAR = \frac{\delta}{\delta t}\left(\frac{\delta W}{\delta m}\right) = \frac{\delta}{\delta t}\left(\frac{\delta W}{\rho \delta V}\right) = \frac{\delta P}{\delta m} = \frac{\delta P}{\rho \delta V}, \tag{6}$$

where W is the electromagnetic energy absorbed in the biological tissue, t is the time, and m denotes mass. P is the power of the electromagnetic wave that spreads the biological tissue, ρ is the density of the tissue, and V is the volume. By introducing the spatial distribution of the intensity of the electric field E (x, y, z), the relationship is as follows:

$$SAR = \frac{\sigma}{\rho} \frac{|E(x,y,z)|^2}{2} \tag{7}$$

Which can be further modified as $SAR = P_a/\rho$. In case of experimental evaluation of the SAR, we can measure the temperature increase ΔT after heating by EM-power in time interval Δt.

$$SAR = c\frac{\delta T(x,y,z,t)}{\delta t} = c\frac{\Delta T(x,y,z,t)}{\Delta t}, \tag{8}$$

where c is the specific heat of the biological tissue or its phantom and SAR very well defines the level of exposure of the biological tissue.

- The 4D—i.e., spatial and time-dependent distribution of temperature $T\ (x,y,z,t)$ in a given biological object, which can be calculated from the so-called Penne's Bioheat Equation, see in part 2.9 of this chapter.

High-frequency electromagnetic fields can penetrate the human body and propagate through. During the propagation of EM waves through biological tissues, their energy is gradually absorbed and converted into heat, thereby increasing the temperature of the irradiated area. To such a wave, biological tissues behave as a lossy dielectric. In such a case, permittivity and permeability become to be complex numbers. The spatial distribution of temperature depends on many factors, the most important of which are:

- The type of the EM wave (i.e., whether it is planar, cylindrical, or spherical).
- Operating frequency determines the EM wavelength (i.e., penetration depth).
- Spatial distribution of the biological tissue in the irradiated volume.
- Dielectric and thermal characteristics of each tissue type in a certain area.
- Blood flowing into the treated area.

2.8 Treatment planning of clinical application of thermotherapy

The term treatment planning for clinical application of the thermotherapy means mathematical and experimental modeling of the effective treatment timing to determine the four-dimensional (4D) distribution of temperature (i.e., 3D in space + temperature behavior with respect to time) during the scheduled treatment (both within the treatment area and in its surroundings).

When preparing a particular type of clinical application, it is necessary to perform a series of experiments and model calculations to create a specific idea about the actual distribution of temperature (with respect to SAR) in the treated area. It is a highly complex problem that is not yet fully resolved. This is due to several factors, of which the most important may be considered:

- highly inhomogeneous nature of the biological object—i.e., mainly irregular and complex definable spatial distribution of different types of biological tissues in the human body,

- in practice, the usually unavailable precise description of the topology of the bloodstream, and particularly its response to external stimuli—i.e., mainly at an increasing temperature in the treatment area,

- very "complicated" nature called near electromagnetic field emitted from the aperture thermotherapy applicators.

In the case of treatment planning, first, we need to do the calculation of SAR 3D distribution and after to do the calculation of the temperature 3D distribution. This distribution inside the treated area (heated by microwave energy q) can be expressed from the well-known Pennes Bioheat Eq. (1948):

$$\rho C_p \frac{\partial T}{\partial t} = \nabla \bullet (k \nabla T) + \omega_b C_{p,b}(T_a - T) + q_m + \rho SAR. \tag{9}$$

where T is tissue temperature (K), t is time (s), ρ tissue density (kg/m^3), C_p specific heat capacity of tissue (J/kg/K), $C_\mathrm{p,b}$ specific heat capacity of blood (J/kg/K), k thermal conductivity (W/m/K), ω_b volumetric blood flow rate (kg/s/m^3) of the specific tissue, T_a arterial blood temperature (usually 37°C), and q_m metabolic heat source rate (W/m^3).

The possibilities of an analytical solution to this equation are limited to a few cases—e.g., the "one-dimensional" case of plane wave penetrating homogeneous phantom. Therefore, computers are to be used to solve this equation to obtain the temperature $T(x, y, z, t)$ time dependence and 3D space distribution. For the treatment planning of microwave thermotherapy, it is possible to use commercially available SW products, e.g., SEMCAD X, Sim4Life, Comsol Multiphysics, and CST Microwave Studio.

In general, it is necessary to solve the time dependence of the temperature $T(x, y, z, t)$ at different points in complex three-dimensional space with the inhomogeneous structure of the biological tissue whose blood supply changes depending on heating. This is not analytically solvable in general, only partially in the case of simplified geometric models.

The versatile option is to apply numerical methods using very powerful computers. The numerical solution then typically uses differential methods or finite element methods. The biggest problem then acts precisely, defining and modeling the bloodstream and its responses to cool or heat certain areas of the human body. The situation is further complicated dramatically by the topology of the heated area. The topology can be a good guess for subsurface treatment in clinical applications. However, the more complicated is the situation for deep regional heating when mapping the treated area requires a CT and/or MRI.

3. Noninvasive temperature monitoring

Noninvasive temperature monitoring of hyperthermia cancer treatment is one of the crucial points for its successful clinical applications. MRI is often discussed to be a prospective solution to this problem. However, it is a costly way (commercially available hyperthermia system controlled by MRI has a price above 1 million EUR).

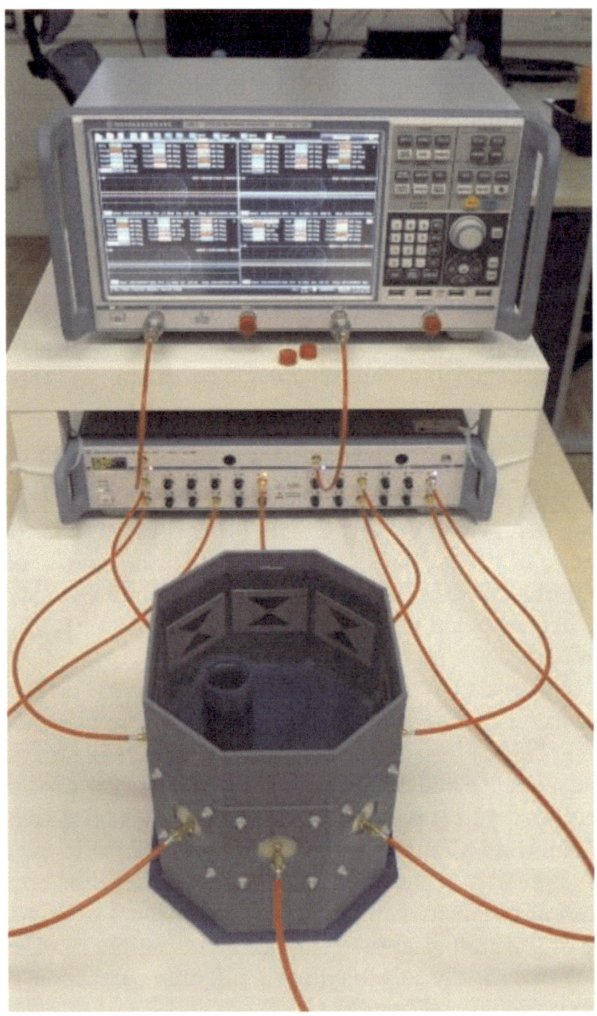

Figure 4.
Photograph of an experimental system for research of MDT.

Because of that, cheaper solutions-based on, e.g., microwave or ultrasound technology, could be a convenient alternative to MRI temperature monitoring. Till now, microwave radiometers have been discussed for this purpose. Radiometers can measure the absolute value of temperature, but their spatial resolution is not sufficient. They integrate thermal noise from certain volumes, and thus, they indicate approximately average temperature inside this volume, so it may happen that the microwave radiometer will not identify existing hot spots or cold spots.

Our theoretical and experimental research work is focused now on microwave differential tomography (MDT). The Department of Biomedical Technology (the Czech Technical University in Prague) developed its own MDT system (see photo in **Figure 4**) in cooperation with Prof. Andrea Massa from Eledia Research Center (Trento, Italy). It seems realistic that the MDT methods can be used for 3D noninvasive temperature monitoring of the treated volume during thermotherapy in oncology. Existing suitable reconstruction algorithms, which allow quasi-real-time monitoring of changes of dielectric properties due to changes of temperature, were implemented. Reconstruction algorithms were tested on different 2D and 3D models. The obtained results using the Distorted Born Algorithm (DBA) and Born Algorithm (BA) were compared in terms of the algorithms' ability to reconstruct the shape and position of the target and flatness of the obtained object function in regions without change in dielectric properties.

4. Microwave applicators

Our research work is oriented toward studies and developments of local external applicators working at 27, 70, 434, and 2450 MHz (see **Figure 2**). These applicators were used to treat deep-seated and/or superficial tumors (treatment depth from 2 up to 8 cm).

Now, following new trends in this field, we continue our research in the important directions of regional applicators (see **Figure 3**). Moreover, we are trying to implement new microwave technologies in the design of new hyperthermia applicators, e.g., applicators based on metamaterial technology. Our BioEM team with Prof. Paul Stauffer from Thomas Jefferson University Hospital in Philadelphia has developed such applicators. Research of MW thermotherapy systems and MW medical diagnostics is in Prague done in cooperation with Dept. of EM Field (Faculty of Electrical Engineering) and Dept. of Biomedical Technique (Faculty of Biomedical Engineering), both are part of the Czech Technical University in Prague. The most important technical activities in this field can be specified as:

- Design of the applicators based on new MW technologies, i.e., development of new type applicators for more effective local, intracavitary, and regional treatment;
- Development of treatment planning, i.e., mathematical and experimental modeling of the effective treatment;
- Feasibility study of noninvasive temperature monitoring, e.g., microwave differential tomography or UWB radar technology;
- MW medical diagnostics (e.g., MW differential tomography).

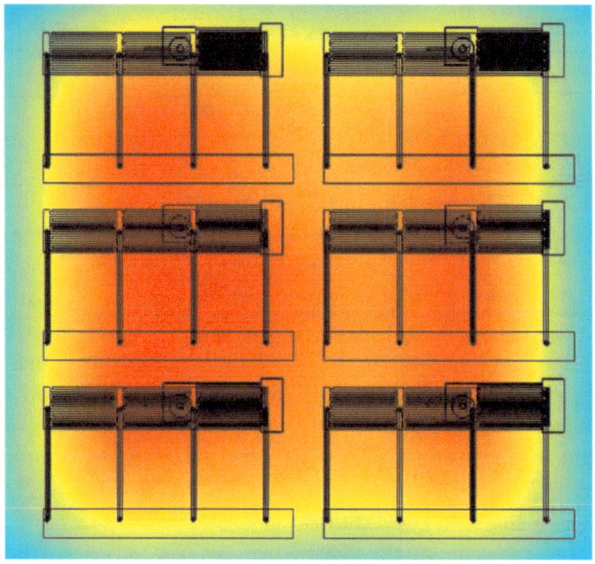

Figure 5.
Calculation of the temperature distribution at 3.5 cm depth under the 3 × 2 MTM elements array. The yellow color denotes the threshold of the treated area.

In **Figure 5**, there is an example of the calculated distribution temperature obtained by a matrix of a 3x2 MTM elements array. The highest temperature level is displayed here in red color, and the yellow color denotes the threshold therapeutic temperature of 41°C.

In actual clinics, we need the treatment planning to create so-called phantoms of the patient body or at least phantoms of the area to be treated, see **Figure 6a** and **b**. In **Figure 6a**, there is an example of a homogeneous phantom; in **Figure 6b** then, there is an example of the anatomical phantom. The first one is suitable for verifying the fundamental behavior of the applicator; the second one then is needed for the 3D SAR and 3D temperature distribution during the treatment of the actual patient.

In **Figure 7**, an example of SAR distribution is calculated for the case of the anatomical phantom. A very strong focus of MW power on a big tumor can be observed here.

5. Microwave medical diagnostics

As mentioned above, recently, there have been strong trends in research to apply microwave technology in medical diagnostics. Significant importance for the future can be identified for above all the following methods: microwave differential tomography, microwave diagnostic UWB radar, and microwave radiometry.

5.1 Microwave differential tomography (MDT)

In Prague, the MDT is developed by a research group from the Dept. of Biomedical Technology in cooperation with Prof. Andrea Massa and his group from ELEDIA Research Center (University of Trento, Italy). Theoretical works are focused on a theory of differential microwave imaging (DMI) in quasi-real-time. Existing suitable

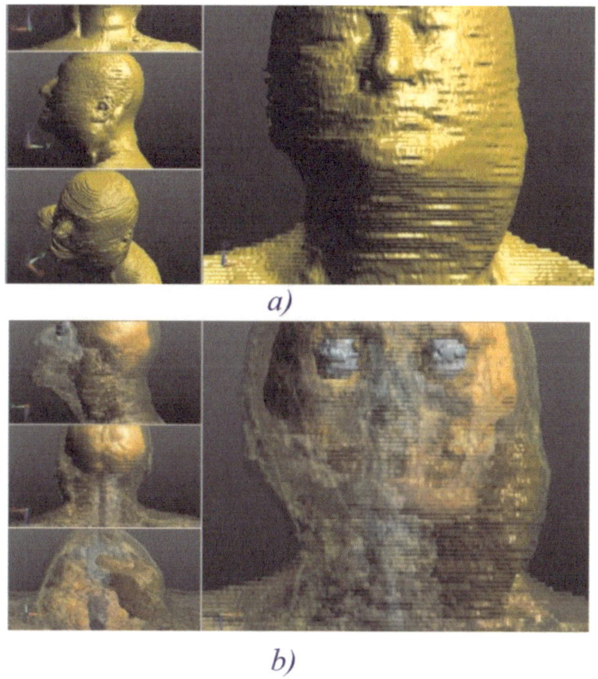

Figure 6.
Example of a homogeneous (a) and the anatomical phantom (b).

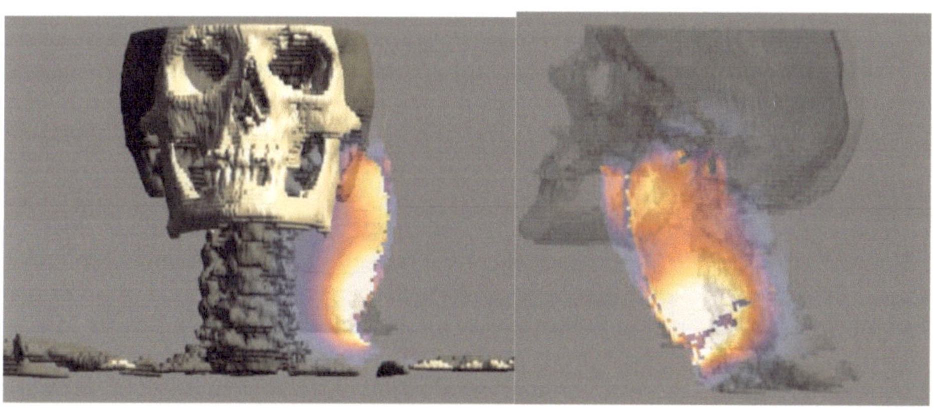

Figure 7.
*Example of SAR distribution calculated for the case of anatomical phantom given in **Figure 6**.*

reconstruction algorithms, namely Distorted Born Algorithm (DBA) and Born Algorithm (BA), which allow quasi-real-time monitoring of changes of dielectric properties due to changes of temperature, were implemented. They were applied and tested both numerically and experimentally within the feasibility studies.

These reconstruction algorithms were tested on numerical data from numerical 2D and 3D simulations; see **Figure 8**.

The below described results based on DBA and BA were compared in terms of the ability to reconstruct the shape and position of the target and flatness of the obtained

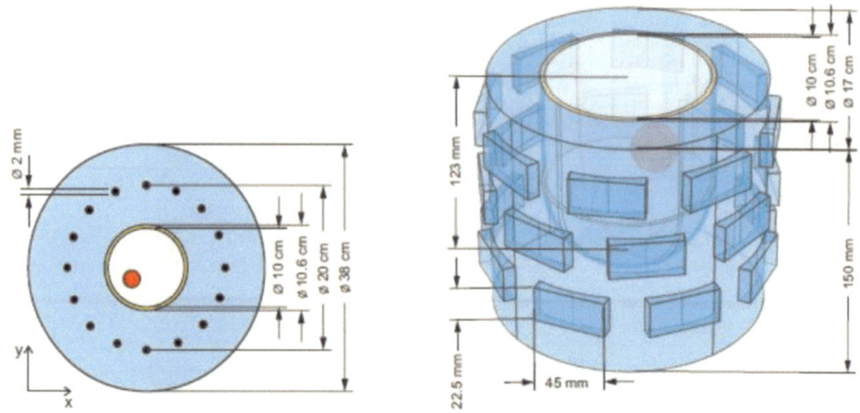

(*a*) Cross-section of a 2D model (*b*) Perspective view of the 3D model

Figure 8.
Numerical models for testing of reconstruction algorithms.

object function in regions without change in dielectric properties. Influences of varying TSVD threshold values, number of voxels, calibration, and normalization were tested. BA with a low TSVD-threshold value leads to clear pictures of the difference in relative permittivity, but we lose information about the difference in conductivity. The described algorithms were tested with a sphere that was virtually homogeneously heated. The resulting pictures were not of the clear boundary of the so-called objective function: the predicted changes of object function are smooth, see **Figures 9** and **10**. Even if the implemented algorithms show several deficits, they represent state of the art and are therefore a suitable starting point in developing the combined MW system. Here described, the principle of noninvasive temperature monitoring, once it is commercially available, would mean a very significant improvement in quality assurance for hyperthermia treatment of oncological patients in actual clinics and for the comfort of their treatment as well.

In **Figure 4**, there is a photograph of the laboratory MDT system built at the Dept. of Biomedical Technique. In this case, it consists of eight bow-tie antennas, but we can go up to 24 antennas in total. Necessary MATLAB scripts for measurements automatization, data acquisition, and image reconstruction were implemented by us. We created numerical models for solving the forward problem, which is necessary for the reconstruction algorithms. A preliminary evaluation of the system based on measurement results was performed at the same time. It seems realistic that the DMI methods can be used for 3D noninvasive temperature monitoring of the treated volume during thermotherapy in oncology.

Currently, we study (by means of numerical simulations) the suitability of different types of antennas, e.g., their EM principle, dimensions, number, and geometrical configuration. We know that the main resolution limit of the described system is a low number of radiating elements. We plan to extend the system to the maximum possible number of antenna elements (i.e., up to 24). We believe there will be considerable improvement in the resolution.

Another prospective possibility of using the principle and technology of DMI is the rapid detection, identification, and classification of strokes (SDI), which would be essential for the quick, qualified decision of what kind of treatment is necessary to

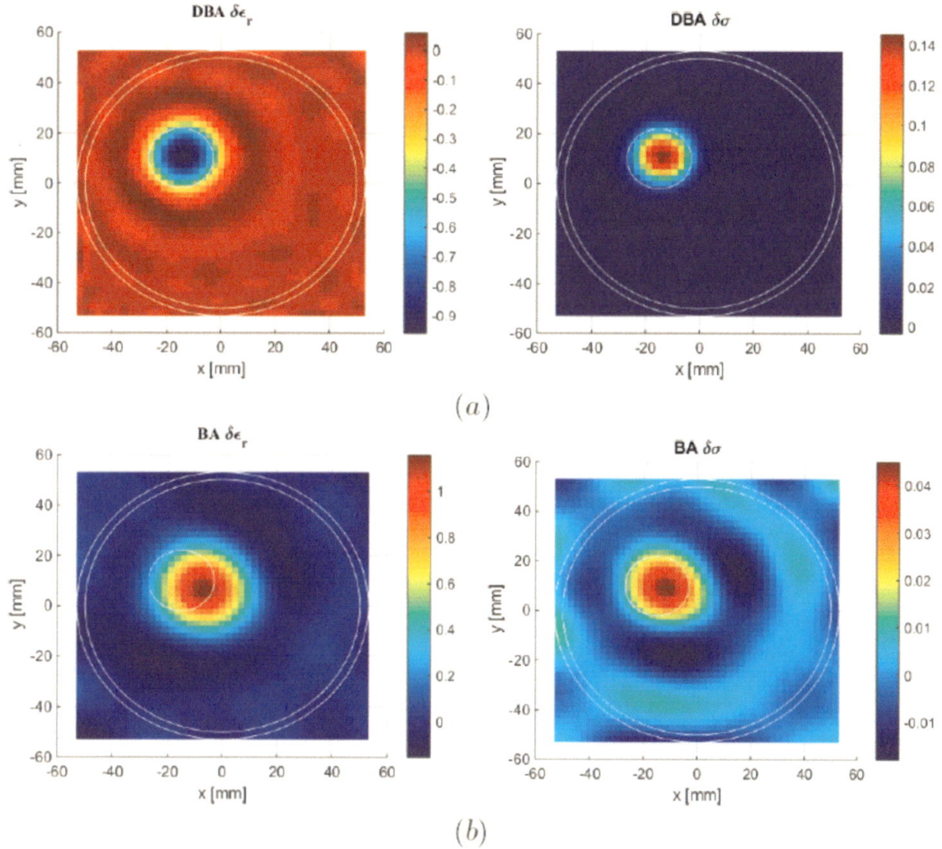

Figure 9.
Results of reconstruction on a 2D model.

give to the stroke patient already in the ambulance when he/she is being transferred to the hospital. The Pioneer research group in this area is a team of Prof. Mikael Persson from Chalmers University in Goeteborg, Sweden.

5.2 UWB radar

Dr. Marko Helbig and Dr. Juergen Sachs from TU Ilmenau in Germany came up with the idea to use microwave UWB radar technology for noninvasive microwave imaging and/or noninvasive temperature monitoring. In Prague, they are followed by people from the Dept. of EM Field.

The detection of temperature change via UWB radar signal is based on the fact that the complex permittivity changes with temperature. We have shown that it is possible to detect these changes by UWB microwave radar. In our case, the antenna array comprises eight dipole antennas (21 x 11 mm). These antennas are excited by the UWB pulse in the frequency band 1–8 GHz. The values of relative permittivity and specific conductivity of all considered tissue temperatures (at starting temperature of 37°C) can be taken, e.g., from the IT'IS Foundation database.

We worked with an experimental antenna setup for UWB temperature change detection to be used in microwave hyperthermia treatment. Our numerical and

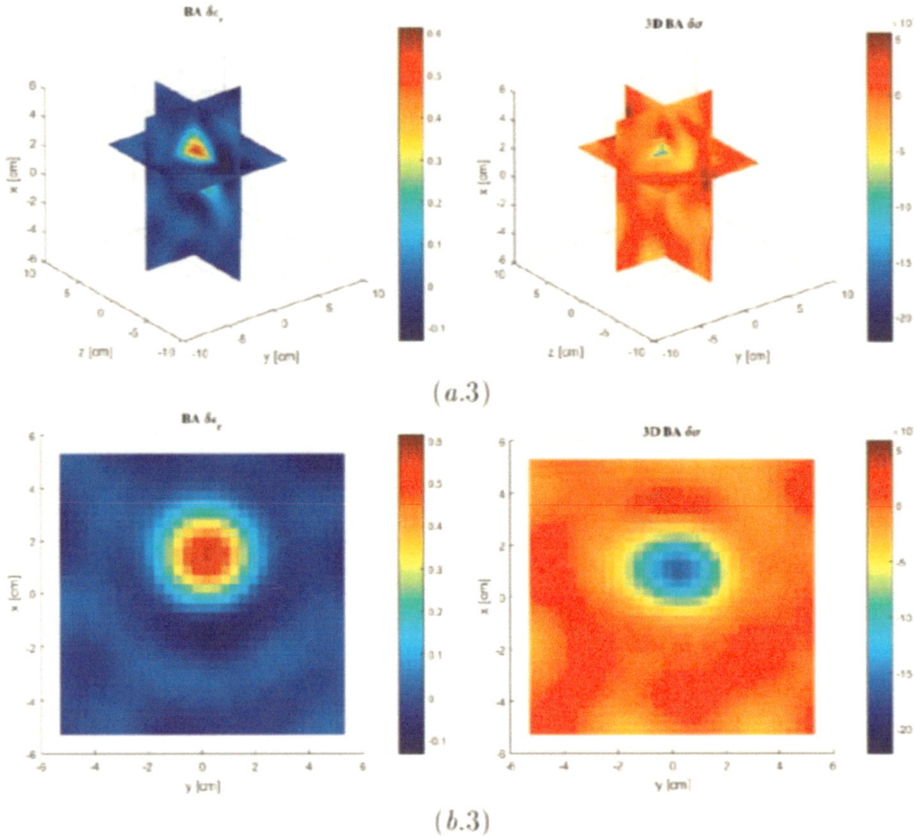

Figure 10.
Results of reconstruction on a 3D model.

laboratory models with implemented frequency and temperature dispersive parameters of biological tissues were used for a series of simulation purposes. The results from our numerical simulations show that it is possible to identify even very low changes in tumor permittivity caused by temperature change.

Our experiments with the homogeneous and nonhomogeneous phantoms have shown that we can detect even different temperature layers. From the reconstructed image, we can partially reconstruct the shape and position of the simulated inhomogeneity. The way to improve the chance for more accurate differential temperature reconstruction is in the higher number of antennas closer to the heated area utilization and in the attenuation correction improvement.

6. Biological effect of microwave power

Research studies on the interactions between the EM field and biological systems have been the subject of high interest during the last decades. Here, we would like to give more details about such kind of research and obtained technical and biological results (i.e., basic description of implemented exposure systems). Two of our recent projects were oriented on the research of thermal effects of EM field (using either waveguide or array applicators). And the third one then on the research of nonthermal effects. Whole-body exposure chamber, operating at 900 MHz, was developed for small animals in the frame of this research project. The setup was designed with respect to homogeneity of induced EM field, elimination of external radiation, and exact determination of absorbed power. Further sufficient space for mice movement was taken into account. The whole-body exposure chamber with an anatomical mouse model was simulated by two different numerical methods, e.g., finite-difference-time-domain method (FDTD) and finite integration technique (FIT), and compared computed SAR values and its dosimetry results.

6.1 Exposure chamber

The major advantage of the system we will describe here is the capability of direct measurement of the whole-body averaged SAR, which is performed by analysis of measured scattering parameters. As the basic idea and principle of the discussed exposure chamber, a circular waveguide was chosen. The advantage of the waveguide structure is a perfect shielding of EM field generated either inside (in order to protect the operators) or generated outside the system (in order to eliminate interference caused by external EM fields). The circularly polarized wave TE_{11} is excited inside the exposure chamber with the aid of two monopoles that have mutually orthogonal orientations, and the distance between them is equal to one-fourth of the wavelength. Such circularly polarized wave provides relatively constant field coupling to each mouse regardless of its position, posture, or movement. The discussed exposure chamber is displayed in **Figure 11**.

EM field distribution and impedance matching of the discussed exposure chamber were optimized and verified by 3D EM field simulators SEMCAD X resp. Sim4Life. Dimensions of the exposure chamber were calculated to use the desired frequency of operation and the volume needed to expose mice. The exposure chamber is made of a copper cylinder with dimensions of 1650 mm in length and 240 mm in diameter. It is terminated by matched loads at both ends (conical shape, 500 mm long, and made of RF absorbers). The reflection loss of the matched load is more than −20 dB at 900 MHz.

The exposed mice are kept in a cylindrical box that is made of Styrofoam. Styrofoam has a dielectric constant of 1.03, i.e., very close to that of air, and thus, the disturbance of exposure and measurements is negligible. The box provides space for two separated mice. Punctured slit-like holes are set on the cover and side of the box for air ventilation. In the study, the mice were held in the chamber only during RF exposures, and therefore, no food or drinking water was necessary.

For the survival of experimental animals inside the exposure chamber, it is important to create efficient ventilation, which will maintain a constant temperature and good air quality in the chamber. The air comes toward mice through the ventilation

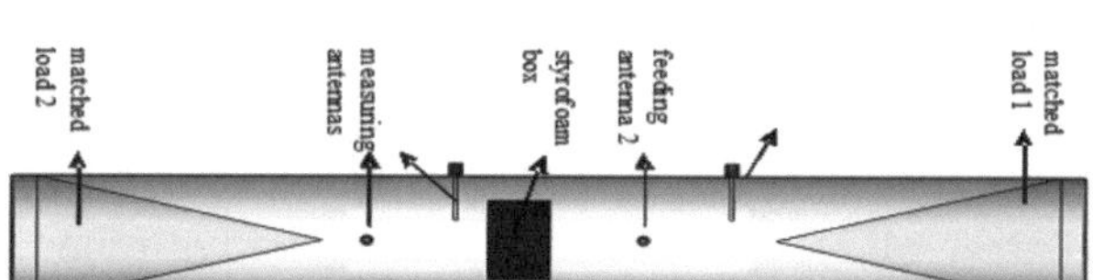

Figure 11.
Waveguide-type exposure chamber for animal experiments.

hole placed below the styrofoam box and flows toward the second opposite ventilation hole placed above the box.

To be able to evaluate the results of experiments with small animals (mice in our case), we need to specify appropriate dosimetry. It is the quantification of the magnitude and distribution of absorbed EM energy within biological objects that are exposed to EM fields. In the case of radiofrequency and microwave frequency bands, there is the dosimetric quantity, which is called SAR (i.e., specific absorption rate). It is defined as the rate at which energy is absorbed per unit mass. The SAR is determined and influenced not only by the incident EM waves but also by the electrical and geometric characteristics of the irradiated subject and objects nearby it. It is strongly related to the internal electric field strength E as well as to the electric conductivity σ and the density of tissues ρ as discussed above and as can be seen, reminded by the following equation.

$$SAR = \sigma.E^2/2\rho \text{ (W/kg)} \tag{10}$$

Therefore, SAR is a suitable dosimetric parameter, even when a studied mechanism is determined to be "athermal." SAR distributions are usually determined from measurements in animal tissues or from numerical calculations. It generally is difficult to measure the SAR directly in a living biological body, and therefore, dosimetry efforts are forced to rely on computer simulations mainly.

An anatomically based dielectric model of an experimental animal is essential for numerical dosimetry. It can be developed commonly from MRI or CT scans. In order to develop it, original gray-scale data must be interpreted into tissue types known as a process of segmentation. In our studies, the CT scans for mouse model development were obtained from the website: http://neuroimage.usc.edu/Digimouse_download.html. The mouse model has the resolution 0.1 mm, meaning voxel size 0.1 x 0.1 x 0.1 mm. Each voxel was assigned to one of 14 different tissue types, such as bone, muscle, brain, etc.

For dosimetry with the numerical voxel models, proper permittivity and conductivity values must be assigned to each tissue. The data from 10 MHz to 6 GHz, derived from 4-Cole-Cole extrapolation based on measurements for small animals, constitute the most widely accepted database for this information. The data are recommended by various international standardization organizations and can be accessed, e.g., from the website http://www.fcc.gov/fcc-bin/dielec.sh.

6.2 Result of biological experiment

In order to verify and rely on numerical dosimetry results, the simulations of the exposure chamber were done in two different EM field simulators (based on two different numerical methods). Our choice was SEMCAD X, which uses the finite difference time domain (FDTD) method, and CST Microwave Studio, which uses the finite integration technique (FIT) method. We used these simulations to the determination of SAR distribution inside the mice during experiments.

Researchers from Medical Faculty in Pilsen, Charles University (Prof. František Vožeh, MD., Jan Barcal, MD.), did biological experiments with the aid of this exposure chamber. With the aim of whether EM exposure can increase the content of free radicals in the exposed tissue, a series of EM exposures to small animals (mice) was done. SAR level was at the level of 0.8 W/kg in the case of these experiments.

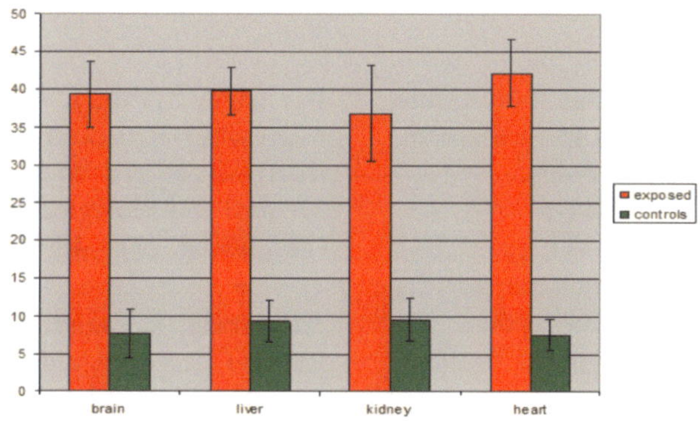

Figure 12.
Results of an animal experiment: in all four studied organ specimens (brain, heart, kidney, and liver), significantly increased content of free radicals was found.

Evaluation of preliminary results is displayed in **Figure 12**. It can be interpreted as a significantly increased content of free radicals being found.

Acknowledgements

This research was funded by Ministry of Education, Youth and Sports of the Czech Republic under Grant LTC19031, and the Student Grant Competition of the CTU, grant number SQS20/203/OHK4/3T/17.

Author details

David Vrba[1], Jan Vrba[1*], Ondrej Fiser[1], Jesus Cumana[2], Milan Babak[2] and Jan Vrba Senior[2]

1 Faculty of Biomedical Engineering, Department of Biomedical Technique, Czech Technical University in Prague, Kladno, Czech Republic

2 Department of EM Field, Faculty of Electrical Engineering, Czech Technical University in Prague, Prague, Czech Republic

*Address all correspondence to: vrba@fel.cvut.cz

References

[1] Vrba J, Boucek J, Vrba J. Waveguide applicators for detection and treatment of cancer. TESLA Electronics, Czechoslovakia. 1984;**2**:44-50

[2] Vrba J, Jiricka K. Radiometer for medical applications. International Science and Colloid. 1984;**29**:105-108

[3] Giannini F, Sorrentino R, Vrba J. “Planar circuit analysis of microstrip radial stub.” IEEE Transactions on MTT. 1984, DOI: 10.1109/tmtt.1984.1132907

[4] Vrba J, Franconi C, Vanucci I. Evanescent mode applicators for subcutaneous hyperthermia. IEEE Transactions on Biomedical Engineering. 1993;**40**(5):397-407

[5] Franconi C, Vrba J, Montecchia F. 27 MHz hybrid evanescent-mode applicators with flexible heating field for deep hyperthermia. International Journal of Hyperthermia. 1993;**9**(5): 655-673

[6] Vrba J, Franconi C, Lapes M. Theoretical limits for the penetration depth of intracavitary applicators. International Journal of Hyperthermia. 1996;**12**(6):737-742

[7] Vrba J. Medical applications of microwaves. Electromagnetic Biology and Medicine. 2005;**24**(3):441-448

[8] Franconi C, Vrba J, Micali F, Pesce F. Prospects for radiofrequency hyperthermia applicator research. International Journal of Hyperthermia. 2011;**27**(2):187-198

[9] Dobsicek-Trefna H, Vrba J, Persson M. Design of a wideband multi-channel system for time reversal hyperthermia. International Journal of Hyperthermia. 2012;**28**(2):175-183

[10] Togni P, Vrba J, Vannucci L. Microwave applicator for hyperthermia treatment on in vivo melanoma model. Medical and Biological Engineering and Computing. 2010;**48**(3):285-292

[11] Drizdal T, Paulides MM, Vrba J, van Rhoon GC. Waveguide applicators for superficial hyperthermia treatment: Is tuning really reaquired? Journal of EM Waves and Applications. 2013;**6**(27): 682-690

[12] Vrba J et al. Technical aspects of microwave thermotherapy. In: RF Interaction with Humans: Mechanisms, Exposures and Medical Applications. IPEM Meeting, Inst. of Physics; London. 2003

[13] Trefna HD, Vrba J, Persson M. Time-reversal focusing in microwave hyperthermia for deep-seated tumors. Physics in Medicine and Biology. 2010; **55**(8):2167-2185

[14] Vrba D, Vrba J. Novel applicators for local microwave hyperthermia based on zeroth-order mode resonator metamaterial. International Journal of Antennas and Propagation. 2014;**2014**: 1-7

[15] Vrba D, Vrba J, Rodrigues DB, Stauffer P. Numerical investigation of novel microwave applicators based on zero-order mode resonance for hyperthermia treatment of cancer. Journal of the Franklin Institute. 2016; **354**(18):8734-8746

[16] Vrba J, Oppl L, Vrba$_{jr}$, J, and Vrba D. Microwave Medical Imaging and Diagnostics. In EuMW 2008, Conference Proceedings. London: Horizon House Publications, 2008. p. 408-411. ISBN 978-1-4244-3794-8

[17] Fiser O, Helbig M, Ley S, Sachs J, Vrba J. Feasibility study of temperature change detection in phantom using M-sequence radar. In: 2016 10th EuCAP. 2016. pp. 1-4

[18] Fiser O, Merunka I, Vrba J. Numerical feasibility study of new combined hyperthermia system for head and neck region. In: European Microwave Conference (EuMC). 2017

[19] Vrba J, Vrba D, Vrba J, Vožeh F, Barcal J, Vannucci L. System for Animal EM Exposure with well Defined Dosimetry. Beijing: URSI GASS; 2014

[20] Ley S, Schilling S, Fiser O, Vrba J, Sachs J, Helbig M. Ultra-wideband temperature dependent dielectric spectroscopy of porcine tissue and blood in the microwave frequency range. Sensors. 2019;**19**(7):1-21

Chapter 11

Wireless Power Transmission on Biomedical Applications

Ting-Wei Wang and Ting-Tse Lin

Abstract

Wireless power transmission (WPT) can provide an alternative for wireless power in implantable medical devices (IMDs). The WPT in implantable medical devices will involve many emerging biomedical topics, such as implantable pacemakers, optogenetic devices, and bio-impedance sensors. To this end, this chapter comprehensively reviews the recent WPT studies for those mentioned above emerging biomedical applications. The specific key components are carried out for those applications. Besides, the operation principle and system design are presented. In conclusion, this chapter's significance can help evolve reliable implantable device development in the future.

Keywords: bio-impedance sensor, electrical field coupling, implantable medical devices, magnetic field coupling, optogenetic device, pacemaker, resonance circuit

1. Introduction

Wireless power transmission (WPT) is an emerging technology to transmit power without wires [1, 2]. It is widely studied in various applications such as consumer electronics charging [3], radio-frequency identification (RFID) system [4], electric vehicle charging [5], biomedical applications [6], etc. Especially in implantable devices, the market benefit of global implantable medical devices achieved a value of US$ 120.5 billion in 2021 by IMARC Group. The market can be expected to reach US$ 168.3 billion by 2027, exhibiting at a compound annual growth rate (CAGR) of 5.54% during 2022–2027. In practice, implantable devices solve today's problems using advanced medical technologies such as a well-known pacemaker and capsule endoscopy [7, 8]. The pacemaker and capsule endoscopy invention brings new perspectives for cardiovascular disease therapy and direct health diagnosis of the whole small bowel. Although several miniatured battery systems have been developed to provide several years of power supply for implantable devices, a battery replacement would induce patients' inconvenient and unexpected surgery risks [9, 10]. Recently, the WPT technique has been widely applied in implantable medical devices that are expected to power supply issues in the future [11, 12].

WPT technique can be mainly divided into the coupling and radiative approaches [13, 14], as shown in **Figure 1**. The WPT coupling can be categorized into magnetic and electric field methods. The radiative approach can be categorized into

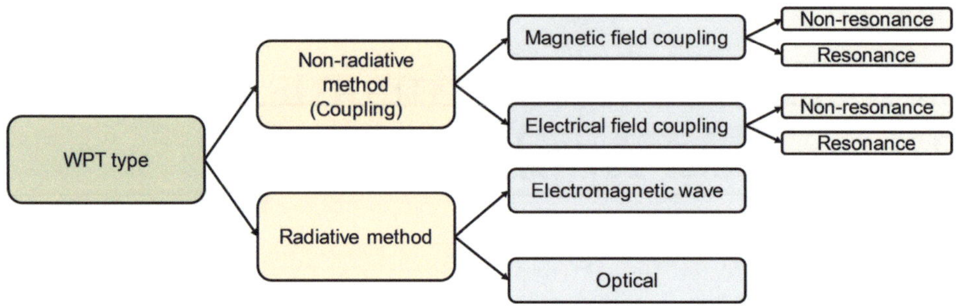

Figure 1.
WPT classification, including nonradiative and radiative methods.

electromagnetic waves and optical power transmission. In the radiative method (far-field), power is transferred by electromagnetic radiation in terms of microwave, laser, and light. Although radiative techniques can transfer power over longer distances, the large environmental loss-induced transfer efficiency is considerable due to poor air medium. The nonradiative method (near field) utilizes magnetic field and electric field coupling to approach wireless power transfer. The main advantage of nonradiative techniques is high transfer efficiency; however, its technical limitation is that the transfer distance is limited.

The nonradiative method is also called coupling-based WPT, which can be described as the power-transmitting side (primary side) and power-receiving side (secondary side). The coupling type can be classified as a magnetic field (**Figure 2a**) and electrical field coupling (**Figure 2b**). Both magnetic and electric field coupling methods can be further categorized as resonance mode, based on LC resonance (**Figure 2c** and **d**).

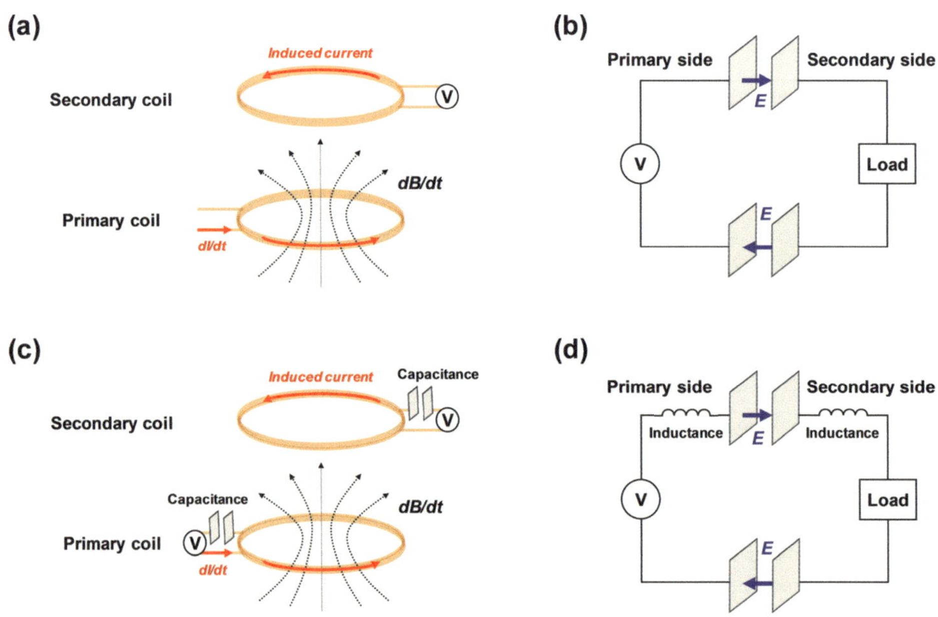

Figure 2.
Magnetic and electric field coupling with and without resonant circuit.

2. Magnetic field coupling without resonance

The magnetic field coupling without resonance type utilizes the principle of electromagnetic induction, as shown in **Figure 3**. According to Ampere's law as Eq. (1), the magnetic pulsing ($d\boldsymbol{B}/dt$) of the primary coil can be generated by time-varying driving current (dI/dt). Where $\boldsymbol{B}$, μ_0, J are magnetic field, permeability, and current density. The electric field $\boldsymbol{E}$ can be induced on the secondary coil, according to Faraday's law as Eq. (2) [9, 15]. Thus, the charge can be stored *via* a secondary coil.

$$\nabla \times \mathbf{B} = \mu_0 \mathbf{J} \tag{1}$$

$$\mathbf{E} = -\frac{\partial \mathbf{B}}{\partial t} \tag{2}$$

To understand the principle of magnetic field coupling without resonance, **Figure 3** shows the topology of the circuit analysis. The mutual inductance, $L_m = k\sqrt{L_1 L_2}$, depends on the distance x between the two coils, where k is the coupling factor ($0 < k < 1$) between the primary and the secondary coil. Based on Kirchhoff's voltage law, Eqs. (3) and (4) can be obtained as below [16]:

$$j\omega L_1 I_1 + I_1 R_1 + j\omega L_m I_2 = U_1 \tag{3}$$

$$j\omega L_2 I_2 + I_2 R_2 + I_2 R_L + j\omega L_m I_1 = 0 \tag{4}$$

I_2 can be expressed as Eq. (5), according to Eq. (4).

$$I_2 = -I_1 \left[\frac{\omega L_m}{\omega L_2 - j(R_2 + R_L)} \right] \tag{5}$$

The relation between I_1 and U_1 can be obtained as Eq. (6), according to Eqs. (3) and (5).

$$I_1 = \left[\frac{R_2 + R_L + j\omega L_2}{(R_1 + j\omega L_1)(j\omega L_2 + R_2 + R_L) + \omega^2 {L_m}^2} \right] U_1 \tag{6}$$

The relation between I_1 and I_2 can be expressed as Eq. (7), according to Eqs. (5) and (6).

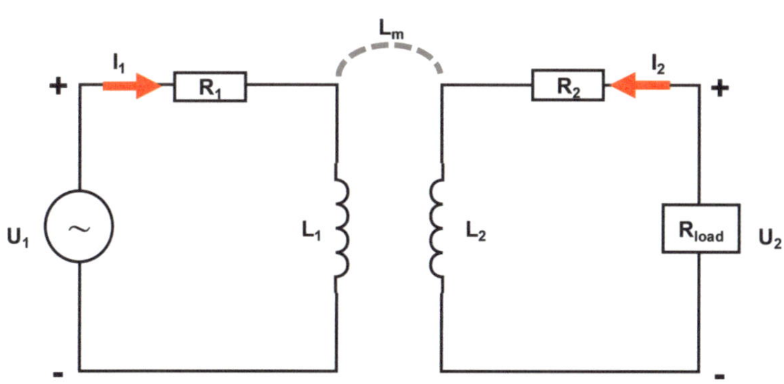

Figure 3.
Circuit topology of electromagnetic induction.

$$I_2 = -\left[\frac{j\omega L_m}{(R_1 + j\omega L_1)(j\omega L_2 + R_2 + R_L) + \omega^2 {L_m}^2}\right] U_1 \tag{7}$$

The ratio of I_1 and I_2 can be expressed as Eq. (8), according to Eqs. (6) and (7).

$$\frac{I_1}{I_2} = -\frac{R_2 + R_L + j\omega L_2}{j\omega L_m} \tag{8}$$

To better understand the input impedance and power efficiency of the circuit, a T-type equivalent circuit is used in **Figure 4**.

$$V_{Lm2} = j\omega L_m I_1 \tag{9}$$

$$V_{Lm2} = \frac{j\omega L_m (R_2 + R_L + j\omega L_2)}{(R_1 + j\omega L_1)(j\omega L_2 + R_2 + R_L) + \omega^2 {L_m}^2} U_1 \tag{10}$$

According to Eq. (11), Z_{in2} can be obtained from Eqs. (8) and (9).

$$Z_{in2} = \frac{V_{Lm2}}{-I_2} = \frac{j\omega L_m I_1}{-I_2} = R_2 + R_L + j\omega L_2 \tag{11}$$

Then, Z_2 can be further obtained, according to Eq. (12).

$$Z_2 = \frac{V_{Lm1}}{I_1} = \frac{j\omega L_m I_2}{I_1} = \frac{\omega^2 {L_m}^2}{R_2 + R_L + j\omega L_2} \tag{12}$$

Finally, the Z_{in1} can be acquired, according to Eq. (13).

$$Z_{in1} = R_1 + j\omega L_1 + Z_2 = R_1 + j\omega L_1 + \frac{\omega^2 {L_m}^2}{R_2 + R_L + j\omega L_2} \tag{13}$$

To understand the power efficiency, the energy loss ratio for the primary-side internal resistance, secondary resistance, and load can be denoted by P_{R1}, P_{R2}, and P_{RL} as Eq. (14).

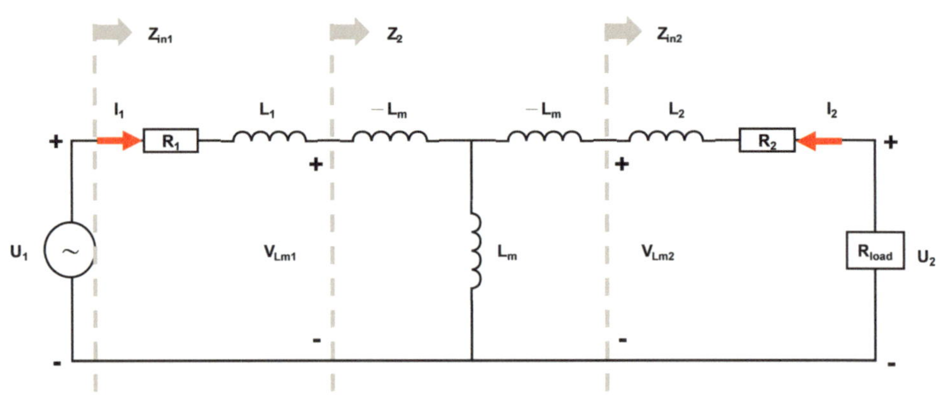

Figure 4.
T-type equivalent circuit for electromagnetic induction.

$$P_{R_1} : P_{R_2} : P_L = |I_1|^2 R_1 : |I_2|^2 R_2 : |I_2|^2 R_L \tag{14}$$

The square ratio of I_2 and I_1 can be expressed as Eq. (15), according to Eq. (8)

$$\left|\frac{I_2}{I_1}\right|^2 = \frac{\omega^2 {L_m}^2}{(R_2 + R_L)^2 + \omega^2 {L_m}^2} \tag{15}$$

The power ratio can be rewritten as Eq. (16), according to Eqs. (14) and (15).

$$P_{R_1} : P_{R_2} : P_L = \left\{(R_2 + R_L)^2 + \omega^2 {L_m}^2\right\} R_1 : \left\{\omega^2 {L_m}^2\right\} R_2 : \left\{\omega^2 {L_m}^2\right\} R_L \tag{16}$$

Finally, the power efficiency can be obtained, according to Eq. (17).

$$\eta = \frac{P_L}{P_{in}} = \frac{P_L}{P_{R_1} + P_{R_2} + P_L} = \frac{\left\{\omega^2 {L_m}^2\right\} R_L}{\left\{(R_2 + R_L)^2 + \omega^2 {L_m}^2\right\} R_1 + \left\{\omega^2 {L_m}^2\right\} R_2 + \left\{\omega^2 {L_m}^2\right\} R_L} \tag{17}$$

3. Magnetic field coupling with resonance

Compared with electromagnetic induction, magnetic resonance coupling utilizes resonance with the series capacitor in primary and secondary sides to approach high efficiency and high power (**Figure 5**). Based on Kirchhoff's voltage law, Eqs. (18) and (19) can be obtained as below [16]:

$$\frac{1}{j\omega C_1} I_1 + R_1 I_1 + j\omega L_1 I_1 + j\omega L_m I_2 = U_1 \tag{18}$$

$$j\omega L_2 I_2 + R_2 I_2 + \frac{1}{j\omega C_2} I_2 + R_L I_2 + j\omega L_m I_1 = 0 \tag{19}$$

I_2 can be expressed as Eq. (20), according to Eq. (18).

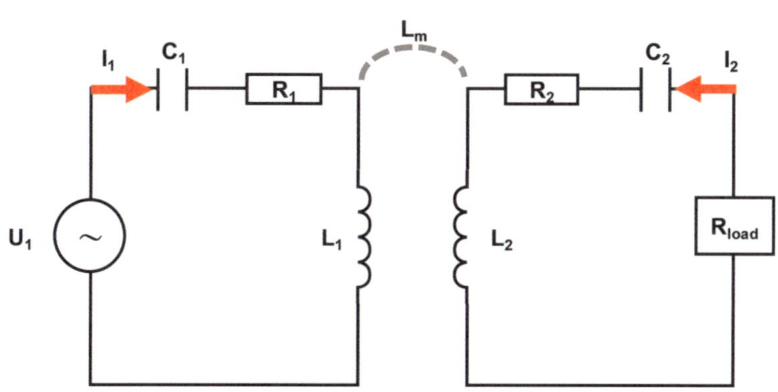

Figure 5.
Circuit topology of magnetic field coupling with resonance.

$$I_2 = \frac{-j\omega L_m}{j\omega L_2 + \frac{1}{j\omega C_2} + R_2 + R_L} I_1 \tag{20}$$

The relation between I_1 and U_1 can be expressed as Eq. (21), according to Eqs. (18) and (20).

$$I_1 = \left\{ \frac{R_2 + R_L + j\left(\omega L_2 - \frac{1}{\omega C_2}\right)}{\left[\left(R_1 + j\left(\omega L_1 - \frac{1}{\omega C_1}\right)\right]\left[R_2 + R_L + j\left(\omega L_2 - \frac{1}{\omega C_2}\right)\right] + \omega^2 {L_m}^2} \right\} U_1 \tag{21}$$

The relation between I_2 and U_1 can be expressed as Eq. (22), according to Eqs. (20) and (21).

$$I_2 = -\left\{ \frac{j\omega L_m}{\left[\left(R_1 + j\left(\omega L_1 - \frac{1}{\omega C_1}\right)\right]\left[R_2 + R_L + j\left(\omega L_2 - \frac{1}{\omega C_2}\right)\right] + \omega^2 {L_m}^2} \right\} U_1 \tag{22}$$

The relation between I_1 and U_1 can be expressed as Eq. (23), according to Eqs. (21) and (22).

$$\frac{I_1}{I_2} = -\frac{R_2 + R_L + j\omega\left(\omega L_2 - \frac{1}{\omega C_2}\right)}{j\omega L_m} \tag{23}$$

When the operating and resonant frequencies are equal on primary and secondary sides, I_1 and I_2 can be expressed as (25) and (26), according to Eqs. (21), (22), and (24).

$$\omega_0 = \omega_1 = \omega_2 = \frac{1}{\sqrt{L_1 C_1}} = \frac{1}{\sqrt{L_2 C_2}} \tag{24}$$

$$I_1 = \frac{R_2 + R_L}{R_1(R_2 + R_L) + \omega^2 {L_m}^2} U_1 \tag{25}$$

$$I_2 = -\frac{j\omega L_m}{R_1(R_2 + R_L) + \omega^2 {L_m}^2} U_1 \tag{26}$$

The ratio of I_1 and I_2 can be rewritten as Eq. (27), according to Eqs. (25) and (26).

$$\frac{I_1}{I_2} = -\frac{R_2 + R_L}{j\omega L_m} \tag{27}$$

To further understand the input impedance and power efficiency of the magnetic field coupling with the resonance circuit, a T-type equivalent circuit is used as **Figure 6**.

$$V_{Lm2} = j\omega L_m I_1 \tag{28}$$

$$V_{Lm2} = j\omega L_m \frac{R_2 + R_L}{R_1(R_2 + R_L) + \omega^2 {L_m}^2} U_1 \tag{29}$$

Z_{in2} can be obtained as Eq. (30), according to Eqs. (26) and (29).

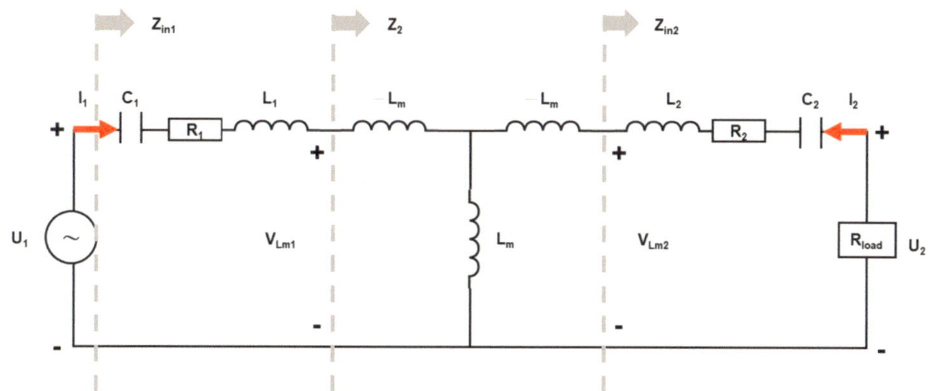

Figure 6.
T-type equivalent circuit for magnetic field coupling with resonance.

$$Z_{in2} = \frac{V_{Lm2}}{-I_2} = \frac{j\omega L_m I_1}{-I_2} = R_2 + R_L \tag{30}$$

Then, Z_2 can be further obtained as Eq. (31).

$$Z_2 = \frac{V_{Lm1}}{I_1} = \frac{j\omega L_m I_2}{I_1} = \frac{\omega^2 {L_m}^2}{R_2 + R_L} \tag{31}$$

Finally, the Z_{in1} can be acquired, according to Eq. (32).

$$Z_{in1} = R_1 + \frac{1}{j\omega C_1} + j\omega L_1 + Z_2 = R_1 + \frac{1}{j\omega C_1} + j\omega L_1 + \frac{\omega^2 {L_m}^2}{R_2 + R_L} \tag{32}$$

To understand the power efficiency, the ratio of energy loss for the primary-side internal resistance, secondary resistance, and load can be denoted by P_{R1}, P_{R2}, and P_{RL} as Eq. (14).

$$P_{R_1} : P_{R_2} : P_L = |I_1|^2 R_1 : |I_2|^2 R_2 : |I_2|^2 R_L \tag{33}$$

The square ratio of I_2 and I_1 can be expressed as Eq. (34), according to Eq. (27).

$$\left|\frac{I_2}{I_1}\right|^2 = \frac{\omega^2 {L_m}^2}{(R_2 + R_L)^2} \tag{34}$$

The power ratio can be rewritten as Eq. (35), according to Eqs. (33) and (34).

$$P_{R_1} : P_{R_2} : P_L = \left\{(R_2 + R_L)^2\right\} R_1 : \left\{\omega^2 {L_m}^2\right\} R_2 : \left\{\omega^2 {L_m}^2\right\} R_L \tag{35}$$

Finally, the power efficiency can be obtained, according to Eq. (36).

$$\eta = \frac{P_L}{P_{in}} = \frac{P_L}{P_{R_1} + P_{R_2} + P_L} = \frac{\left\{\omega^2 {L_m}^2\right\} R_L}{\left\{(R_2 + R_L)^2\right\} R_1 + \left\{\omega^2 {L_m}^2\right\} R_2 + \left\{\omega^2 {L_m}^2\right\} R_L} \tag{36}$$

	Efficiency	Power at R_{load}
Magnetic field coupling without resonance	$\frac{\{\omega^2 L_m{}^2\}R_L}{\{(R_2+R_L)^2+\omega^2 L_m{}^2\}R_1+\{\omega^2 L_m{}^2\}R_2+\{\omega^2 L_m{}^2\}R_L}$	$\left\lvert\frac{j\omega L_m}{(R_1+j\omega L_1)(j\omega L_2+R_2+R_L)+\omega^2 L_m{}^2}\right\rvert^2 U_1{}^2 R_L$
Magnetic field coupling With resonance	$\frac{\{\omega^2 L_m{}^2\}R_L}{\{(R_2+R_L)^2\}R_1+\{\omega^2 L_m{}^2\}R_2+\{\omega^2 L_m{}^2\}R_L}$	$\left\lvert\frac{j\omega L_m}{R_1(R_2+R_L)+\omega^2 L_m{}^2}\right\rvert^2 U_1{}^2 R_L$

Table 1.
Comparison between magnetic field coupling under either no or with resonance.

Table 1 summarizes the results from the previous derivation in terms of the efficiency and power of the two types of magnetic field coupling circuits. With C_1 and C_2 inserted on the primary and secondary sides, the power efficiency and power at R_{load} will increase, which is why resonance is important.

4. Electric field coupling

Compared with magnetic field coupling, the electric field coupling without resonance uses a displacement current instead of an electromagnetic induction (**Figure** 7), as Eq. (37). Where ε, **E**, and **D** are permittivity, electric field intensity, electric flux density, respectively [17–19].

$$I = \frac{\partial \mathbf{D}}{\partial t} \bullet S = \varepsilon \frac{\partial \mathbf{E}}{\partial t} \bullet S \tag{37}$$

In **Figure 8**, the mutual inductance of C_m between two capacitances on the distance x between the two coils can be written as $C_m = k\sqrt{C_1 C_2}$.

The electrical field coupling schematic diagram can be equivalent to π-type coupling, as shown in **Figure 9**. Based on the mesh-analysis method, the basic mathematical expression can be described as a matrix, according to Eq. (38).

$$\begin{bmatrix} U \\ 0 \\ 0 \end{bmatrix} = \begin{bmatrix} R_1 + j\omega L_1 + \frac{1}{j\omega(C_1 - C_m)} & -\frac{1}{j\omega(C_1 - C_m)} & 0 \\ -\frac{1}{j\omega(C_1 - C_m)} & \frac{1}{j\omega(C_1 - C_m)} + \frac{1}{j\omega C_m} + \frac{1}{j\omega(C_2 - C_m)} & -\frac{1}{j\omega(C_2 - C_m)} \\ 0 & -\frac{1}{j\omega(C_2 - C_m)} & R_2 + R_L + j\omega L_2 + \frac{1}{j\omega(C_2 - C_m)} \end{bmatrix} \begin{bmatrix} I_{l1} \\ I_m \\ I_{l2} \end{bmatrix} \tag{38}$$

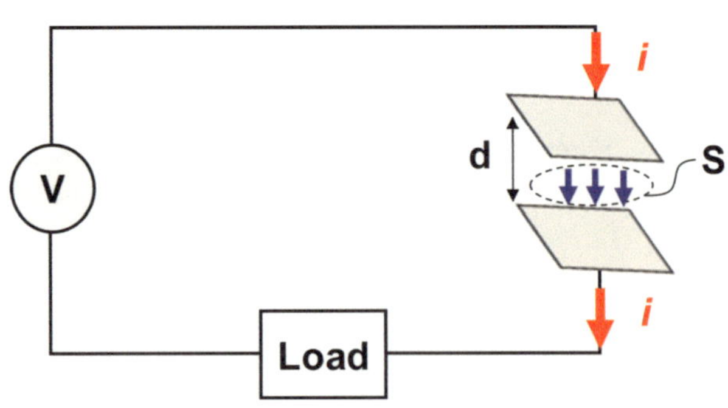

Figure 7.
Diagram of displacement current.

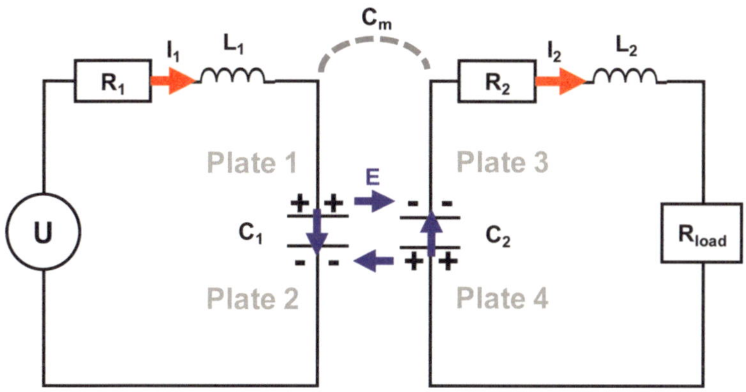

Figure 8.
Circuit topology of electric field coupling with resonance.

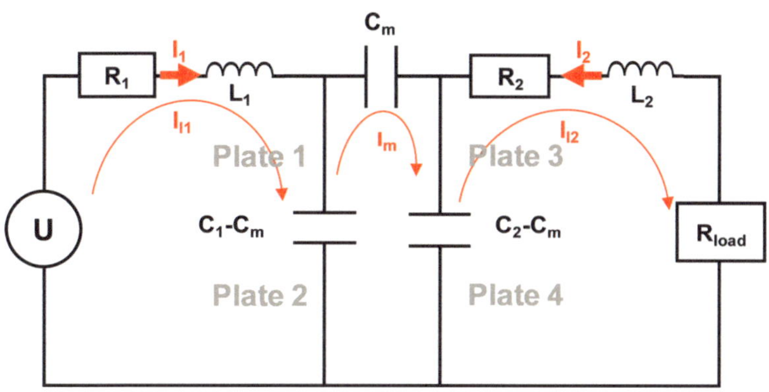

Figure 9.
π-type equivalent circuit for electric field coupling with resonance.

$$Z_{in2} = \frac{1}{j\omega C_2 + \frac{1}{R_2 + j\omega L_2 + R_L}} \tag{39}$$

$$Z_2 = \frac{1}{\omega^2 C_m^2 Z_{in2}} \tag{40}$$

$$Z_{in1} = \frac{1}{j\omega C_1 + \frac{1}{Z_2}} + j\omega L_1 + R_1 \tag{41}$$

To analyze the input impedance, the extended π-type circuit with a separate capacitance is shown in **Figure 10**. The input impedance of each stage can be expressed as Eqs. (38), (39), and (40).

When operating frequency is equal to the resonant frequency on the secondary side, the imaginary part of the Z_{in2} becomes zero. Then, the resonant frequency can be acquired as Eq. (42).

$$\omega_2 = \sqrt{\frac{1}{L_2 C_2} - \left(\frac{R_2 + R_L}{L_2}\right)^2} \tag{42}$$

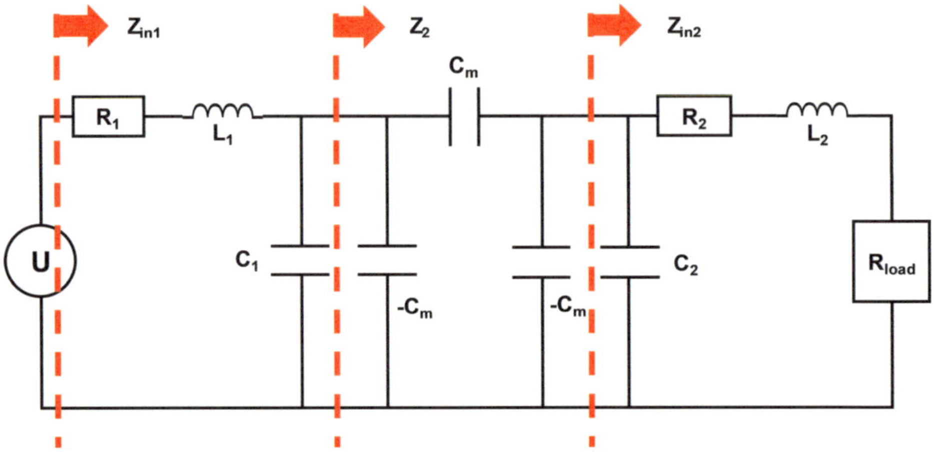

Figure 10.
π-type circuit with a separate capacitance for electric field coupling with resonance.

In the same way, when the operating frequency is equal to the resonant frequency on the primary side, the imaginary part of the Z_{in1} becomes zero. The L_1 can be obtained as Eq. (43).

$$L_1 = \frac{C_1(R_2+R_L)^2 - \omega_1^2\left\{C_2\left[(R_2+R_L)^2+(\omega L_2)^2\right]-L_2\right\}\left\{L_2C_1-\left(C_1C_2-C_m^2\right)\left[(R_2+R_L)^2+(\omega L_2)^2\right]\right\}}{\omega_1^4\left\{L_2C_1-\left(C_1C_2-C_m^2\right)\left[(R_2+R_L)^2+(\omega L_2)^2\right]\right\}^2+\omega_1^2C_1^2(R_2+R_L)^2} \tag{43}$$

To understand the power efficiency, the ratio of energy loss for the primary-side internal resistance, secondary-side internal resistance, and load can be denoted by P_{R1}, P_{R2}, and P_{RL} as Eq. (44).

$$P_{R_1} : P_{R_2} : P_{R_L} = |I_1|^2R_1 : |I_2|^2R_2 : |I_2|^2R_L \tag{44}$$

The square ratio of I_2 and I_1 can be calculated as Eq. (45), according to Eq. (38) [16].

$$\left|\frac{I_1}{I_2}\right|^2 = \frac{\omega^2C_1C_2\left[(R_2+R_L)^2+(\omega L_2)^2\right]+\mathbf{C}_1^2(1-\omega^2L_2C_2)}{\mathbf{C}_m^2} \tag{45}$$

The power ratio can be rewritten as Eq. (46), according to Eqs. (44) and (45).

$$P_{R_1} : P_{R_2} : P_L = \left\{\omega^2C_1C_2\left[(R_2+R_L)^2+(\omega L_2)^2\right]+\mathbf{C}_1^2(1-\omega^2L_2C_2)\right\}R_1 : \{\mathbf{C}_m^2\}R_2 : \{\mathbf{C}_m^2\}R_L \tag{46}$$

Finally, the power efficiency can be expressed as Eq. (47).

$$\eta(\omega) = \frac{P_L}{P_{in}} = \frac{P_L}{P_{R_1} + P_{R_2} + P_L}$$
$$= \frac{\{C_m^2\}R_L}{\left\{\omega^2 C_1 C_2\left[(R_2+R_L)^2 + (\omega L_2)^2\right] + C_1^2(1-\omega^2 L_2 C_2)\right\}R_1 + \{C_m^2\}R_2 + \{C_m^2\}R_L} \quad (47)$$

5. WPT system overview

The basic topology of magnetic field resonant coupling is shown in **Figure 11** [16, 20, 21]. The transmitter and receiver sides are coupled through the magnetic field (**Figure 11a**) and electric field (**Figure 11b**). The DC power supply generates the source power. The inverter is responsible for DC/AC conversion to produce the high-frequency AC power. The AC power is transferred to the receiver side and converted from AC to DC power by a rectifier. Then, the power is transferred to the load. In the WPT operating frequency for implantable devices, the common frequency is a at low frequency because of the reduction of tissue heating effect [22]. The typical frequency is 13.56 MHz within the license-free industrial, scientific, and medical (ISM) band [23].

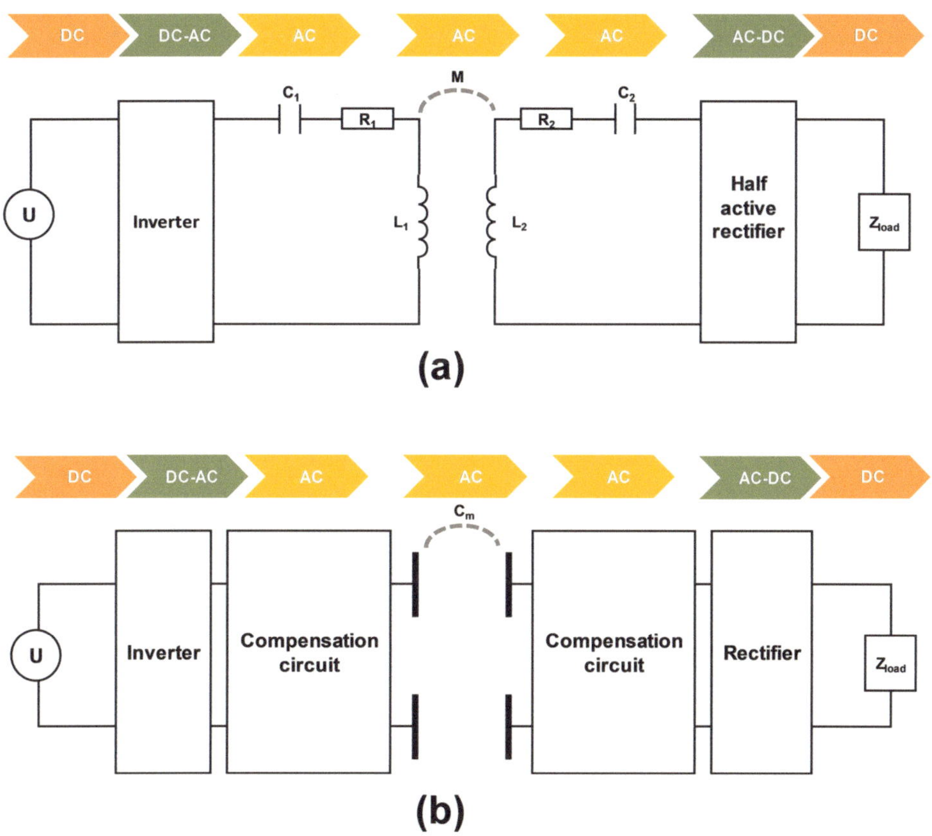

Figure 11.
System overview for (a) magnetic field coupling and (b) electric field coupling methods.

The class E power amplifier is commonly applied in MHz-range WPT applications because of its excellent power efficiency from the properties of zero voltage switching and zero voltage derivative switching [24]. The class-E power amplifier can be depicted in **Figure 12**. In theory, it has a maximum efficiency of 100% [25], according to Eq. (48). The overall circuit consists of the power supply V_{dc}, RF-chuck, and passive load network. The RF-chuck with high reactance ensures DC voltage pass through the system. The passive load network is responsible for minimizing the overlap between drain voltage and drain current and further reducing the output power loss [26].

$$P_{dissipation} = \frac{1}{T_0}\int_{T_0} I_{DS} * V_{DS}\, dt = 0 \tag{48}$$

Some group has proposed a class E power amplifier-based WPT system for implantable biomedical in practical applications. Mutashar *et al.* [27] designed a wireless magnetic resonant coupling system using a class-E power amplifier and inductive power links via two spiral transmitters and receiver coils (**Figure 13a**). The 13.56 MHz carrier frequency within the ISM band range was selected to avoid tissue damage. The proposed class-E power amplifier can reach up to 87.2% power efficiency. The overall inductive coupling system achieves a power transmission efficiency of 73%. The compact structure could be implemented in bio-implants in the future. In practical applications for the electrical field coupling method, Narayanamoorthi *et al.* [24] developed a class-E-based capacitive coupled wireless power transfer system for biomedical implants (**Figure 13b**). They optimize the class E-based capacitive coupling WPT systems at an operating frequency of 13.56 MHz. Moreover, the matching circuit consisted of inductive-capacitive-inductive (LCL) impedance was used to improve the power efficiency of the power amplifier up to 96.34%.

6. WPT in emerging biomedical topics

WPT systems are rapidly evolving for biomedical implants, including emerging topics of implantable pacemaker devices, implantable optogenetic devices, and implantable impedance plethysmography (IPG).

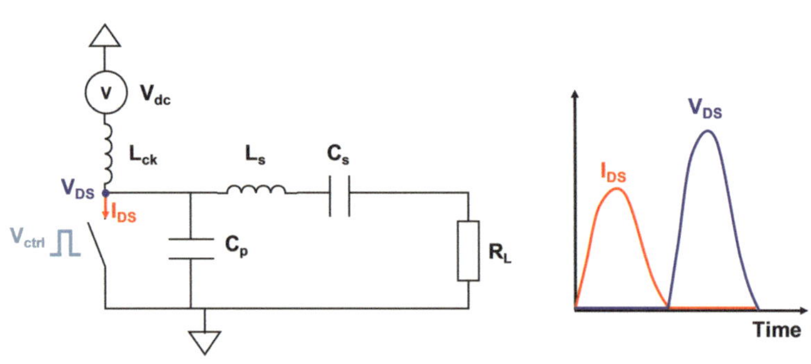

Figure 12.
Circuit diagram of a class E power amplifier.

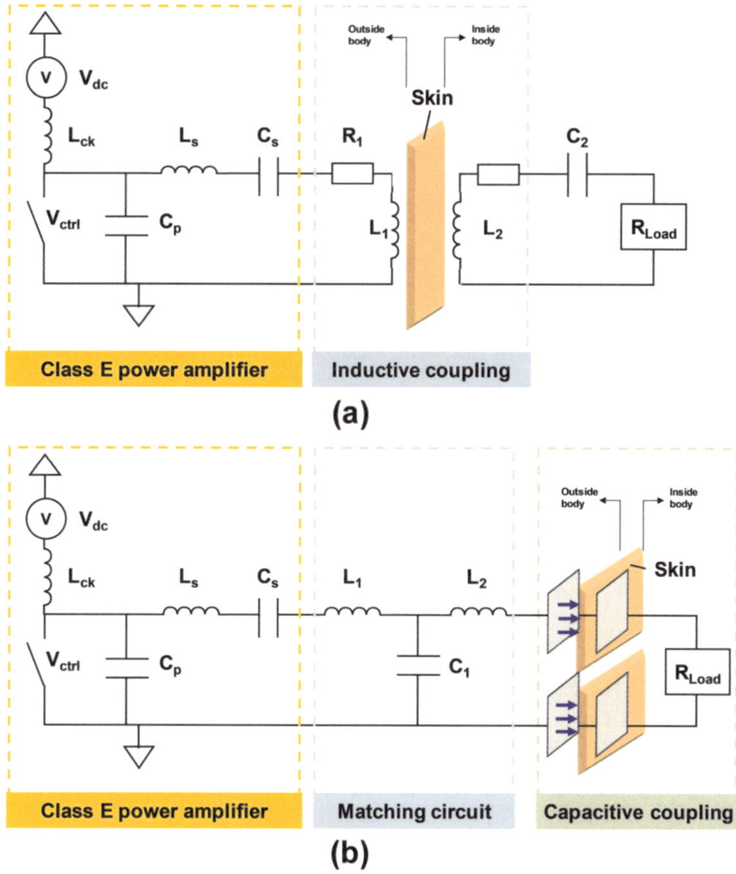

Figure 13.
WPT system by (a) magnetic field coupling [27] and (b) electric field coupling for implantable applications [24].

6.1 Wireless charge-based implantable pacemaker

Heart failure is one of the most killers worldwide. It occupies one-third of all global death. Nowadays, the implantable cardioverter pacemaker is a common approach to directly suppress heart failure [28, 29]. However, the battery life could cause health risks and patient inconvenience. Parinaz Abiri *et al.* [30] developed an inductively powered wireless pacing device to realize remote cardiac stimulation. They designed a Class E-based magnetic field resonant coupling system with a resonant frequency of 13.56 MHz to pace rhythm on Yucatan miniature pig.

6.2 WPT-based implantable optogenetic device

Optogenetics is an emerging field that provides an alternative for direct electrical stimulation to modulate membrane voltage. By genetic modification, the ion channels in the cell can be sensitive to certain wavelengths of light [31]. Nowadays, several advanced biomedical topics have involved the optogenetic topic in different applications, including neuroscience to control animal behavior [32, 33] and cardiology to find out a solution for cardiovascular disease [34, 35]. Yu *et al.* [36] implement an implantable optogenetic device implanted into the dog's chest. WPT-based optogenetic modulation of cardiac sympathetic nerve activity can prevent ventricular arrhythmias.

6.3 WPT-based implantable bio-impedance sensor

Bio-impedance sensing is a noninvasive sensing technology based on Ohm's law. It was commonly applied to cell structural depiction, human composition, and physiological measurement [37–40]. The main advantage of bio-impedance sensing is providing wearable and low-cost applications in biomedical engineering. The bio-impedance technique commonly uses the two-pair electrodes to detect arterial pulsation by Ohm's law. One pair of electrodes is responsible for current excitation in our body. Furthermore, the other pair electrodes are responsible for sensing voltage signals. Theodor *et al.* [41] developed an implantable bio-impedance sensor to measure the artery activity from the domestic pig for additional cardiovascular parameters extraction such as pulse rate and blood pressure [42].

7. Conclusion

This chapter has presented the WPT methodology for implantable medical devices. The common configuration of WPT and its principle were demonstrated, including magnetic field coupling and electric field coupling. Most of all, the recent emerging biomedical topics such as wireless-charging pacemakers, implantable optogenetic devices, and implantable bio-impedance sensors were demonstrated. Overall, this chapter thoroughly presented the electric principle of WPT and extended it to emerging biomedical applications.

Author details

Ting-Wei Wang[1,2] and Ting-Tse Lin[3,4,5,6*]

1 Department of Electrical Engineering, California Institute of Technology, Pasadena, CA, USA

2 Department of Medical Engineering, California Institute of Technology, Pasadena, CA, USA

3 Division of Cardiology, Department of Internal Medicine, National Taiwan University Hospital, Taipei, Taiwan

4 Cardiovascular Center, National Taiwan University Hospital Hsinchu Branch, Hsinchu, Taiwan

5 College of Medicine, National Taiwan University, Taipei, Taiwan

6 Division of Cardiology, Department of Internal Medicine, National Taiwan University Hospital Hsinchu Branch, Hsinchu, Taiwan

*Address all correspondence to: ttlin111@gmail.com

References

[1] Sun L, Ma D, Tang H. A review of recent trends in wireless power transfer technology and its applications in electric vehicle wireless charging. Renewable and Sustainable Energy Reviews. 2018;**91**:490-503. DOI: 10.1016/j.rser.2018.04.016

[2] Bi Z, Kan T, Mi CC, Zhang Y, Zhao Z, Keoleian GA. A review of wireless power transfer for electric vehicles: Prospects to enhance sustainable mobility. Applied Energy. 2016;**179**:413-425. DOI: 10.1016/j.apenergy.2016.07.003

[3] Hoang H, Lee S, Kim Y, Choi Y, Bien F. An adaptive technique to improve wireless power transfer for consumer electronics. IEEE Transactions on Consumer Electronics. 2012;**58**(2): 327-332. DOI: 10.1109/TCE.2012.6227430

[4] Kiani M, Ghovanloo M. An RFID-based closed-loop wireless power transmission system for biomedical applications. IEEE Transactions on Circuits and Systems II: Express Briefs. 2010;**57**(4):260-264. DOI: 10.1109/TCSII.2010.2043470

[5] Iqteit N, Yahya K, Khan SA. Wireless power charging in electrical vehicles. In: Zellagui M, editor. Wireless Power Transfer. Recent Development, Applications and New Perspectives. London, UK: IntechOpen; 2021

[6] Meng E, Sheybani R. Insight: Implantable medical devices. Lab on a Chip. 2014;**14**(17):3233-3240. DOI: 10.1039/C4LC00127C

[7] Basar MR, Ahmad MY, Cho J, Ibrahim F. Application of wireless power transmission systems in wireless capsule endoscopy: An overview, Sensors (Basel, Switzerland). 2014;**14**(6):10929-10951. DOI: 10.3390/s140610929

[8] DeForge WF. Cardiac pacemakers: A basic review of the history and current technology. Journal of Veterinary Cardiology: the Official Journal of the European Society of Veterinary Cardiology. 2019;**22**:40-50. DOI: 10.1016/j.jvc.2019.01.001

[9] Wang T-W, Lin T-T. Electromagnetic compatibility issues in medical devices. In: Kishk A, editor. Electromagnetic Compatibility. London, UK: IntechOpen; 2021

[10] Kanaan AI, Sabaawi AM. Implantable wireless systems: A review of potentials and challenges. In: Al-Rizzo H, Abushamleh S, editors. Antenna Systems. London, UK: IntechOpen; 2021

[11] Shadid R, Noghanian S. A literature survey on wireless power transfer for biomedical devices. International Journal of Antennas and Propagation. 2018;**2018**:4382841. DOI: 10.1155/2018/4382841

[12] Khan SR, Pavuluri SK, Cummins G, Desmulliez MPY. Wireless power transfer techniques for implantable medical devices: A review. Sensors. 2020; **20**(12):3487. DOI: 10.3390/s20123487

[13] Etemadrezaei M. Wireless power transfer. In: Rashid MH, editor. Power Electronics Handbook. 4th ed. Oxford, UK: Butterworth-Heinemann; 2018. pp. 711-722

[14] Garnica J, Chinga RA, Lin J. Wireless power transmission: From far field to near field. Proceedings of the IEEE. 2013; **101**(6):1321-1331. DOI: 10.1109/JPROC.2013.2251411

[15] Wang TW, Sung YL, Lin SF. Cardiac influence of repetitive transcranial magnetic stimulation in small animals.

IEEE Journal of Electromagnetics, RF and Microwaves in Medicine and Biology. 2020;**4**(4):279-285. DOI: 10.1109/JERM.2019.2958686

[16] Imura T. Wireless Power Transfer: Using Magnetic and Electric Resonance Coupling Techniques. Singapore: Springer Nature; 2020

[17] Wang TW, Zhang H, Lin SF. Influence of capacitive coupling on high-fidelity non-contact ECG measurement. IEEE Sensors Journal. 2020;**20**(16): 9265-9273. DOI: 10.1109/JSEN.2020. 2986723

[18] Wang TW, Lin SF. Negative impedance capacitive electrode for ECG sensing through fabric layer. IEEE Transactions on Instrumentation and Measurement. 2021;**70**:1-8. DOI: 10.1109/TIM.2020.3045187

[19] Wang T-W, Lin S-F. Non-contact capacitive sensing for ECG recording in small animals. Measurement Science and Technology. 2020;**31**(12):125703. DOI: 10.1088/1361-6501/ab8cfc

[20] Kindl V, Zavrel M, Drabek P, Kavalir T. High efficiency and power tracking method for wireless charging system based on phase-shift control. Energies. 2018;**11**(8):2065. Available from: https://www.mdpi.com/1996-1073/11/8/2065

[21] Yi K. Capacitive coupling wireless power transfer with quasi-LLC resonant converter using electric vehicles' windows. Electronics. 2020;**9**(4):676. DOI: 10.3390/electronics9040676

[22] Freeman DK, Byrnes SJ. Optimal frequency for wireless power transmission into the body: Efficiency versus received power. IEEE Transactions on Antennas and Propagation. 2019;**67**(6):4073-4083. DOI: 10.1109/TAP.2019.2905672

[23] Sanchez-Montero R, Lopez-Espi P-L, Alen-Cordero C, Martinez-Rojas J-A. Bend and moisture effects on the performance of a U-shaped slotted wearable antenna for off-body communications in an Industrial Scientific Medical (ISM) 2.4 GHz band. Sensors. 2019;**19**(8):1804. DOI: 10.3390/s19081804

[24] Narayanamoorthi R, Vimala Juliet A, Chokkalingam B, Padmanaban S, Leonowicz ZM. Class E power amplifier design and optimization for the capacitive coupled wireless power transfer system in biomedical implants. Energies. 2017;**10**(9): 1409. DOI: 10.3390/en10091409

[25] Lim A, Tan A, Kong Z-H, Ma K. A design methodology and analysis for transformer-based Class-E power amplifier. Electronics. 2019;**8**(5):494. DOI: 10.3390/electronics8050494

[26] Sokal NO, Sokal AD. Class E-A new class of high-efficiency tuned single-ended switching power amplifiers. IEEE Journal of Solid-State Circuits. 1975; **10**(3):168-176. DOI: 10.1109/JSSC.1975.1050582

[27] Mutashar S. Efficiency improvement of wireless power transmission for bio-implanted devices. International Journal of Biomedical Engineering. 2013;7: 721-724

[28] Kirk JA et al, Pacemaker-induced transient asynchrony suppresses heart failure progression, Science Translational Medicine. 2015;7(319): 319ra207. DOI: 10.1126/scitranslmed.aad2899

[29] Khazanie P, Hellkamp AS, Fonarow GC, Curtis LH, Al-Khatib SM,

Hernandez AF. Permanent pacemaker use among patients with heart failure and preserved ejection fraction: Findings from the Acute Decompensated Heart Failure National Registry (ADHERE) National Registry. American Heart Journal. 2018;**198**:123-128. DOI: 10.1016/j.ahj.2017.12.020

[30] Abiri P et al. Inductively powered wireless pacing via a miniature pacemaker and remote stimulation control system. Scientific Reports. 2017;**7**(1):6180. DOI: 10.1038/s41598-017-06493-5

[31] Deisseroth K. Optogenetics. Nature Methods. 2011;**8**(1):26-29. DOI: 10.1038/nmeth.f.324

[32] Mahmoudi P, Veladi H, Pakdel FG. Optogenetics, tools and applications in neurobiology. Journal of Medical Signals and Sensors. 2017;7(2):71-79

[33] Boyden ES. Optogenetics and the future of neuroscience. Nature Neuroscience. 2015;**18**(9):1200-1201. DOI: 10.1038/nn.4094

[34] Williams JC, Entcheva E. Optogenetic versus electrical stimulation of human cardiomyocytes: Modeling insights. Biophysical Journal. 2015; **108**(8):1934-1945. DOI: 10.1016/j.bpj.2015.03.032

[35] Entcheva E, Kay MW. Cardiac optogenetics: A decade of enlightenment. Nature Reviews Cardiology. 2021;**18**(5):349-367. DOI: 10.1038/s41569-020-00478-0

[36] Yu L et al. Optogenetic modulation of cardiac sympathetic nerve activity to prevent ventricular arrhythmias. Journal of the American College of Cardiology. 2017;**70**(22):2778-2790. DOI: 10.1016/j.jacc.2017.09.1107

[37] Wang TW, Chen WX, Chu HW, Lin SF. Single-channel bioimpedance measurement for wearable continuous blood pressure monitoring. IEEE Transactions on Instrumentation and Measurement. 2021;**70**:1-9. DOI: 10.1109/TIM.2020.3035578

[38] Wang TW et al. Single-channel impedance plethysmography neck patch device for unobtrusive wearable cardiovascular monitoring. IEEE Access. 2020;**8**:184909-184919. DOI: 10.1109/ACCESS.2020.3029604

[39] Wang T-W et al. Bio-impedance measurement optimization for high-resolution carotid pulse sensing. Sensors. 2021;**21**(5):1600. DOI: 10.3390/s21051600

[40] Wang TW, Sung YL, Chu HW, Lin SF. IPG-based field potential measurement of cultured cardiomyocytes for optogenetic applications. Biosensors & Bioelectronics. 2021;**179**:113060. DOI: 10.1016/j.bios.2021.113060

[41] Theodor M et al. Implantable impedance plethysmography. Sensors. 2014;**14**(8):14858-14872. DOI: 10.3390/s140814858

[42] Wang TW, Lin SF. Wearable piezoelectric-based system for continuous beat-to-beat blood pressure measurement. Sensors (Basel, Switzerland). 2020;**20**(3):851. DOI: 10.3390/s20030851

Chapter 12

Theoretical Premises and Contemporary Optimizations of Microwave Tomography

Paul M. Meaney and Keith D. Paulsen

Abstract

Microwave imaging has long been proposed as an effective means for biomedical applications—breast cancer detection and therapy monitoring being the most prominent because of the endogenous dielectric property contrast between malignant and normal breast tissue. While numerous numerical simulations have been presented demonstrating feasibility, translation to actual physical and clinical implementations have been lacking. In contrast, the Dartmouth team has taken somewhat counterintuitive but fundamentals-based approaches to the problem—primarily addressing the confounding multipath signal corruption problem and exploiting core concepts from the parameter estimation community. In so doing, we have configured a unique system design that is a synergism of both the hardware and software worlds. In this paper, we describe our approaches in the context of competing strategies and suggest rationales for why these techniques work—especially in 2D. Finally, we present data from actual neoadjuvant chemotherapy exams that confirm that our technique is capable of imaging the tumor and also visualizing its progression during treatment.

Keywords: microwave, tomography, breast, log transform, multi-path, clinical

1. Introduction

The primary basis for microwave breast imaging is that there is considerable dielectric property contrast between malignant and normal tissue. This contrast is a complicated issue and its understanding has evolved substantially over time. At the most simplistic level, microwaves are especially useful as "water detectors." At low microwave frequencies, water typically has relative permittivity values between 75 and 80 while that for fat ranges between 5 to 10. The earliest assumptions about breast tissue were that it was primarily fatty and tumors contained far more water on account of their rapid replication and proliferation [1]. However, recent studies have provided a more nuanced appreciation of normal breast tissue composition. In a study by Woodard and White [2], in the context of assessing the physical content of tissue for radiographic purposes, they found that the two constituents of breast tissue—adipose and mammary tissue—varied considerably from woman to woman, and concomitantly, so did their water content. In fact, the water content variations were:

adipose—11.4–30.5%, and mammary—30.2–72.6%. More recent studies have tried to provide a more complex presentation of the properties. Studies by Lazebnik et al. [3], Sugitani et al. [4], Martellosio et al. [5], and Cheng and Fu et al. [6] have all presented values for all three tissue types referred to as adipose, fibroglandular and tumor, respectively. Relatively consistently, the studies have typically shown the adipose properties to be quite low, those for the fibroglandular to be substantially elevated and those for the tumors to be the highest.

However, there is considerable variation between results. As has been pointed out by Meaney et al. and [7, 8] Salahuddin et al. [9], there are important weaknesses and even flaws in the methodologies of these studies which could easily skew the desired results. Two of the more critical problems revolve around the use of the ubiquitous open-ended dielectric probe and the frequency sampling regimen used in algorithms for fitting the data to established Cole-Cole curves for broadband parameter estimation [3]. With regards to the former, it is well known that the penetration depth or sampling volume in front of the dielectric probe is on the order of 1/6th that of the probe diameter [10, 11]. Given that the probe diameters of the most common commercial probe (Keysight Technologies, Santa Clara, CA) are on the order of about 2 mm diameter, this suggests that the penetration depth is on the order of 0.3 mm. However, various reports explicitly state the assumed sampling volume to be 3 mm deep or more [12]. The significance is that it is relatively straightforward for a pathologist to perform a tissue analysis over a sample that is 3 mm thick. For a volume as small as 0.3 mm, the analysis is substantially less informative. In these cases, the probe measurements are really only assessing superficial dielectric properties. In addition to compositional issues near the surface, this is also where the greatest temperature (cooling after excision) and moisture (drying after excision) gradients appear especially during measurements of excised tissue. Both temperature and water content can have dramatic influence over the tissue properties. As a whole, these factors are not adequately addressed in recent reports and open the door for substantial variation.

With respect to the frequency sampling regimens, this is primarily a problem related to how the properties are fitted to the Cole-Cole equation to determine broadband coefficients. The norm for this is to acquire data in a logarithmic frequency fashion and was established by Cole and Cole [13]. The reason for this is that when the dielectric property data is plotted in the Cole-Cole plane, the sample points form nearly a circle and the coefficients are derived from the circle's features such as, the radius and center location. When the data is collected in a logarithmic fashion, the points end up being spaced nearly evenly around the arc of the circle, while for the linearly spaced frequencies, the data points are highly concentrated in a localized corner of the plot. For fitting purposes, the evenly spread out data points produce more accurate findings. The problem becomes further compounded when the data spans multiple relaxation zones (the most prominent cross over point for tissue properties occurs within the range of about 1.5–2.5 GHz) [14]. When a linear sampling regimen is used, the lower relaxation zone is dramatically under-sampled compared to that for the higher zone such that the Cole-Cole fitting is grossly skewed to bias the results for the higher zone. While many of the reports properly sampled the data logarithmically with respect to frequency, some have not [3].

More recently there have been several clinical breast imaging studies [15–17]. The Preece study was based on the University of Bristol radar imaging technique which only produces intensity maps, where localized hot spots are indicative of tumors. Even so, the clinical data suggests that their system is capable of distinguishing tumor from normal tissue, even in dense breasts suggesting that there must be property contrast between the tumor and fibroglandular tissue. The reports by Poplack et al. [16] and Meaney et al. [17] suggest that there is contrast in both the permittivity and conductivity between malignant and normal tissue. The more interesting aspect is that the contrast is only statistically significant for the conductivity images. Given that there are questions related to the absolute dielectric property levels of the different tissue types for all ex vivo studies, we have attempted to develop a hypothesis based on previous literature. In particular, extensive studies by Foster et al. showed distinctly different tissue permittivity and conductivity relationships as functions of water content [18]. In these situations, the fat and fibroglandular tissue would have either no water or low levels of free water (i.e. the bound water in the fibroglandular tissue is primarily bound to long sugars and proteins). In these situations, the ionic content of the solutions is not able to exhibit higher conductivity because there is little to no free water necessary for conduction. However, the tumor has considerable free water and exhibits high conductivity. This mechanism is one means for explaining the relatively high conductivity contrast between the tumor and normal tissue. Conversely, the permittivity acts more as a measure of bulk water. In this situation, the fibroglandular tissue has relatively high water content compared with fat, but not as high as that for tumor. But the percentage difference between that for tumor and fibroglandular is typically in the range of 10–20% compared with what could amount to factors of 2× or more for the conductivity. This can explain the more subdued permittivity contrast we observe for the imaging studies between normal and malignant tumor. These features need to be explored further—preferably in clinical studies.

As was alluded to above, there are multiple near field microwave imaging techniques. These consist primarily of (a) radar techniques, (b) tomography or inverse problem techniques, (c) thermoacoustic imaging, and (d) holography. The radar techniques are generally a type of synthetic aperture radar (SAR) methods that typically utilize either backscatter or transmission data. For the backscatter approaches, a large amount of wideband data is acquired for many positions around the object surface and then time delays are synthetically added or subtracted from each measurement to focus sequentially at each pixel within the imaging domain. The contributions from each measurement are summed at each pixel and the resulting intensity maps are displayed as the images [19]. Multiple simulation efforts have been developed with the most advanced phantom and patient experiments performed by the Fear group at the University of Calgary [20]. Transmission techniques have also been developed—primarily by the group at the University of Bristol [21]. Theirs utilizes a fixed array of wideband antennas which directly contact the breast—albeit they utilize a coupling gel to enhance coupling and minimize unwanted contributions from multipath signals. This has advanced from simulation and phantom experiments to more extensive clinical trials [15].

Tomographic and inverse problems have been studied extensively in simulation with only very few translating to phantom and clinical work. These approaches typically utilize mostly transmission data and require nonlinear inverse algorithms to produce actual maps of the tissue permittivity and conductivity [22]. Simulation work includes studies by Rocca et al. [23], Fhager and Persson [24], Catapano et al. [25] and Shea et al. [26]. Extensive phantom and ex vivo animal studies have been performed by Semenov et al. [27]. The most comprehensive clinical work has been performed by Meaney et al. which includes studies of normal patients [28, 29], a diagnostic study comparing images of patients with and without tumors [16], and a study monitoring the progression of tumors during neoadjuvant chemotherapy [17]. For these studies, the imaging technique can distinguish tumors from normal tissue and benign lesions to a level of significance for lesions 1 cm and larger. For the therapy monitoring study, the technique was able to determine whether a tumor was responding to treatment or not within the first 30 days of the chemotherapy regimen.

The thermoacoustic techniques generally apply a low duty cycle, high power microwave pulse which is selectively absorbed by the malignant tissue and subsequently causes a mechanical vibration which can be detected by ultrasound transducers. The images are produced by synthetically combining the signals from the different ultrasound transducers. The technique has seen limited success in both phantom and clinical work [30]. The holography approach is being studied primarily at McMaster University by Dr. Nikolova. To date, they have developed an initial prototype which has produced promising phantom results [31].

For this chapter, we focus on tomographic or inverse problem approaches. Microwave tomography and/or microwave inverse problems have now been studied at great length for several decades [25, 26, 32]. The preponderance of efforts has been in simulation with very few advancing to actual implementation and/or clinical exams. While factors such as cost, exam time and image reconstruction complexity/costs are often cited as prime reasons for failure, our experience has led us to focus on two factors that inhibit progress. These are the need to explicitly contend with the problems of multi-path signal corruption [33] and the need for variance stabilizing transformations in the reconstruction process [34]. Viewing microwave imaging from the context of these two challenges clearly illustrates flaws in conventional approaches and inhibits overall progress.

Multi-path signal corruption has been acknowledged for decades. In the context of classic radar and telecommunications applications, signals such as ground clutter constructively and destructively add to the desired signals and excite unwanted artifacts such as ghosting [35]. For far-field applications such as radar, these artifacts can often be as little as a minor nuisance. However, for near field imaging situations, the corruption can be extreme—to the degree that the multi-path signal can easily completely overwhelm the desired one [33]. For far-field situations, the primary mechanism for multi-paths are reflections off of neighboring structures or surfaces and recombining with the original signal [36]. However, for near field cases, they very often propagate as surface waves along interfaces of support structures and the coupling medium or along the outside of antenna feedlines. One likely reason why this phenomenon goes unconsidered is the fact that these structures are simply not included in the models for numerical simulations [26, 37]. For many implementations, the computational costs are already enormous when simply including just the antennas, coupling medium and the target. Adding complexities such as the feedlines, support structure and the coupling medium tank would simply overwhelm the capabilities of modern computers. Consequently, for most simulation efforts, these structures are simply ignored—in fact, even neighboring antennas are usually also eliminated in the name of computational speed and cost. The unfortunate result is essentially a precise rendering of an unrealistic scenario.

There are few options in compensating for this challenge. The primary factor is that the multi-path signals originate from the same desired signal—i.e. it is the same frequency. Because of this, sophisticated filtering approaches are not effective. Techniques such as time-gating have been proposed in different implementations but have not resulted in any published results for microwave tomography [38]. One of the more challenging aspects of time gating is the need for a very broad band signal with fine sampling between frequencies. Given that the measured microwave data is most often acquired in the frequency domain, this can add dramatically to the acquisition time. In addition, many of the proposed antennas simply do not operate over sufficient bandwidth to make this possible.

One technique employed by the Dartmouth group is the use of a lossy coupling medium [39]. This poses unique challenges, but when considering the imaging problem in a comprehensive manner, there is considerable merit to it. The main drawback is that propagating across even a short span in a lossy imaging medium can easily exceed the measurement dynamic range of high quality commercial vector network analyzers (VNA). This is not a trivial concern, but when properly understood, it is possible to devise systems to adequately accommodate this. Dynamic range considerations are discussed in more detail in the Methods section. The primary benefit is that the highly attenuating medium dramatically suppresses the unwanted surface waves. **Figure 1** shows a set of simulations for a monopole antenna radiating into a coupling medium where the active part of the antenna is positioned several centimeters above a Plexiglas plate [40]. For the low attenuating liquids (low conductivity), the surface waves easily reach the Plexiglas via coaxial modes traveling on the outsides of the feedlines. Once sufficiently powerful signals reach the low loss plate, planar modes are excited and the waves propagate unimpeded everywhere. However, as the conductivity of the liquid is increased, the coaxial modes are dramatically reduced to the point that for the last example, no signal reaches the Plexiglas. In this instance, the desired signal still propagates in a well-behaved beam pattern from the active part of the monopole antenna. In effect, we have traded a nearly impossible problem—i.e. uncontrolled propagation from damaging multi-path signals—for the need for a high dynamic range VNA which is just a difficult problem. It should be noted that multi-port VNA's with the necessary dynamic range are now commercially available from some vendors—albeit at a significant cost.

As a side note, in explicitly dealing with multi-path issue, we have been able to realize several opportunities that would not have been possible had we taken a more conventionally intuitive approach. For instance, we have found that the monopole antennas are ideal in this setting. First, the naturally occurring resistive loading of the lossy bath dramatically improves the monopole antenna bandwidth [41]. These can be used from roughly 500 MHz to 3 GHz with a −10 dB return loss across the band—well in excess of that for most conventional antennas. While these antennas are essentially isotropic radiators, their low profile allows them to be packed tightly around the target which dramatically reduces the propagation distances compared with more

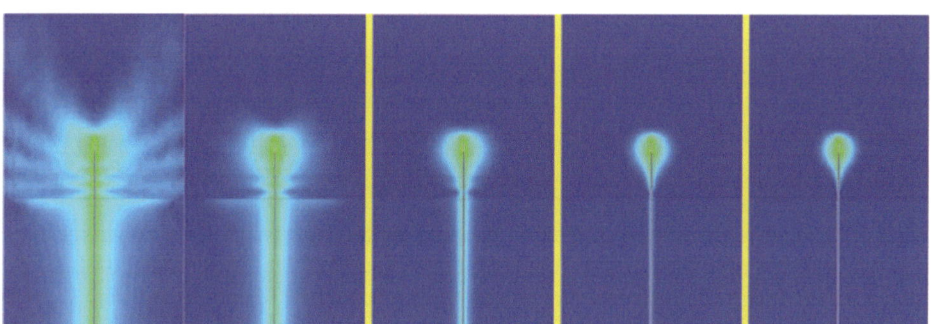

Figure 1.
Simulations of the field patterns and associated surface waves for differing coupling bath conductivities (S/m) (Permission granted to re-print images by Human Press [40]).

conventional aperture antennas. This shortening of the propagation distance easily compensates for the signal loss suffered due to having lower directivity. Ultimately, the close packing feature is a substantial advantage over the loss of directivity. In addition, because of the lossy bath, antenna low physical profile and the isotropic radiation pattern, there is essentially no mutual coupling between antennas even when spaced as closely as 2 cm apart [42]. Mutual coupling can be significantly debilitating for antenna array performance. Finally, the combination of the low loss medium and the low profile antennas makes the broadcast waves appear to be propagating in a mostly dielectric medium. This allows us to exploit the discrete dipole approximation (DDA) as an efficient means for simulating the signal forward solution which is recognized as the largest time cost in the reconstruction process [43]. We are now able to recover 2D images in 6 s and estimate that fully 3D images can be reconstructed in a few minutes—all without the aid of parallel processors and graphical processing units (GPUs). Overall, the advantages of using a lossy coupling medium have led to dramatic innovations.

Finally, while much has been written about the mathematics and algorithms regarding microwave tomography and/or inverse problems, the most common approach is a non-linear, iterative one which broadly falls under the category of multi-parameter estimation problems. These problems have been studied extensively within the probability and statistics community for which a host of definitions and techniques have been developed to optimize algorithmic performance. In particular, work by Box and Cox in the 1950s and 1960s devised ways to assess the performance of different single-step and iterative algorithms and derived a wide range of suitable transformations to improve performance [44]. Their focus centered around characterizing problems where the data was inherently heteroscedastic and developing transformations to make the data more homoscedastic—i.e. amenable to standard, least squares multi-parameter estimation techniques. Two of the more ubiquitous examples include the log transforms used in X-ray CT and optical coherence tomography (OCT) [45, 46]. The basic assumptions are that the error function difference between the measured and computed field values should have a zero mean and a normal distribution. A convenient way to test this is to simply examine the residual vector after the reconstruction [47]. For the X-ray CT case, the image reconstructions are simply not possible without the transformation. This is also the case with OCT, for which the log transform is now widely adopted [48].

In earlier work, we demonstrated that the residual data for the microwave case was highly heteroscedastic when applied to an algorithm operating directly on the complex field data (**Figure 2a**) [34]. Our interpretation was that it was largely due to the wide signal strength dynamic range of the field values. However, once we applied the log transformation, the residuals were significantly more normal with a near zero mean (**Figure 2b**). The main challenge here is that when taking the log transform of a complex number, the result is the log magnitude and its phase [49]. The phase term immediately implies that there could be some form of unwrapping necessary. For the X-ray CT case, the detected signals are all real numbers, so there is no phase term. In OCT, the governing equation for the light is the transport equation and the phase is generated by harmonically modulating the light with a 100 MHz signal [46]. The major point here is that the wavelength associated with the 100 MHz is very large such that there is never enough scattering to generate phase changes greater than +/− 180 degrees. In effect, the data is always unwrapped. However, for the microwave breast imaging case, the dielectric scatterers are often physically on the same order size as the wavelengths and the scattered fields can quite frequently change phase by values well

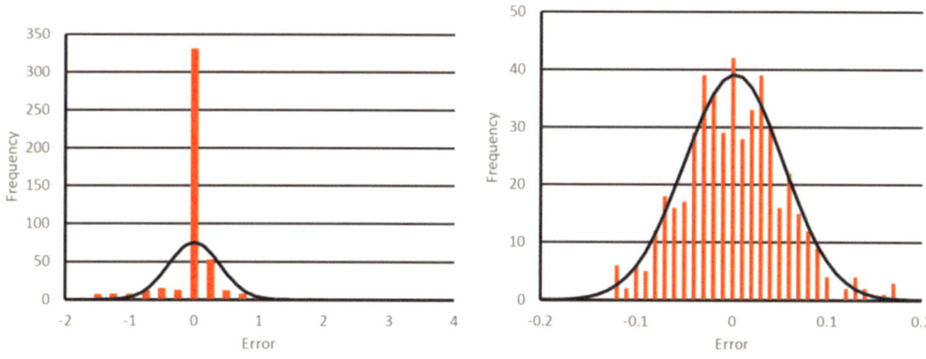

Figure 2.
Histograms of the residual error vector after reconstruction for the (a) non-transformed and (b) log transformed algorithms (Permission granted by Wiley Publishers to re-print graphs [34]).

in excess of +/− 180 degrees. For the parameter estimation problems utilizing complex data, it is critical that all of the measured and computed phase pairs be on the same Riemann sheet for the algorithms to work properly. We briefly describe methods for unwrapping the phase in Section 2.2 which are covered in more detail in Meaney et al. [50].

Operationally, the key is that when starting at a baseline of a homogeneous medium, adding a contrasting target to the imaging zone can change the phases measured or computed for the different transmit/receive pairs substantially—often easily exceeding the single Riemann sheet bounds of −180 to +180 degrees. Depending on the antenna orientations, there are nominally some cases where the measurement and computed phases are already on the same Riemann sheet at the start of the algorithm, but there can also be many that are not. In cases where they are not, by mapping all of the measured and computed phases to the baseline Riemann sheet—i.e. −180 to +180 degrees—(which is effectively what happens when using the non-transformed algorithm), it essentially transforms that element into a "bad" data point. While not universally viewed in this context, the various solutions that have been proposed generally work with the net effect of forcing the starting measured and computed phase values to be on the same Riemann sheet. These include: (1) introducing a priori information [24], (2) frequency hopping [51], and (3) simply adding more data [25, 27]. A priori information works because it effectively assumes a sample target at the start of the image reconstruction process that has similar characteristics to the actual target. The desired result is that it generally positions the measured and computed phases for all or most transmit/receive pairs on the same Riemann sheet. There are several reports utilizing this technique in simulation that appear promising; however, these techniques have generally not advanced beyond simulation studies. These results unfortunately end up being biased based on the quality of the initial guess. Frequency hopping acts in a similar manner as for applying a priori information in that images are reconstructed at progressively increasing frequencies with the results from lower frequencies used as starting guesses for the subsequent higher frequency reconstructions. At the lowest frequency, assumptions can be made that the phase changes are modest and it can be assumed that all of the data is unwrapped. While the lower frequency images can be quite blurred because of the associated larger wavelengths, the algorithms converge more reliably. The property images that are transposed to the next higher frequency can be close enough to the actual images at the increased frequency such that it essentially positions the phases for all transmit/receive pairs on the appropriate Riemann sheet. The notion is that the resolution of each successively higher frequency increases until one reaches the highest frequency. Finally, a number of groups have advocated utilizing substantially more measurement data than that prescribed by the Dartmouth team [25, 27]. While these often advance the notion that there needs to be as much measurement data as number of unknowns, our own experience based on a wealth of literature suggests that this may not be necessary [52]. An equally valid interpretation in the context of this discussion is that they hope to increase the amount of "good" data such that the amount of "good" data simply outweighs the amount of "bad" data for which a least squares image process can achieve a reasonable image. For many real situations, each of these solutions may be unrealistic. Accurate a priori information may be difficult to generate at the time of or before the actual imaging session. Frequency hopping may be unrealistic because it is difficult to devise antennas with sufficient bandwidth to accommodate the algorithm. Finally, adding more data inevitably implies that the measurement system will require more channels which inherently leads to increases in the algorithm complexity and hardware costs.

The Dartmouth team has devised robust unwrapping techniques for both the measured and computed phases which are briefly summarized in Section 2.2 [50]. These generally exploit the wide operating bandwidth of the monopole antennas and even the nature of the algorithm convergence. One important consequence of these developments is that the amount of measurement data can be kept to a minimum (typically 16 antennas configured in a circle for 2D images) which dramatically reduces overall system cost and complexity [52]. The algorithm is fast and does not suffer from convergence to unwanted local minima even when starting from an initial estimate of the coupling bath properties. These approaches have been developed in the context of utilizing actual measurement data to maximize the benefit from each piece of measurement data while not imposing Riemann sheet criteria that essentially transform "good" data into "bad". This approach is summarized in Section 2.2 and demonstrates excellent convergence behavior using real measurement data.

2. Methods

2.1 Hardware

The heart of the system concept is the monopole antenna array submerged in a lossy glycerin: water coupling bath. As discussed in the Introduction, the monopole antennas offer important advantages such as small size, ease of modeling and broad bandwidth while also presenting different opportunities. The major trade-off is that the signal attenuation is great, even for a typical physical distance of 15 cm across the lossy medium in the illumination chamber, such that most commercial vector network analyzers are inadequate. This has changed somewhat in that there are now commercial 24 channel systems from Rohde & Schwartz (Munich, Germany) and Keysight Technologies (Santa Clara, CA); however, their costs can easily exceed $100 K depending on the options necessary. For realizing an actual system, the most critical trade-offs are cost versus performance. In our case, this discussion involved numerical modeling concerns along with critical aspects of the microwave technology. For instance, as mentioned above, there is considerable debate regarding the required number of antennas. Our approach has been to configure just enough antennas for a single-plane array and then exploit mechanical motion to move the array vertically to achieve complete 3D coverage of the target—in this case the breast. We have demonstrated that 16 antennas is sufficient for the single array [52], which has kept our costs to modest levels. **Figure 3** shows the antennas mounted to a horizontal plate which is supported by opposing motors that can bring the antennas to different levels surrounding the breast to enable full volumetric coverage of the breast. Below we summarize the more significant issues for the microwave electronics.

2.1.1 Large dynamic range

In general, the dynamic range is limited on the low end by the noise floor which is defined as $N = kTB$ where k is Boltzman's constant, T is the temperature and B is the sampling bandwidth. The primary option for decreasing the noise floor is increasing the sampling time (i.e. additional averaging) which is the inverse of the bandwidth. Theoretically the noise floor could be made infinitely low if one allowed for the data acquisition to take forever. However, for practical systems, the lower end is also limited by the discretization resolution of the A/D converter. The most common way to address both is to add increased amplification in front of the receiver while also increasing the sampling time. The typical VNA is configured so that the lowest possible noise floor is roughly −100 dBm. Given that the transmit signals typically can range on the order of 1–10 mW, the maximum dynamic range is roughly 102 dB (assuming that we can realistically pick out a signal that is roughly 8 dB greater than the noise floor). In building our custom system, we have employed the Ettus B210 agile transceiver which by themselves can only see down to −110 dBm. By adding 20 dB gain, increasing the sampling time, and adding a low noise amplifier, we have simultaneously improved the overall noise figure of the system while also lowering the noise floor and boosting the signal. In this way, we are able to achieve substantially higher dynamic range capabilities in a low cost configuration. The Ettus boards cost a fraction of that of typical VNA's and are programmable using general purpose instrumentation software including Matlab (Mathworks, Natick, MA) and LabView (National Instruments, Austin, TX) [53].

(a) (b) (c)

Figure 3.
Photographs of the antenna array, support structure and motors used for vertical, mechanical array motion: (a) close-up of mounting plate, (b) side view of tank, antennas and support structure, and (c) top view of the antennas in the tank.

2.1.2 Channel-to-channel isolation

One hypothetical way to achieve a large dynamic range would be to use a 2-channel VNA coupled to a double-pole, 16 throw (DP16T) switch matrix to achieve coverage for all 16 channels. We have already discussed the challenges in achieving the dynamic range for a 2-channel system. Unfortunately, the multi-pole switch matrix introduces its own set of problems. In cases where the 2-port VNA already has a dynamic range approaching 140 dB, it is necessary for the switch matrix to have channel-to-channel isolation on the order of 140 dB. In practice it is especially difficult to achieve this and very rarely do switch manufacturers specify isolation greater than roughly 80–100 dB. There are ways to achieve this, but it generally requires substantial shielding and the addition of extra single-pole, single-throw (SPST) switches which introduces substantial insertion loss and inevitably detracts from the overall dynamic range.

Our latest concept is to exploit new software defined radio (SDR) technology as the key building block of our system [53]. In this case, we utilize the Ettus Research B210 (Austin, TX) which has two channels, each having a transmit/receive port and a dedicated receive port. Our design involves a physical separation of the boards used for transmit and receive (extra isolation), along with a set of switch/amplifier modules which add to the overall channel-to-channel isolation along with improving the dynamic range with a 20 dB gain amplifier in front of the receiver channel. One of the signals from the Transmit B210 is used as a reference signal to synchronize the transmit oscillators with those of the receive modules. **Figure 4** shows a schematic of a representative 4-channel system along with an initial prototype showing the separate transmit and receive modules along with the switch and amplifiers. This design can be readily scaled up to a full 16 channel system.

With respect to isolation, **Figure 5a** shows the shielded housings for the B210 boards. The primary feature is the ridge in the cover which lightly rests on a gold strap running across the board that physically separates the RF portion of the board from the lower frequency electronics and power supply circuitry. Tests have shown that the "one-way" leakage from these housings is on the order of −110 dB (from a nominally

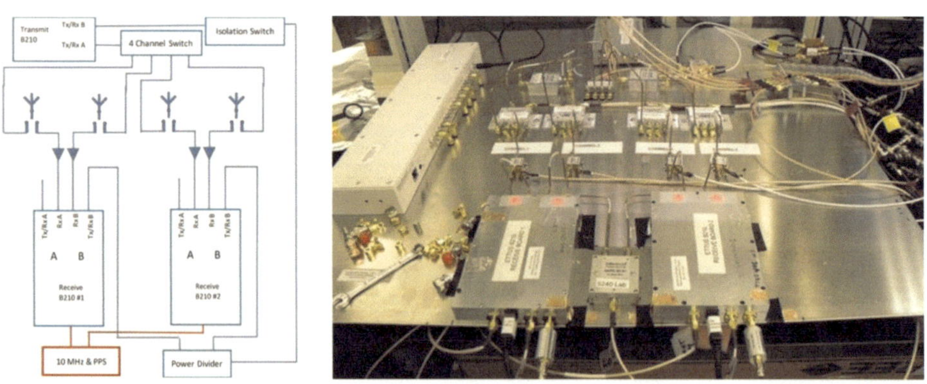

(a) (b)

Figure 4.
(a) Schematic diagram of a 4-channel measurement system and (b) a photograph of a first prototype (a is reproduced from "Meaney PM, Hartov A, Bulumulla S, Raynolds T, Davis C, Schoenberger F, Richter S, Paulsen KD, "4-channel, vector network analyzer microwave imaging prototype based on software defined radio technology," Review of Scientific Instruments, *vol. 90, pp. 044708-1—044708-14, 2019," with permission of AIP Publishing [53]).*

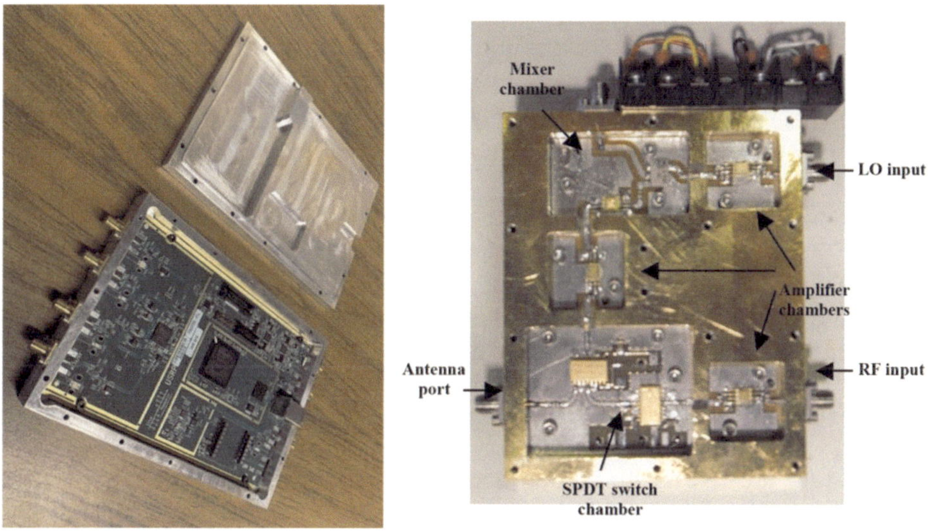

(a) (b)

Figure 5.
Photograph of the (a) shielded housings for the Ettus B210 SDR circuit boards, and (b) the shielded housing for an earlier transceiver module (b is reproduced from "Li D, Meaney PM, Raynolds T, Pendergrass SA, Fanning MW, Paulsen KD, "A parallel-detection microwave spectroscopy system for breast imaging," Review of Scientific Instruments, *vol. 75, pp. 2305–2313, 2004," with permission of AIP Publishing [54]).*

transmitted signal of 1 mW). While a signal of this amplitude may propagate along the outer surface of the electronics, because all of the other components of the system have comparable isolation specification, the greatest signal level that could leak back into the system would be on the order of −220 dB because of the "two-trip" attenuation in escaping the isolated board and then re-entering a companion board. **Figure 5b** shows the internal features of an earlier, custom transceiver module which included several amplifiers and switches along with a downconverting mixer. All feedthrus for the bias and control lines were coaxial low pass filters to restrict RF signals from escaping outside of the housings. Each component within the housing was sequestered to its own channelized compartment to limit internal leakage and the covers were configured with raised surfaces for tight fits of the two housing portions to further limit leakage to the outside. While somewhat cumbersome, in large quantity the housings can be fabricated at relatively low cost and have proven to be effective in limiting leakage, even to the extreme levels of −140 dBm.

2.1.3 Synchronization

Commercial VNA's are attractive because they produce measurements that guarantee built-in synchronization of the transmit and receive signals. However, as was discussed earlier and also in the subsequent sections, these network analyzers are unattractive because of their costs. There are a host of means for achieving synchronization of signals in different hardware configurations. For our SDR-based system, we exploit the phase locking of the transmit and receive oscillators with respect to a 10 MHz reference signal from their Octoclock (Ettus Research, Austin, TX). In addition, accurate and consistent triggering of the measurements is achieved via the PPS signal which is also supplied by the Octoclock. In the case with separate transmit and receive modules, we exploit the fact that the second output signal from the dedicated transmit board is synchronized to the output of the channel that feeds the antennas. This is subsequently used as a reference signal and is fed into an unused receive port of the receive channels for sampling and synchronization with the receive boards [53, 54].

2.2 Software

While much has been made of a range of non-linear iterative schemes and various stochastic approaches for image reconstruction approaches, we have opted to use the robust and well-regarded Gauss-Newton algorithm. **Figure 6** shows examples of the magnitude contour plots for 2D wave propagation from different antennas surrounding the medium with a high contrast object. The differences between the values measured at the receive antenna locations for the situation where an object is present minus that for the case for the homogeneous bath becomes the measurement data for the algorithm. Note that the techniques are completely translatable to 3D but are more easily understood in the context of 2D.

For the untransformed algorithm, the minimization statement examines the differences between the measured field values and the associated computed values at each receive antenna and for all views of the object from each iteration:

$$\min \left\| E^{m} - E^{c}\left(k^{2}\right) \right\|^{2} \tag{1}$$

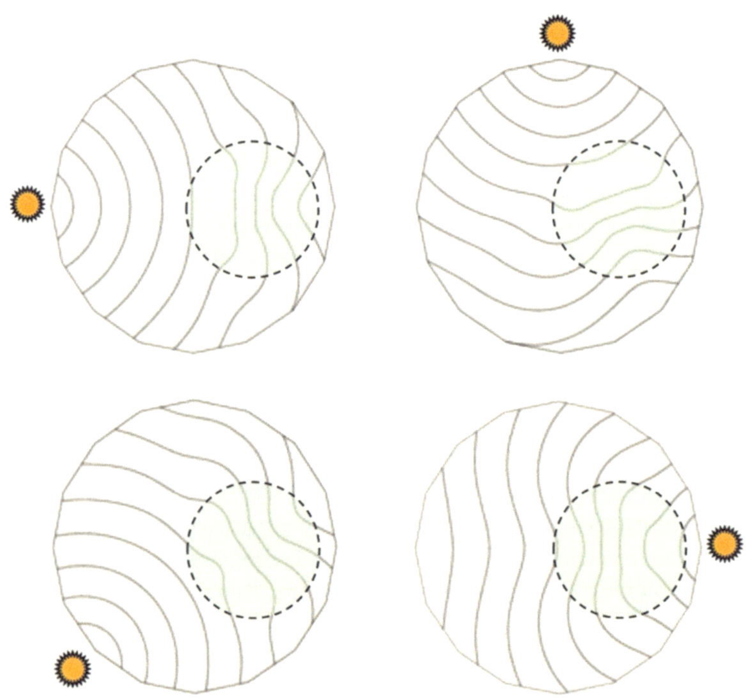

Figure 6.
2D magnitude contour plots within an imaging zone containing a high contrast scatterer for point source illuminations from different directions.

where E^m and E^c are the complex measured and computed electric field vectors, respectively, and k^2 is the wavenumber squared which is essentially the image and embeds both the permittivity and conductivity, respectively. The associated log-transformed equation is [49]:

$$\min \left\|\Gamma^m - \Gamma^c\left(k^2\right)\right\|^2 + \left\|\Phi^m - \Phi^c\left(k^2\right)\right\|^2 \tag{2}$$

where Γ^m, Γ^c, Φ^m, and Φ^c are the measured and computed log-magnitude and phase vectors, respectively. The two equations are essentially the same except for the fact that the phase and magnitude need to be accounted for explicitly. In addition, both the measured and computed phases need to be properly unwrapped. The image is then updated at each iteration by:

$$k_i^2 = k_{i-1}^2 + S\Delta k^2 \tag{3}$$

where k_i^2 and k_{i-1}^2 are the wavenumbers squared vectors at iterations i and i + 1, Δk^2 is the image update at each iteration and S is the iteration step size. In all cases, the initial property estimate of the image is that of the homogeneous coupling bath which is the least biased estimate possible. The stopping criteria is generally met when the changes in the image (Δk^2) are small—determined empirically. For all cases, regularization is necessary to stabilize the solution. For this situation, we have found both Tikhonov and Marquardt–Levenberg techniques to be suitable [55, 56].

For applications such as MR, phase unwrapping is normally performed as a function of position—i.e. the phase at a specific location is set as the reference phase and the phases at neighboring locations are compared with it [57]. If the differences between the values at new locations differ from the reference ones by more than 180°, the values for the new ones are adjusted by multiples of 360° until the differences are minimized. This process is continued until the entire spatial domain has been unwrapped. For the microwave measured values, this is impossible since the phases can only be known at the receive antenna locations—i.e. the phase can vary more than 180° between antennas. Our alternative is to unwrap the phases as a function of frequency. A baseline assumption implies that the scattered phases should be within the bounds of +/− 180° for the lowest frequency. The measured phases are forced to be within the baseline Riemann sheet, i.e. −180 to +180°. The values at the next frequency are unwrapped against these, followed by those at the next frequency being unwrapped with respect to the second set and so on. This is described in more detail in Meaney et al. [50]. For the computed phases, it would be possible to unwrap the values as a function of position. However, this is tedious, complicated and slow. We previously introduced a novel technique to unwrap the phases as a function of image reconstruction iteration [50]. As long as the phases do not change dramatically between iterations of the reconstruction process, it can be assumed that their differences will be less than 180°. To ensure this criterion, we deliberately force the step size (Eq. (3)) to be small during the earliest iterations so that the image, and hence the phases, differ only slightly between iterations. In this way, the phases only need to be computed at the antenna locations, and the values for the previous iteration need to be stored for comparison purposes. This technique is described in more detail in Meaney et al. [50].

Finally, the most substantial computational cost for these types of algorithms are for the forward field solutions which need to be computed for each transmit antenna at each image reconstruction iteration. Different techniques have been proposed to reduce this time, most notably finite element (FEM) and finite difference time domain (FDTD) modeling [26, 58]. However, these can be quite slow, especially as the forward problem grows in size, and often requires multi-processor computers and GPU processors to keep the computation times modest. While our 2D FEM-based approach can usually produce images in roughly 5 minutes, we have explored ways to further reduce the time. In particular, we have been experimenting with the discrete dipole approximation (DDA) to improve efficiency [59]. This technique is generally not suitable when there are large, high contrast scatterers in the field of view—especially other bulky antennas; however, because our monopole antennas only slightly perturb the fields, and almost not at all when submerged in a lossy bath, this algorithm is ideal for our approach. The primary notion is that it utilizes Gauss-iterative based forward solver techniques whose most substantial time cost is a matrix–vector multiplication that needs to be performed repeatedly ($O(N^2)$) where N is the vector length). However, it can be easily shown that the associated matrix can be formulated to be a symmetric, block Toeplitz matrix. Each block matrix can then be easily converted to a circulant matrix. Circulant matrix–vector multiplication can be performed efficiently ($O(N\log N)$) by use of the convolution theorem and the use of the fast Fourier transform (FFT). We have shown that this can reduce the forward solution time by a factor of 20–50 times with respect to efficient commercial software—i.e. full reconstructions in 6 seconds or less. This technique is described in detail in Hosseinzadegan et al. [59, 60].

3. Results

3.1 Algorithm convergence

While the final desired result of the iterative image reconstruction algorithm is the associated property images, the best way to assess the quality of the process is by analyzing the computed electric fields as compared with the measurement data [47]. As can be seen from Eq. (2), the minimization statement should converge to a point where measured and computed magnitudes and phases achieve a reasonable match. In essence, the final image can really be considered a bi-product of this process. A key reason why this distinction is important is that there are ample situations where researchers employ a range of a priori information to achieve reasonable results. In a laboratory or experimental setting this might appear reasonable; however, these ad hoc measures often introduce biases which can artificially skew the final results. The best unbiased way to examine the process is to focus on the actual terms being compared.

Figure 7 shows the normalized (a) magnitude and (b) phase values for the receiver positions associated with a single transmitter for both the measured values (heavy black lines) and computed values (narrower colored lines) at selected iterations [50]. The data was collected at 1300 GHz and both the magnitude and phase values have been normalized to the cases where there is no target in the bath—i.e. that the reconstruction process starts with an assumption of an empty bath which is reasonable since it is possible to exactly know the bath properties in all situations. Note that the measured phases exceed the +/− 180 degree bounds in some cases, implying that our algorithm is robust even for large, high contrast targets which often generate some measurements outside of the primary Riemann sheet. As a useful analogy, we refer to

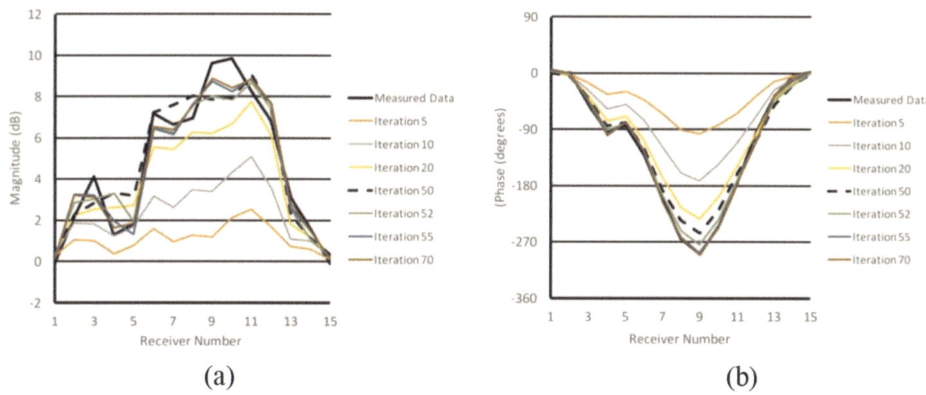

Figure 7.
1300 MHz (a) Magnitude and (b) phase projections for the measured data and multiple computed values at representative iterations. Note that the dashed black lines represent the projections at the end of the first step (Permission granted by Wiley Publishers to re-print graphs [50]).

the curves for each measurement set as a projection, much like the nomenclature used in X-ray CT. It should be noted that in this example, a two-step algorithm was used whereby the first 50 iterations used a standard Levenberg–Marquardt technique to produce a good, albeit smoothed image. After which, a Tikhonov algorithm was used for 20 iterations to more carefully refine the images. The intermediate projection is a black, dashed line for both the magnitude and phase.

For both the magnitude and phase, the projections monotonically approach those for the measurement data. The projections for the intermediate image have not yet matched the measurement data, but their curves retain many of the characteristics of the measurements but with a reduced overall strength and are also somewhat smoothed. Once the second step is employed, the computed magnitudes and phases rapidly approach the measured values. The convergence of the phase values is almost exact, while those for the amplitude are good but not as precise. The discussion in Section 4 provides a possible rationale for this mismatch. Regardless, the match is quite good and gives confidence that the process works well. It should be noted that this is for an actual breast exam where the measurement data inherently has 3D characteristics while the algorithm is a 2D one. In addition, these plots are for only one of the 16 antennas but is reasonably representative of all sets.

3.2 Therapy monitoring

While microwave imaging is poised to play a substantial role in the detection and diagnosis of breast cancer, it may be that its most important role will be in the area of neoadjuvant chemotherapy monitoring [17]. In this setting, it is important to be able to tell if the tumor is responding adequately to treatment at a sufficiently early time point so that treatment can be altered appropriately. Measures such as a physical exam present morphological information regarding the tumor response—i.e. size—however, these changes are generally quite slow to manifest compared to physiological changes. Early results suggest that the microwave images may be more physiological in nature [17].

Figure 8a shows representative sagittal MR scans of the ipsilateral breast for a woman with heterogeneously dense breasts and a 6.5 x 3.7 x 7.1 cm^3 tumor in the right breast just before treatment started: (i) T2, (ii) T1 with gadolinium, and (iii) the T1

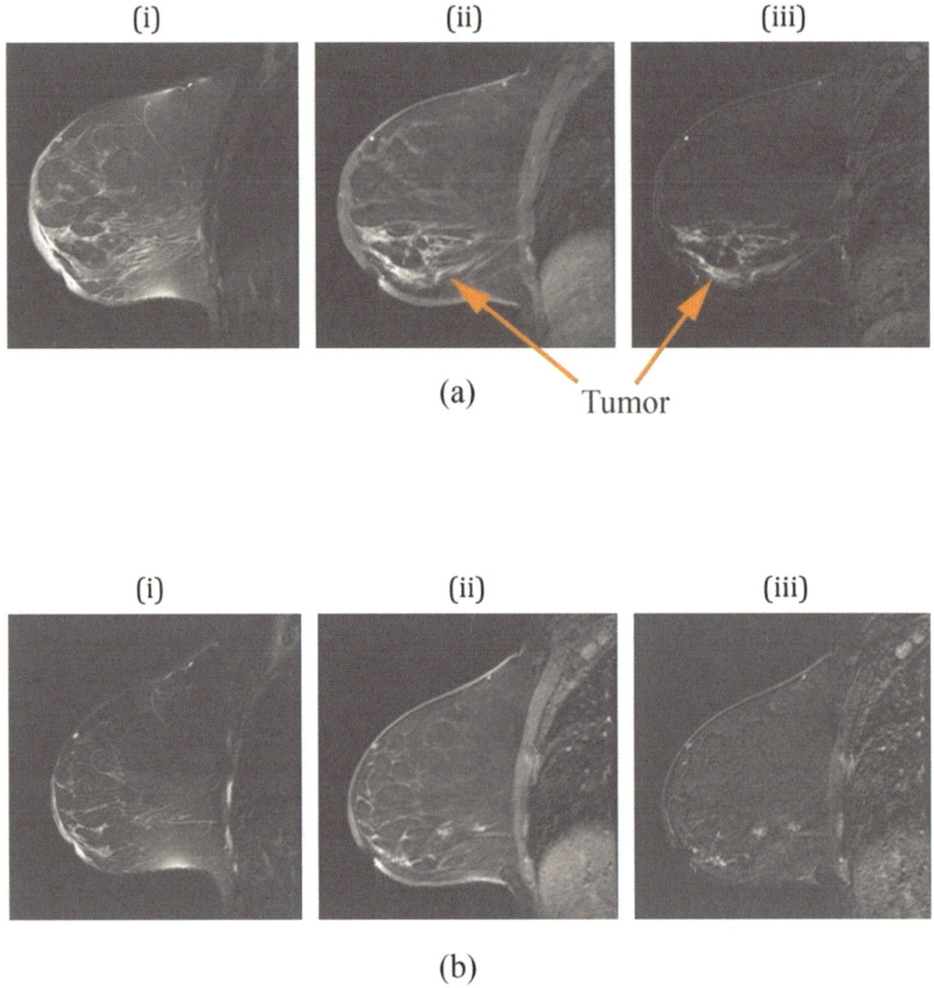

Figure 8.
Sagittal MR images of the right breast of a patient with a complete pathologic response: (a) images prior to therapy and (b) at day 85. For each set, (i) is a T2-weighted image, (ii) is a contrast-enhanced image from a spoiled gradient recalled (SPGR) sequence, and (iii) is a subtraction image between (ii) and a pre-contrast baseline, respectively (Reproduced from [17], with the permission of Springer Nature—License link: https://creativecommons.org/licenses/by/4.0/).

subtraction image for the cases with and without gadolinium. (i) Shows a large thickening of the skin surrounding most of the breast, while (ii) and (iii) show the extent of the tumor. **Figure 8b** shows the corresponding images after 85 days of treatment. The skin thickening had diminished substantially and the subtraction image indicates only small, minor enhancements.

Figure 9 shows the 1300 MHz permittivity and conductivity images for planes 5, 6, and 7, respectively, where plane 1 is closest to the chestwall and plane 7 is closest to the nipple. Set (a) is for the contralateral breast (normal) and sets (b), (c), and (d) are for the ipsilateral breast at different time points during the treatment—prior to treatment, after 44 days of treatment, and just prior to surgery, respectively. In all cases, the outline of the breast is readily visible separating the tissue from the surrounding bath. For the contralateral breast, the majority of the tissue exhibits quite low

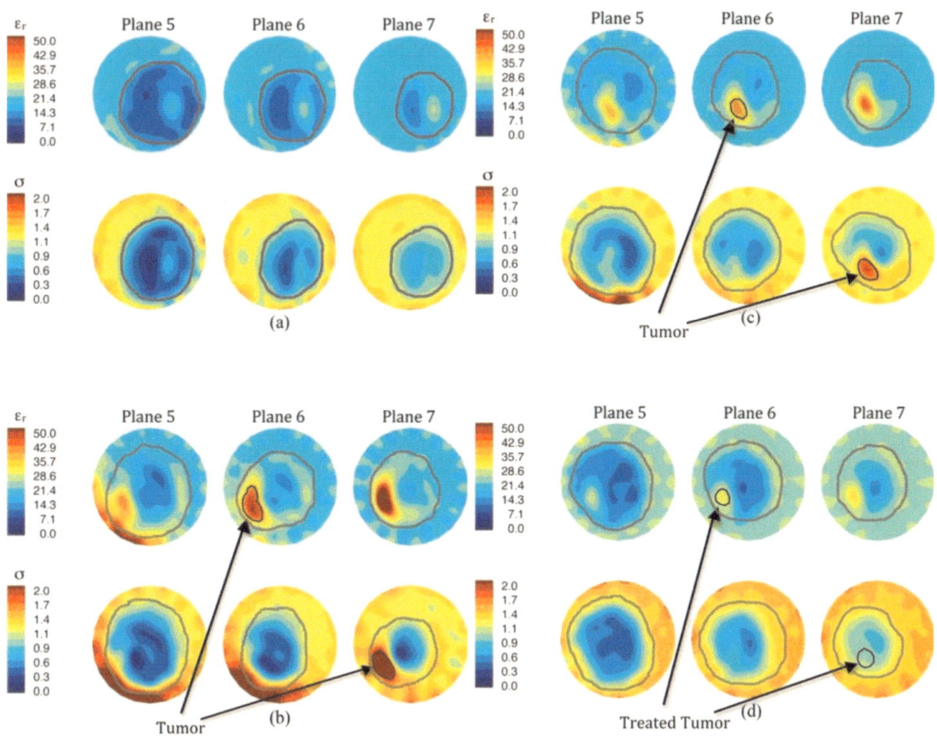

Figure 9.
1300 MHz microwave tomographic images. Imaging planes five to seven are shown corresponding to the three closest planes to the nipple with the permittivity on the top row and conductivity on the bottom row. (a) Left (contralateral) breast prior to treatment, (b) right (ipsilateral) breast prior to treatment, (c) right breast 44 days into treatment, and (d) right breast immediately prior to surgery (Reproduced from [17], with the permission of Springer Nature—License link: https://creativecommons.org/licenses/by/4.0/).

permittivity and conductivity which is consistent with adipose tissue, while there are also centralized zones of slightly elevated tissue corresponding to the fibroglandular tissue. For the diseased breast at the first two time points, there is a pronounced ring of highly elevated tissue properties—primarily to the lower left quadrant. This feature is consistent with the edema that would be present from the skin thickening observed in the MR images. For these same time points, the localized, elevated properties of the tumor are also visible, albeit slightly diminished at the 44 day mark. By the final set, both the microwave observable skin thickening and tumor have essentially disappeared in agreement with the MR images shown in **Figure 8b**. It is noteworthy that the tumor has shrunk considerably with respect to size and property intensity by the midway point suggesting that the microwave images are capable of recognizing early tumor response.

4. Discussion

One of the more interesting aspects of our imaging system revolves around the notion of why a 2D algorithm works in the first place. While there have been efforts by others exploring 2D imaging, most have reverted to 3D implementations when 2D efforts failed. Notably, Semenov et al. [61] concluded that the reason their 2D system did not work was because of the inherent mismatch between the actual 3D wave propagation and the assumed 2D propagation in their algorithm. However, we contend that there are scientifically valid reasons why our 2D algorithm works well. The most important is that by choosing the monopole antennas, we are able to position them on a quite small diameter circle surrounding the target. Given the vertical orientation of the antennas, in the limit of a decreasing array circle diameter, the aspect ratio of the antenna length to the diameter of the circle naturally increases. This can be theoretically extrapolated to the point where the domain closely mimics that of cylindrical geometry. For competing systems, the motivation has often centered about acquiring as much measurement data as possible. This often implies that the antenna array must be configured on a larger diameter circle for the simple reason relating to space limitations associated with the physical size and the increased number of radiating elements. In Semenov et al. [62], thousands of antennas (effective number based partially on mechanical motion) were positioned on a 60 cm diameter circle for a roughly 8 hour, 3D data acquisition, single frequency exam time, while the diameter for the Dartmouth system is only on a 15.2 cm diameter where the broadband data acquisition takes roughly 10 minutes. For the latter, the field patterns have simply not devolved into their spherical radiation behavior within the small zone while they are considerably further along that process for the larger diameter system.

In Meaney et al. [63], we compared the magnitude and phase values for the scattering from a simple phantom illuminated in our imaging chamber. It is noteworthy that there was very little difference between the phase calculations using both 2D and 3D models. There was some deviation for the magnitude, but the differences never exceeded 0.5 dB. It is worth examining the far field behavior of the 2D and 3D waves, in spite of the fact that it is a simple estimation in this situation. For the phase in the main beam, the phases propagate as a function of R (distance from the antenna) for both the 2D and 3D cases. This presumably accounts for the close match of the previous calculations. However, for the magnitude, they decay as a function of $1/R$ and $1/R^2$ for the 2D and 3D cases. Likewise, this likely partially accounts for the previously mentioned differences. As hypothesized above, these differences are likely mitigated because we operate within such a physically constrained space. As a final note, the lossy coupling bath most likely assists for this comparison by substantially attenuating (and almost eliminating) signals that might propagate out of the plane of interest only to be reflected back into it during the propagation process. Regardless, our 2D configuration in conjunction with our 2D algorithm has provided good images in a variety of settings from simulations, phantom experiments, animal experiments and even clinical trials.

5. Conclusions

The Dartmouth microwave imaging system is quite robust because of its synergism of hardware and numerical modeling expertise. Competing implementations of microwave imaging approaches generally ignore multipath signals. In fact, most numerical models preclude their existence by not incorporating structures which would excite surface wave propagation. The Dartmouth approach directly addresses this phenomenon which has led to a range of hardware and software innovations. Most notably, by being able to pack the antennas in such a small circle about the target, various factors have aligned to make 2D imaging a reality.

In addition, by incorporating fundamental concepts from parameter estimation theory, we have been able to implement a version of the image reconstruction process that essentially eliminates the a priori information requirement. While the log transform is most often used for purely real mathematical functions, there is precedent for its use with complex functions—notably for optical coherence tomography. The concept takes on an added level of complexity for microwave imaging because of the phase unwrapping criteria. By developing simple and effective ways to unwrap the phases, we now have an algorithm which does not require a priori information and does not converge to local minima or unwanted solutions. This type of robustness is essential when working in actual clinical situations.

Finally, the results bear out that microwave images can be diagnostically useful in the breast neoadjuvant chemotherapy monitoring setting. Initial pilot results suggest that the property changes observed as early as 30 days after treatment correlate well with final treatment response. Essentially the microwave properties provide a physiological assessment of the response which is known to occur earlier than measures based on morphological phenomena. While physiological models have been proposed that may explain the behavior observed in the imaging trials, more data is required to confirm this phenomenon. Regardless, these results suggest that microwave imaging may be a viable approach and may provide additional tools for clinicians in their fight to treat breast cancer.

Acknowledgements

This work was supported by NIH/NCI grant # R01-CA240760.

Conflict of interest

Drs. Meaney and Paulsen are co-inventors on several patents related to microwave imaging.

Author details

Paul M. Meaney* and Keith D. Paulsen
Dartmouth College, Hanover, USA

*Address all correspondence to: paul.meaney@dartmouth.edu

References

[1] Chaudhary SS, Mishra RK, Swarup A, Thomas JM. Dielectric properties of normal and malignant human breast tissues at radiowave and microwave frequencies. Indian Journal of Biochemistry & Biophysics. 1984;**21**: 76-79

[2] Woodard HQ, White DR. The composition of body tissues. British Journal of Radiology. 1986;**59**:1209-1219. DOI: 10.1259/0007-1285-59-708-1209

[3] Lazebnik M, Popovic D, McCartney L, Watkins CB, Lindstrom MJ, Harter J, et al. A large-scale study of the ultrawideband microwave dielectric properties of normal, benign and malignant breast tissues obtained from cancer surgeries. Physics in Medicine and Biology. 2007; **52**:6093-6115. DOI: 10.1088/0031-9155/ 52/20/002

[4] Sugitani T, Kubota SI, Kuroki SI, Sogo K, Arihiro K, Okada M, et al. Complex permittivities of breast tumor tissues obtained from cancer surgeries. Applied Physics Letters. 2014;**104**: 253702

[5] Martellosio A, Pasian M, Bozzi M, Perregrini L, Mazzanti A, Svelto F, et al. Dielectric properties characterization from 0.5 to 50 GHz of breast cancer tissues. IEEE Transactions on Microwave Theory and Techniques. 2017;**65**:998-1011. DOI: 10.1109/TMTT.2016.2631162

[6] Cheng Y, Fu M. Dielectric properties for non-invasive detection of normal, benign, and malignant breast tissues using microwave theories. Thoracic Cancer. 2018;**9**:459-465. DOI: 10.1111/ 1759-7714.12605

[7] Meaney PM, Gregory A, Epstein N, Paulsen KD. Microwave open-ended coaxial dielectric probe: Interpretation of the sensing volume re-visited. BMC Medical Physics. 2014;**14**:1756-6649

[8] Meaney PM, Gregory AP, Seppälä J, Lahtinen T. Open-ended coaxial dielectric probe effective penetration depth determination. IEEE Transactions on Microwave Theory and Techniques. 2016;**64**:915-923. DOI: 10.1109/ TMTT.2016.2519027

[9] Salah-Ud-Din S, Meaney PM, Porter E, O'Halloran M. Investigation of abscissa scales for dielectric measurements of biological tissues. Biomedical Physical & Engineering Express. 2017;**3**:015020

[10] Alanen E, Lahtinen T, Nuutinen J. Variational formulation of open-ended coaxial line in contact with layered biological medium. IEEE Transactions on Biomedical Engineering. 1998;**45**: 1241-1248. DOI: 10.1109/10.720202

[11] Gregory AP, Clarke RN, Hodgetts TE, Symm GT. RF and dielectric measurements upon layered materials using coaxial sensors. National Physical Laboratory, Report MAT 13. 2008

[12] Hagl DM, Popovic D, Hagness SC, Booske JH, Okoniewski M. Sensing volume of open-ended coaxial probes for dielectric characterization of breast tissue at microwave frequencies. IEEE Transactions on Microwave Theory and Techniques. 2003;**51**: 1194-1206. DOI: 10.1109/ TMTT.2003.809626

[13] Cole KS, Cole RH. Dispersion and absorption in dielectrics: I. alternating current characteristics. The Journal of Chemical Physics. 1941;**9**:341-351. DOI: 10.1063/1.1750906

[14] Foster KR, Schepps JL, Schwan HP. Microwave dielectric relaxation in muscle: A second look. Biophysical Journal. 1980;**29**:271-281. DOI: 10.1016/S0006-3495(80)85131-9

[15] Preece AW, Craddock IJ, Shere M, Jones L, Winton HL. MARIA M4: Clinical evaluation of a prototype ultrawideband radar scanner for breast cancer detection. J Med Imag. 2016;**3**: 033502

[16] Poplack SP, Paulsen KD, Hartov A, Meaney PM, Pogue B, Tosteson T, et al. Electromagnetic breast imaging: Pilot results in women with abnormal mammography. Radiology. 2007;**243**: 350-359. DOI: 10.1148/radiol.2432060286

[17] Meaney PM, Kaufman PA, Muffly LS, Click M, Wells WA, Schwartz GN, et al. Microwave imaging for neoadjuvant chemotherapy monitoring: Initial clinical experience. Breast Cancer Research. 2013;**15**:35. DOI: 10.1186/bcr3418

[18] Schepps JL, Foster KR. The UHF and microwave dielectric properties of normal and tumour tissues: Variation in dielectric properties with tissue water content. Physics in Medicine and Biology. 1980;**25**:1149-1159. DOI: 10.1088/0031-9155/25/6/012

[19] Hagness SC, Taflove A, Bridges JE. Two-dimensional FDTD analysis of a pulsed microwave confocal system for breast cancer detection: Fixed-focus and antenna-array sensors. IEEE Transactions on Biomedical Engineering. 1998;**45**:1470-1479. DOI: 10.1109/10.730440

[20] Fear EC, Sill J, Stuchly MA. Experimental feasibility study of confocal microwave imaging for breast tumor detection. IEEE Transactions on Microwave Theory and Techniques. 2003;**51**:887-892. DOI: 10.1109/TMTT.2003.808630

[21] Klemm M, Craddock IJ, Leendertz JA, Preece A, Benjamin R. Radar-based breast cancer detection using a hemispherical antenna array—Experimental results. IEEE Transactions on Antennas and Propagation. 2009;**57**: 1692-1704. DOI: 10.1109/TAP.2009.2019856

[22] Paulsen KD, Meaney PM, Gilman L. Alternative Breast Imaging: Four Model-Based Approaches. Boston, MA: Springer Publishers; 2005

[23] Rocca P, Benedetti M, Donelli M, Franceschini D, Massa A. Evolutionary optimization as applied to inverse scattering problems. Inverse Problems. 2009;**25**:123003. DOI: 10.1088/0266-5611/25/12/123003

[24] Fhager A, Persson M. Using a priori data to improve the reconstruction of small objects in microwave tomography. IEEE Transactions on Microwave Theory and Techniques. 2007;**55**:2454-2462. DOI: 10.1109/TMTT.2007.908670

[25] Catapano I, Di Donato L, Crocco L, Bucci OM, Morabito AF, Isernia T, et al. On quantitative microwave tomography of female breast. Progress In Electromagnetics Research. 2009;**97**: 75-93. DOI: 10.2528/PIER09080604

[26] Shea JD, Kosmas P, Hagness SC, Van Veen BD. Three-dimensional microwave imaging of realistic numerical breast phantoms via a multi-frequency inverse scattering technique. Medical Physics. 2010;**37**:4210-4226

[27] Semenov SY, Bulyshev AE, Abubakar A, Posukh VG, Sizov YE, Souvorov AE, et al. Microwave-tomographic imaging of high contrast

objects using different reconstruction approaches. IEEE Transactions on Microwave Theory and Techniques. 2005;**53**:2284-2294. DOI: 10.1109/TMTT.2005.850459

[28] Poplack SP, Paulsen KD, Hartov A, Meaney PM, Pogue B, Tosteson T, et al. Electromagnetic breast imaging—Average tissue property values in women with negative clinical findings. Radiology. 2004;**231**:571-580. DOI: 10.1148/radiol.2312030606

[29] Meaney PM, Fanning MW, Raynolds T, Fox CJ, Fang Q, Kogel CA, et al. Initial clinical experience with microwave breast imaging in women with normal mammography. Academic Radiology. 2007;**14**:207-218. DOI: 10.1016/j.acra.2006.10.016

[30] Kruger RA, Reinecke DR, Kruger GA. Thermoacoustic computed tomography—Technical considerations. Medical Physics. 1999;**26**:1832-1837. DOI: 10.1118/1.598688

[31] Amineh RK, Khalatpour A, Nikolova NK. Three-dimensional microwave holographic imaging using co- and cross-polarized data. IEEE Transactions on Antennas and Propagation. 2012;**60**:3526-3531. DOI: 10.1109/TAP.2012.2196932

[32] Joachimowicz N, Pichot C, Hugonin JP. Inverse scattering: Iterative numerical method for electromagnetic imaging. IEEE Transactions on Antennas and Propagation. 1991;**39**:1742-1753. DOI: 10.1109/8.121595

[33] Meaney PM, Schubitidze F, Fanning MW, Kmiec M, Epstein N, Paulsen KD. Surface wave multi-path signals in near-field microwave imaging. International Journal of Biomedical Imaging. 2012;**2012**:697253

[34] Meaney PM, Fang Q, Rubaek T, Demidenko E, Paulsen KD. Log transformation benefits parameter estimation in microwave tomographic imaging. Medical Physics. 2007;**34**:2014-2023. DOI: 10.1118/1.2737264

[35] Woods AJ, Rourke T. Ghosting in anaglyphic stereoscopic images. Proc SPIE 5291, Stereoscopic Displays and Virtual Reality Systems XI. 2004;354-365. DOI: 10.1117/12.537424

[36] Talbi L, Delisle GY. Experimental characterization of EHF multipath indoor radio channels. IEEE Journal on Selected Areas in Communications. 1996;**14**:431-440. DOI: 10.1109/49.490228

[37] Gilmore C, Mojabi P, LoVetri J. Comparison of an enhanced distorted Born iterative method and multiplicative-regularized contrast source inversion method. IEEE Transactions on Antennas and Propagation. 2009;**57**:2341-2351. DOI: 10.1109/TAP.2009.2024478

[38] Sun C, Zhao H, Feng W, Du S. A frequency-domain multipath parameter estimation and mitigation method for BOC-modulated GNSS signals. Sensors. 2018;**18**:721. DOI: 10.3390/s18030721

[39] Meaney PM, Fox CJ, Geimer SD, Paulsen KD. Electrical characterization of glycerin: Water mixtures and the implications for use as a coupling medium in microwave tomography. IEEE Transactions on Microwave Theory and Techniques. 2017;**65**:1471-1478. DOI: 10.1109/TMTT.2016.2638423

[40] Meaney PM, Paulsen KD. Addressing multipath signal corruption in microwave tomography and the influence on system design and algorithm development. Open Access Journal of Biomedical Engineering and Biosciences. 2018;**1**:1-7

[41] Fox CJ, Meaney PM, Shubitidze F, Potwin L, Paulsen KD. Characterization of a monopole antenna in a lossy medium for microwave breast computed tomography. In: Ibrahim TS, Crozier S, Fear E, editors. New Electromagnetic Methods and Applications of Antennas in Biomedicine. 2008. p. 5

[42] Meaney PM, Geimer SD, Paulsen KD. Mutual coupling in a tomographic imaging system. Eur Conf Antenn Propag, Berlin, DE. 2009; 2948-2949.

[43] Grzegorczyk TM, Meaney PM, Kaufman PA, diFlorio-Alexander RM, Paulsen KD. Fast 3-D tomographic microwave imaging for breast cancer detection. IEEE Transactions on Medical Imaging. 2012;**31**:1584-1592. DOI: 10.1109/TMI.2012.2197218

[44] Box GEP, Cox DR. An analysis of transformations. Journal of the Royal Statistical Society B. 1963;**35**: 211-252. DOI: 10.1111/j.2517-6161.1964.tb00553.x

[45] Curry TS III, Dowdey JE. Christensen's Physics of Diagnostic Radiology. 4th ed. Philadelphia, PA: Lippincott, Williams and Wilkins; 1990

[46] Pogue BW, Poplack SP, McBride TO, Wells WA, Osterman KS, Osterberg UL, et al. Quantitative hemoglobin tomography with diffuse near-infrared spectroscopy: Pilot results in the breast. Radiology. 2001;**218**: 261-266. DOI: 10.1148/radiology.218.1.r01ja51261

[47] Beck JV, Arnold KJ. Parameter Estimation in Engineering and Science. New York: Wiley; 1977

[48] Arridge SR. Optical tomography in medical imaging. Inverse Problems. 1999;**15**:R41-R93

[49] Meaney PM, Paulsen KD, Pogue BW, Miga MI. Microwave image reconstruction utilizing log-magnitude and unwrapped phase to improve high-contrast object recovery. IEEE Transactions on Medical Imaging. 2001; **20**:104-116. DOI: 10.1109/42.913177

[50] Meaney PM, Geimer SD, Paulsen KD. Two-step inversion in microwave imaging with a logarithmic transformation. Medical Physics. 2017;**44**:4239-4251. DOI: 10.1002/mp.12384

[51] Chew WC, Lin JH. A frequency-hopping approach for microwave imaging of large inhomogeneous bodies. IEEE Microwave Guided Wave Letters. 1995;**5**:439-441. DOI: 10.1109/75.481854

[52] Fang Q, Meaney PM, Paulsen KD. Singular value analysis of the Jacobian matrix in microwave image reconstructions. IEEE Transactions on Antennas and Propagation. 2006;**54**: 2371-2380. DOI: 10.1109/TAP.2006.879192

[53] Meaney PM, Hartov A, Bulumulla S, Raynolds T, Davis C, Schoenberger F, et al. 4-channel, vector network analyzer microwave imaging prototype based on software defined radio technology. The Review of Scientific Instruments. 2019; **90**:044708. DOI: 10.1063/1.5083842

[54] Li D, Meaney PM, Raynolds T, Pendergrass SA, Fanning MW, Paulsen KD. A parallel-detection microwave spectroscopy system for breast imaging. The Review of Scientific Instruments. 2004;**75**:2305-2313. DOI: 10.1063/1.1764609

[55] Gockenbach MS. Inverse problems and Tikhonov regularization. In: Carus Mathematical Monographs. Providence, RI: American Mathematical Society; 2016

[56] Marquardt D. An algorithm for least-squares estimation of nonlinear parameters. SIAM Journal on Applied Mathematics. 1963;**11**:431-441. DOI: 10.1137/0111030

[57] Wang Y, Li D, Haacke EM, Brown JJ. A three-point Dixon method for water and fat separation using 2D and 3D gradient-echo techniques. Journal of Magnetic Resonance Imaging. 1998;**8**: 703-710. DOI: 10.1002/jmri.1880080329

[58] Meaney PM, Paulsen KD, Ryan TP. Two-dimensional hybrid element image reconstruction for TM illumination. IEEE Transactions on Antennas and Propagation. 1995;**43**:239-247

[59] Hosseinzadegan S, Fhager A, Persson M, Meaney PM. A discrete dipole approximation solver based on the COCG-FFT algorithm and its application to microwave breast imaging. International Journal of Antenna Propagation. 2019;**2019**: 9014969. DOI: 10.1155/2019/9014969

[60] Hosseinzadegan S, Fhager A, Persson M, Geimer SD, Meaney PM. Discrete dipole approximation-based microwave tomography for fast breast cancer imaging. IEEE Transactions on Microwave Theory and Techniques. 2021;**69**:2741-2752. DOI: 10.1109/TMTT.2021.3060597

[61] Semenov SY, Svenson RH, Boulyshev AE, Souvorov AE, Borisov VY, Sizov Y, et al. Microwave tomography: Two-dimensional system for biological imaging. IEEE Transactions on Biomedical Engineering. 1996;**43**:869-877. DOI: 10.1109/10.532121

[62] Semenov SY, Svenson RH, Boulyshev AE, Souvorov AE, Nazarov AG, Sizov Y, et al. Three-dimensional microwave tomography: Experimental prototype of the system and vector Born reconstruction method. IEEE Transactions on Biomedical Engineering. 1999;**46**:937-946. DOI: 10.1109/10.775403

[63] Meaney PM, Grzegorczyk T, Paulsen KD, Jeon SI. Log transformation with Gauss-Newton microwave image reconstruction reduces incidence of local minima convergence. IEEE Int Symp Antenn Propag, Charlestown, SC. 2009

Section 5

Industry Applications

Chapter 13

Additive Manufacturing of RF Waveguide Components

Mauro Lumia, Giuseppe Addamo, Oscar Antonio Peverini, Flaviana Calignano, Giuseppe Virone and Diego Manfredi

Abstract

The exponential growth of publications, in the last years, on the use of additive manufacturing (AM) technologies in the microwave field proves the increasing interest of research institutions and industries in these techniques. Some advantages of AM with respect to conventional machining are weight reduction, design flexibility, and integration of different functionalities (electromagnetic, thermal, and structural) in a single part. This chapter presents the most employed AM technologies for the manufacturing of RF waveguide components. First, an overview of the AM processes is discussed with particular care on material properties and post-processing. Then, an extensive survey on microwave-guided components fabricated by AM processes published in literature is shown.

Keywords: additive manufacturing, 3D printing, waveguide components, microwave components, SLM, SLA, FDM

1. Introduction

According to the American Society of Testing and Materials (ASTM), additive manufacturing (AM) is defined as "the process of joining materials to make objects from 3D model data, usually layer upon layer, as opposed to subtractive manufacturing methodologies." Often this term is substituted by 3D Printing (3D Printing is typically associated with people printing at home or in the community; additive manufacturing is typically associated with production technologies and supply chains, but they both produce parts by the addition of layers). AM technologies can be classified into seven categories, namely, binder jetting, material jetting, direct energy deposition, sheet laminations, material extrusion, powder bed fusion, and vat photopolymerization. Each category includes several processes that share the same principle used for layer modeling and different materials that can be processed (**Table 1**).

AM has been first applied for rapid prototyping of visualization models and tooling. Recently, the improvement in the process's accuracy and material properties of the manufactured objects have expanded the field of applications. Indeed, AM is currently used to manufacture personalized prostheses, replacement organs, and implants in the medical sector and produce complex lightweight components for the aerospace, automotive, and sports industries. Recently, AM has been applied by

Process	Technologies	Materials
Binder jetting	Ink-jetting 3D printing	Metal Polymer Ceramic
Material jetting	Polyjet Ink-jetting	Photopolymer Wax
Direct energy deposition	Direct Metal Deposition Electron Beam Direct Melting	Metal
Sheet laminations	Laminated Object Manufacture Ultrasonic Consolidation	Metal Polymer Ceramic Paper
Material extrusion	Fused Deposition Modeling	Polymer
Powder bed fusion	Selective Laser Sintering Selective Laser Melting Electron Beam Melting	Metal Polymer Ceramic
Vat-photopolymerization	Stereolithography Digital Light Processing	Photopolymer Ceramic

Table 1.
Seven categories of AM technologies [1].

RF industries for the development of next generations of microwave and millimeter-wave components for sensors, imaging systems, and satellite communication (SATCOM) [1].

A generic AM process starts with a model generated using a three-dimensional Computer-Aided Design (3D CAD) system. Then the model is converted into the STL file format that approximates the 3D model with a mesh of triangles. Then, this file is transferred to the AM machine to set the process parameters. Such settings are defined according to the geometry of the model (e.g., position and orientation of the components, design of support structures) and to the building process (e.g., energy source, material constraints, and layer thickness). At the end of the printing process, the part is removed from the building platform and prepared for the post-processing operations, for example, cleaning, sandblasting or shot-peening, thermal treatments, and plating [2].

The main advantage of AM process is manufacturing lightweight components with complex internal surfaces in a single part. Moreover, by eliminating tools, the design flexibility is increased. On the other hand, the main concerns are the manufacturing accuracy and the surface roughness that are worse than standard manufacturing processes and strongly depend on the material and process parameters.

In the microwave area, Selective Laser Melting (SLM)[1], Stereolithography (SLA), and Fused Deposition Modeling (FDM™) are the most investigated technologies. The application of AM processes in manufacturing microwave and millimeter-wave components strictly depends on the accuracy, cost, and performance requirements. From this point of view, basic knowledge of the characteristics of a single process is necessary. This overview is reported in Sections 2–4. Section 5 is, instead, devoted to giving a survey of the principal results (in terms of realized components) actually achieved in the specialized literature. In this summary, all the important aspects for microwave

[1] The process Selective Laser Melting is, nowadays, sometimes called Laser Powder Bed Fusion (LPBF).

engineering are reported, i.e., operative band, measured results versus the expected ones, and an explanation of this difference. This section is split into subsections for reader convenience according to the component category.

2. Selective laser melting

SLM is an AM process that allows the manufacturing of all metal parts. This technology can easily realize complex geometries with interior features and channels. Regarding **Figure 1**, the manufacturing process starts with a thin layer of metal powder spread by a recoater along with the building platform. Then, a high-energy laser beam selectively fuses the deposited powder layer. The laser follows the contour defined in the STL file. Once a layer is completed, the building platform is lowered, the new powder is spread, and the laser melts this new layer. The process is repeated until the parts are completely manufactured. Due to the high temperatures necessary for melting, the process takes place in a protected atmosphere, normally argon, to prevent oxidation of the parts [1]. At the end of the process, the excess powder is removed; the parts still attached to the building platform undergo a stress-relieving job in an oven. This thermal treatment is necessary to reduce deformations of the parts caused by the high thermal stresses arising during the manufacturing process. Finally, the components are detached from the building platform and eventually subjected to surface finish treatment as polishing and shot peening [3].

Metal powder properties are important in the final quality and cost of the part built via SLM. The main properties influencing the process can be subdivided into three categories, which are as follows: [4]:

- geometrical (size and shape);
- metallurgical (microstructure, composition);
- mechanical/physical (flowability, absorption of light).

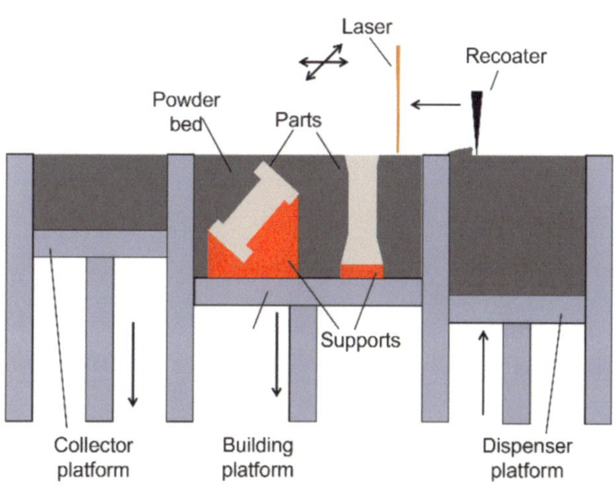

Figure 1.
Selective laser melting process.

As far as the first point is concerned, fine particles enable high-density parts with good surface quality, while the spherical shape improves flowability and, hence, mechanical properties [5]. Irregular powder particles can lead to poor surface finish, low density, and increased defects [6].

Extremely important steps in the SLM process are the orientation of the part in the building platform and the design of the supporting structures. Supports have mainly three purposes, which are as follows:

- to fix the part to the building platform;
- to conduct excess heat away;
- to prevent deformation or collapse of the part.

Typical drawbacks are as follows:

- an increment of the building and the post-processing time;
- their removal is not, in general, an easy task, and the risk of damage is quite high.

A possible solution to reduce the number of supports consists of choosing an optimal building orientation. It is worth noticing that the staircase effect has to be also well-considered for specific applications. The generation of the staircase effect is described in **Figure 2**. The STL file format is a triangular approximation of the nominal CAD. If the layer thickness is too high or the inclination angle is too small, the staircase effect becomes more remarkable. On the other hand, overhanging surfaces is another important aspect to consider. These surfaces are areas not supported by solidified material during the building process. The heat-conduction rate of powder-supported zones is lower than the solid-supported zones, while the absorbed energy is higher. The melt pool created by the laser becomes too large and sinks into the powder. Therefore,

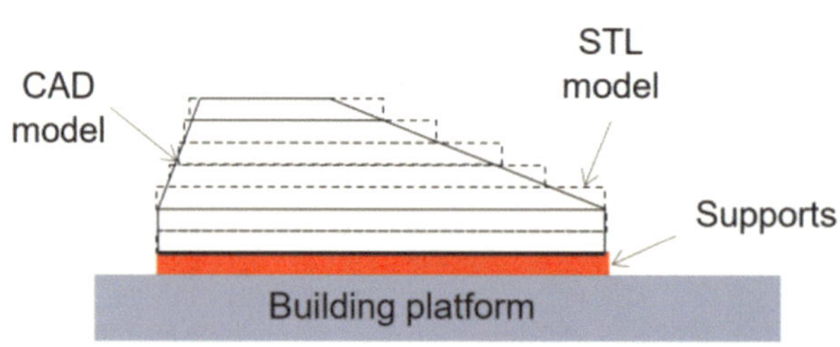

Figure 2.
Generation of the staircase effect.

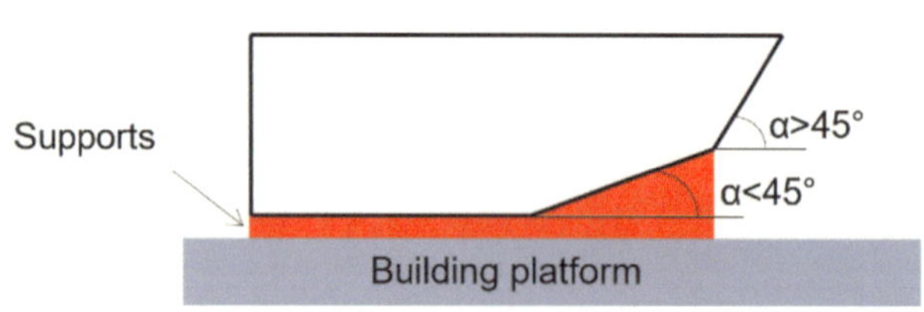

Figure 3.
Self-supporting angles.

deformation occurs if these surfaces are not supported. Supporting structures are usually built with a low density during the manufacturing of the part, and they must be manually removed at the end of the manufacturing process. A clever solution is represented by self-supporting angles (**Figure 3**). Based on experimental results, downward sloping faces with angles $\alpha > 45°$ are self-supporting. At the same time, staircase effects can be reduced by increasing sloping angles. Moreover, in this way, the value of surface roughness decreases. On the contrary, angles lower than 30° should quickly be avoided since the staircase effect increases [7].

Materials commonly used in the SLM process are aluminum alloys, titanium alloys, stainless steel, Ni-based alloys, and cobalt-chromium alloys [1]. From an RF point of view, the most interesting ones are the aluminum ones as the AlSi10Mg alloy. This material exhibits high electrical conductivity, low-specific weight, high corrosion resistance, and good mechanical properties. The typical achievable accuracy guaranteed for this aluminum alloy is in the order of ±0.1 mm [3]. However, better manufacturing accuracy has been observed in literature for components designed with an AM-oriented approach.

3. Stereolithography

Stereolithography (SLA) was developed in 1984 by Charles Hull and was the first available commercial AM process. SLA is a vat photopolymerization process based on the solidification of a liquid resin using a UV laser. Since the process takes place in a liquid, support structures are necessary during the building phase. These are made by the same material of the parts and are specified in the machine parameter settings [1].

Concerning **Figure 4**, the manufacturing process starts with the building platform lowered from the top of the vat by a layer thickness. Then, a recoater blade smooths the surface of the vat, and a UV laser cures the material. Then, the platform is lowered by a layer thickness, and the process is repeated until the part is completed. At the end of the process, the platform is lifted, the part is drained and removed from the platform [8]. Then, it is placed in a UV oven to complete the curing. During the solidification process,

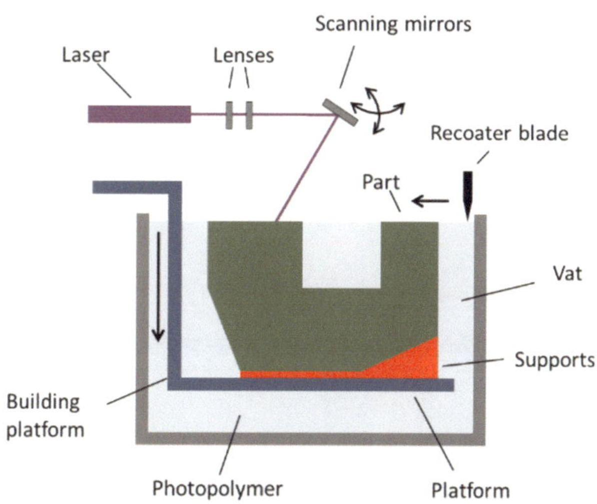

Figure 4.
Stereolithography process.

usually, the photopolymer shrinks. This shrinkage induces compression stresses on the previous layers that may cause curling and distortion. These effects can be reduced by adopting clever scan strategies, such as Star Weave or ACES [1].

Acrylates resins were the first photopolymers developed. These resins had high reactivity but produced inaccurate parts due to a significant shrinkage (5–20%) and a tendency to warp and curl. They are low viscosity resins used for visual or anatomic models, with a low accuracy but high-speed manufacturing. On the contrary, epoxy resins present high viscosity and are used for functional parts. They have slow photo speed but allow more accurate, harder, and stronger parts than the acrylate ones.

Furthermore, they exhibit low levels of shrinkage (1–2%), reducing the risk of warp and curl. Most of the commercially available resins are epoxides with acrylate content to combine the advantages of both materials [2]. Moreover, it is possible to suspend ceramic particles in a resin to obtain a ceramic–polymer composite material and improve mechanical and thermal properties.

SLA allows the manufacturing of parts with good accuracy ranges from 25 to 50 μm and smooth surfaces; the typical average surface roughness Ra is lower than 10 μm. However, the metal plating of the internal channels can be critical in terms of adhesion, uniformity of the metal coating, and long-term stability [9], limiting the applicability of SLA in the manufacturing of some RF applications (e.g., waveguide components for space).

4. Fused deposition modeling

FDM, also known as Fused Filament Fabrication (FFF), is a material extrusion process in which material is heated and then dispensed through a nozzle layer by layer. The process was invented and patented by Scott Trump in 1989, who then founded Stratasys Inc. The basic process is based on a robot arm that moves two nozzles where a filament of polymeric material is fused and then deposited layer by layer on a platform (**Figure 5**). The filament is supplied by an unrolled spool and

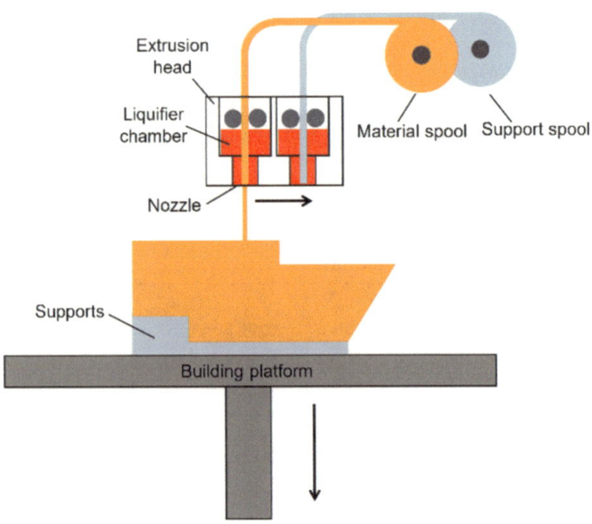

Figure 5.
FDM process.

pressed into the extrusion head. The material is then heated utilizing electrical resistance [1]. Heat is conducted to the liquefier chamber to obtain a liquid state. The material inside the chamber should be maintained in a molten state, with particular care on the temperature since some polymers degrade at high temperatures and could burn [2]. The filament is softened, the molten material gets off through a nozzle of reduced diameter and deposits on the building platform [1].

The nozzle diameter determines both the shape and size of the extruded filament and the minimum feature size that can be printed—the larger the nozzle diameter, the faster the process, but with lower precision. Material extrusion is controlled by the pressure difference between the chamber and the atmosphere. If the pressure is maintained constantly, the material flows at a constant rate with a constant cross section. The same happens if the nozzle speed is kept constant. Any change in the direction of the extrusion head must result in a change in the corresponding material flow rate. Otherwise, a different amount of material will be deposited in that region. Once the material is extruded, gravity and surface tension can cause a change in the shape, while the cooling and drying effects can change the size of the material deposited. This effect can be reduced by minimizing the differential temperature between the chamber and the atmosphere. Bonding with the previous layer and the adjacent region can be ensured by residual heat energy or solvents and wetting agents in the extruded filament [2].

FDM machines can be equipped with one to three nozzles. In this case, two nozzles deposit two different building materials, while the third one is filled with a soluble material for supporting structures. The former case is employed for a low-cost solution for large industrial machines. In this way, the support material is not removed manually but is dissolved in a chemical bath after the fabrication process.

Many thermoplastic materials are available for FDM and FFF processes; the most commonly used for RF applications are listed below [1].

- Acrylonitrile Butadiene Styrene (ABS) is a generic thermoplastic material used in injection molding processes. The main properties are hardness, strength, and heat resistance. Companies developed different kinds of ABS, each with specific properties, and many colors are available for FDM printers. ABS exhibits good adhesion so that it can be used with high printing speed.

- PolyActic Acid (PLA) is a common biodegradable plastic obtained from corn or sugar cane. It is used for shopping bags and packaging, but it can also be used for biocompatible medical implants. PLA is one of the materials most used for low-cost printers because of its environmental sustainability. However, it is not ideal for high-temperature environments or outdoor applications. Like ABS, different colors and different types of PLA are available. ABS exhibits good adhesion so that it can be used with high printing speed

- Polycarbonate (PC) is a polymer with good temperature resistance and high impact resistance. However, PC is very hygroscopic and absorbs moisture from the air. Moreover, it must be used with a heated bed to avoid warping problems. PC is available in black or transparent colors.

- Polyamide (PA) is commonly known as Nylon. PA is the most common nylon material used in FDM printers because it has good strength, a low-friction coefficient, and can take up vibrations or impacts. However, it tends to warp more than ABS and PLA. For this reason, the use of a heated building volume is recommended.

- ULTEM is a thermoplastic material developed by Saudic and commercialized by Stratasys. It has high heat and chemical resistance, good strength, flexibility, and impact resistance. It is used for aerospace and medical parts.

- PEEK is an organic thermoplastic material developed by Victrex. It has high heat and chemical resistance and good strength. It is a biocompatible material used for manufacturing medical, aerospace, and automotive parts.

FDM allows the manufacturing of multi-material parts. The use of a soluble support material offers good design flexibility. The main concern with this technology is that the nozzle dimensions limit the accuracy. The typical resolution of the process is about 100–200 μm. As in the case of SLA, a metal plating of the internal channels is mandatory for waveguide RF application.

5. RF applications

The manufacturing of RF components by AM process is particularly challenging for different reasons. The internal channels of the parts have to be designed considering the typical AM post-processing operations, for example, powder removal in SLM or metal plating in SLA and FDM. Moreover, due to the typical accuracy of the AM processes, an electromagnetic robust design approach is mandatory in the case of high RF-performance components.

The simplest strategy consists of splitting the component into two or more parts. This way is usually employed in the case of SLA and FDM to ease electroless metal plating processes. Sometimes this is also used in SLM products, particularly for complex internal structures. The main drawback is misaligning the different parts with consequent leakage effect and, subsequently, higher losses. Moreover, the intrinsic features of AM processes are not employed, making their employment, therefore, less effective and convenient.

The designed part should be realized monolithically to take full advantage of the AM process. To achieve this goal, the electromagnetic design has to be adapted to the process in a so-called "*AM-oriented design*." The most important is aligning the propagation axis with the building direction, avoiding, at the same time, the presence of overhanging structure in the internal channel (for instance, stubs or corrugations). These criteria, then, force a complete reconsideration of well-known RF structures. One of the main advantages is better manufacturing accuracy and symmetry of the cross section (essential characteristic for dual-polarization components).

The following subsections summarize the most common 3D-printed RF components, namely—waveguides, filters, horn antennas, dual-polarization components (OMTs and septum polarizers). Finally, the last subsection describes some examples of the integration of different RF functionalities into a single component.

5.1 Waveguides

Waveguide lines are the simplest candidates for evaluating the best achievable accuracy of the AM technologies for microwave components. An important parameter is the realized metal loss per wavelength/cm. Different factors influence these values:

- the material conductivity;
- the surface roughness;
- the dimensional accuracy;

Deep research has been done considering different AM technologies and a frequency range that spans from Ku- to E- band (10–170 GHz). A summary of these works with relevant results is reported in **Table 2**. As observed, SLM and SLA are the most commonly exploited technologies since they ensure good accuracy and low roughness compared to other AM processes.

Going more into detail, an interesting example of a waveguide manufactured by FDM is reported in ref. [10]. Firstly, a WR42 waveguide has been printed using ABS. Then, the same printer was customized to deposit a low-cost conductive silver ink. The manufactured waveguide exhibits a measured loss of 0.11 dB/cm for the entire K-band.

As far as the SLA is concerned, mainly the W- and D-band (75–170 GHz) have been considered. In Refs. [11–13], the copper plating has been applied, showing a W-band loss mean value of 0.06 dB/cm. An interesting comparison between commercial and SLA waveguides is reported in [12], showing that SLA components present a measured loss that is almost double of the commercial ones: 0.06 dB/cm versus 0.03 dB/cm and 0.26 dB/cm and 0.15 dB/cm in W- and D-band, respectively.

As far as the SLM is concerned, the main test refers to the frequency range from 18 to 110 GHz. Aluminum is the material most used. However, few examples have been carried out by using copper and nickel alloys [11]. An interesting comparison is presented in Ref. [11], where the authors compare the losses of a commercial WR42 waveguide with SLM (in aluminum) and a CNC machined realization. As it could be expected, the lowest attenuation occurs in the commercial waveguide (0.004 dB/cm), while the SLM and CNC parts exhibit a loss of about 0.02 dB/cm and 0.03 dB/cm, respectively. Still considering SLM, in Ref. [15], a study on the circular waveguide,

Ref.	Frequency (GHz)	Waveguide	Technology	Material	Loss (dB/cm)
[10]	18–26	WR42	FDM	Silver plated	0.11
[14]	18–26	WR42	SLM	AlSi10Mg	0.02
[15]	26–38	Circular	SLM	AlSi10Mg	0.01
[11]	75–110	WR10	SLM	GRCOP-84 Inconnel 625 AlSi10Mg	0.141 0.369 0.103
[12]	75–110	WR10	SLA	Copper plasted	0.055
[12]	75–110	WR10	SLA (DLP)	Copper plated	0.063
[13]	75–110	WR10	SLA	Copper plated	0.06
[12]	120–170	WR6	SLA (DLP)	Copper plated	0.26

Table 2.
Losses of AM waveguides.

operating in Ka-band, is carried out. The measured aluminum prototypes exhibit a loss of 0.01 dB/cm. The co-polar and cross-polar transmission coefficients have been measured to understand the SLM's feasibility for dual-polarization systems, thanks to the double symmetry shape. The measurements show a spurious cross-polarization term of −25 dB/−40 dB.

A comparison, in W-band, between parts SLM realized with different alloys (aluminum, copper, and nickel alloys) is reported in Ref. [11], showing the better behavior of the aluminum prototype.

5.2 Waveguide filters

Filters are one of the most demanding elements from a manufacturing point of view. The high-standing waves developing inside the components lead to high sensitivity to mechanical tolerances and high susceptibility to multifactor discharge and passive intermodulation products (PIM). For this reason, they are an important benchmark for the AM manufacturing of passive waveguide components [3].

A detailed review of 3D-printed microwave filters has been recently reported [16]. The frequency range considered in literature is mainly 7–22 GHz, although a few examples at higher frequency bands (up to 110 GHz) have also been discussed. Metal, plastic, and ceramic materials have been employed in published works.

FDM has been used in Ref. [17] to manufacture two X-band filters with PLA filament. Each component has been manufactured in two halves with an E-plane cut. The parts are coated with a copper spray to create a conductive surface for the galvanization process with further copper. Despite the accuracy of the printing and painting process, good results have been obtained for both filters. In Ref. [18], a two-pole X-band filter in the WR90 waveguide insert has been manufactured using PC filament. The printed insert has been coated with a nickel spray and then electroplated with copper.

As far as the SLA process is concerned, in Ref. [19], a four-pole quasi-elliptic filter, working at 810 MHz ± 16 MHz, is presented. The filter is composed of two mushroom-shaped resonators. The filter has been realized in two blocks, metalized with silver painting and then electroplated with copper. The same manufacturing approach has been reported in Ref. [20] for a two-pole filter.

Considering SLM realizations, an example of a W-band filter is presented in Ref. [21]. The filter is based on five rectangular resonators coupled using inductive irises. Two stainless steel prototypes have been manufactured, one coated with 5 μm of copper showing different measured electrical conductivity: 1.25 × 107 S/m and 5.96 × 107 S/m, respectively. As far as, the scattering parameters are concerned, the stainless steel filter shows a center frequency shift down by 1.66 GHz and a minimum return loss of 24.41 dB in the passband, while the copper-plated presents a frequency shift of 0.9 GHz with a return loss is 26.56 dB.

Some examples of monolithic realization and comparison between materials and processes are reported in Refs. [9, 22]. In Ref. [9], a comparison for Ku/K filter prototypes realized both in SLM (in aluminum, titanium, and maraging steel alloys) and SLA (copper plated) is shown. The study proves that an AM-oriented architecture provides excellent results in both roughness and mechanical accuracy; at the same time, the metallization of the SLA component is extremely complicated in the inner surfaces of the filter. A similar comparison is reported in Ref. [22], where two prototypes, operating in X-band, have been manufactured—one by SLM with an aluminum-copper alloy, the other by SLA, and then copper electroplated. The most interesting contribution of this paper was the excellent performance, even for the SLA filter.

Ceramic materials have also been used for the 3D printing of waveguide filters with two different approaches. The first method consists of manufacturing ceramic components, subsequently, metal plating. For example, in Ref. [23], a ceramic-filled resin has manufactured hemispherical resonator waveguide filters. Four Ka-band filters have been manufactured, metal plated, and measured with good RF results. The electroless plating process consists of a deposition of palladium, nickel, and copper, followed by a passivation layer of silver. In the second approach, the materials' dielectric properties are exploited to reduce the overall envelope.

For instance, Ref. [24] manufactured a dielectric perturber for a third-order filter with a bandpass shifting from 10 to 12 GHz. The filter was manufactured using copper with a standard manufacturing process, while the perturber was manufactured by SLA using zirconia (εr = 32, tan δ = 0.002). In Ref. [25], the authors used alumina (εr = 9.1, tan δ = 0.0001) to manufacture the dielectric resonators in a sixth-order quasi-elliptic bandpass filter. The housing has been 3D printing in two blocks using plastic material and then metal plated. An important aspect of this work is that no supports or glue are needed. The dielectric resonators are placed in the middle of their cavities to maximize their Q-factor.

5.3 Horn antennas

Waveguide horns are commonly used in high-performance antenna feed chains in SATCOM applications. This class of antennas, indeed, can provide excellent performances in terms of gain, bandwidth, return loss, and cross-polarization [26]. Therefore, much effort has been made to study the 3D printing of waveguide horns. Most of the works are from X-band to K-band (7–27 GHz), although some examples can be found up to 300 GHz [27].

As far as the SLA process is concerned, three works are particularly interesting. In the first one [28], a 2–12 GHz double-ridge horn antenna is presented. The internal surface has been coated with a layer of silver ink with a conductivity of 4 × 105 S/m.

In Ref. [29], a comparison between SLA and standard process (lathe in aluminum) has been considered to realize a spline horn operating in the Ku-band. Both prototypes exhibit good agreement between the simulated and measured value of the co-polar pattern, while worse performances in terms of losses and cross-polarization level can be noticed in the AM one. The latter is mainly related to the higher surface roughness, which has been estimated equal to 1.79 μm. In Ref. [30], a metal-plated 240 GHz choke horn antenna has been reported. The antenna has been designed to obtain 12 dBi gain from 200 to 280 GHz. The measured radiation pattern is in good agreement with the simulation.

Considering the SLM process, six works are particularly significant [26, 27, 31–34]. The first four refer to the manufacturing of classical architecture to understand the applicability of the SLM technique. The last two show some interesting work in the exploration of AM features.

In Ref. [26], the manufacturing of three smooth-wall horns, working in Ku-, Ku/K-band, and Q/V-band, is presented. The smooth-wall design allows the alignment of the building direction with the propagation axis, ensuring a good cross-sectional symmetry and low cross-polarization level. The horns have been manufactured using aluminum alloy. All the antennas exhibit good agreements between measured and simulated values. In particular, return-loss higher than 33 dB, cross-polarization lower than −28 dB, and peak gain of 25 dBi have been obtained. Moreover, a 3D scan of the outer section of the horns reveals an accuracy between 0.06–0.08 mm.

In Ref. [27], the manufacturing of 300 GHz corrugated horn antennas. Four prototypes have been manufactured using a tin-bronze alloy with an electrical conductivity of 7 MS/m. The dimensions of the prototypes differ by 5% with respect to the theoretical model. Nevertheless, the measurement results have observed good symmetry between E-plane and H-plane. However, the antenna efficiency is relatively low due to the conductivity of the material used in manufacturing. To improve the RF performances, all the prototypes have been plated with 3 μm of gold (conductivity of 44 MS/m).

In Ref. [31], an X/Ku-band (8–18 GHz) chocked horn SLM manufactured in the aluminum alloy is presented. To investigate process manufacturing accuracy and repeatability, 15 prototypes have been manufactured, showing an overall accuracy of 0.1–0.3 mm and a roughness Ra of 3.5 μm. The feeding gap and the radiating aperture have been milled after the manufacturing process, with an accuracy of ±0.02 mm.

In Ref. [32], a Ku-band spline horns SLM manufactured in aluminum and titanium alloys are presented. The manufactured components' roughness is about 16 μm for both the materials, but it reduces to 3.4 μm for the Al one after the post-processing. The radiation pattern of the two prototypes is in good correlation with the simulation, but the titanium horn exhibits a worse cross-polarization level due to a slight ellipticity of the aperture. In the same paper, the manufacturing of a cluster of four spline horns is presented, showing a mass reduction of 30% with respect to classical manufacturing.

In Ref. [33], a perforated X-band (8-12GHz) horn antenna has been presented to lighten the component. The holes' dimensions are lower than 1/15th of the wavelength at 12 GHz. Not considering the SMA connector, the antenna weight is just 8 g. The good agreement between simulated and measured RF results demonstrates an efficient method to reduce weight without affecting performance. A similar idea has been applied in Ref. [34] for a steel pyramidal horn antenna.

FDM process has been used to manufacture waveguide horns up to 15 GHz. In Ref. [35], the authors presented the first additively manufactured dielectric-loaded profiled conical horn antenna in the frequency range from 9 to 15 GHz. The component has been manufactured using polylactic acid (PLA) with relative permittivity $\epsilon_r = 2.72$. Two concentric dielectric cores with different infill factors have been printed to obtain the desired relative permittivity inside the horn. After the manufacturing process, the outer surfaces of the horn have been plated with 10 μm of copper. Surface roughness of 8.050 μm has been measured after the plating process. The measured radiation pattern is in good agreement with the simulation, with a cross-polarization level less than −21 dB up to 13 GHz. A measured S11 lower than −10 dB has been obtained from 10 to 14 GHz. The antenna exhibits a gain of 16.0–20.0 dBi and total efficiency of 60–91% over the frequency range from 9 to 15 GHz. In Ref. [36], three X-Ku pyramidal horn antennas have been 3D printed using ABS material and then metal plated using copper or chromium wand nickel, respectively. As expected, the copper prototype exhibits the best performance.

5.4 Dual-polarization components

Orthomode transducers (OMT) and septum polarizers are classical components used in dual-polarization antenna feed-chain systems. Separate the two linear polarization of the incoming signal routing them to two different rectangular waveguides [37]. Septum polarizers are employed to route the two circular polarizations in the common waveguide (typically circular or square) to the TE10 modes in two different rectangular waveguides and vice versa. The most important parameters are insertion loss, isolation between the rectangular ports, and cross-polarization in transmission [38].

Starting with SLM realizations, a C-band septum polarizer operating in the frequency band 3.6–4.4 GHz is shown in Ref. [39]. The component has been manufactured of Scalmalloy® with a measured dimensional error of 0.6 mm. The return loss is better than 25 dB and the isolation greater than 23 dB, while the deviation between measured and simulated axial ratio is about 1 dB mainly caused by a not perfectly circular common section. Moving up in frequency, in Ref. [40], two single sidearm OMTs are presented. The components are designed in the frequency band 10–15 GHz. The two OMTs have been compared with a realization by conventional manufacturing techniques. In this work, the OMT geometry realized was not figured out for AM process.

Nevertheless, both components exhibit RF performances in good accordance with the traditionally machined components. An example of an OMT realization of an AM-oriented geometry is reported in Ref. [41], where the RF measured performances well match the simulated one. Isolation over 50 dB and return loss better than 20 dB have been achieved in the operative frequency range. Considering the Ka-band, an asymmetric side-coupling OMT has been presented in Ref. [37]. The OMT geometry has been conceived to be AM-oriented by using a multi-slope junction. The manufactured prototype exhibits good accordance with measured and simulated RF results. In particular, the insertion loss is lower than 0.25 dB, and the return loss is better than 27 dB. Then, in Ref. [42], the OMT design has been integrated with a twist in the coupled arm. The measured results are in good agreement with the simulation. In particular, the return loss for both polarizations is better than 27 dB, and the insertion loss is lower than 0.15 dB and 0.2 dB for the inline and coupled channels, respectively, which corresponds to an equivalent resistivity of 16 μΩcm. The prototype has been controlled by a computer tomography scanner showing a mechanical accuracy in the range 0.02–0.04 mm.

By considering the SLA process, in Ref. [43], a V-Band (50–75 GHz) OMT manufactured is presented. The OMT has been metal plated with a process developed by SWISSto12. This consists of copper plating and passivation using a thin layer of gold or silver. The novelty of this OMT is the multi-step conical post on the turnstile junction. This permits easy printing by SLA.

Septum polarizers have also been realized. In Ref. [38], a Ka-band prototype realized in SLM is presented. Two 45° bends have been integrated to accommodate standard WR28 flanges. The device has been designed to align the propagation axis with the building direction to guarantee the best cross-sectional symmetry. Thanks to this choice, the measured cross-polarization level is better than 30 dB and the isolation better than 28 dB.

In Ref. [44], a broadband septum polarizer has been manufactured in SLA and copper plated. The paper shows a novel design with a triangular common port design that allows a higher bandwidth than circular or square waveguide polarizers. The polarizer has been firstly designed in W-band and manufactured with a standard machining process. Then, the design was scaled in K-band and manufactured with an SLA printer. The prototype has been coated with conductive ink and then subjected to galvanic copper plating. Both the components exhibit isolation greater than 17 dB and a return loss better than 14 dB over 37.8% of bandwidth. Moreover, the insertion loss is lower than 0.4 dB for the 3D-printed prototype.

5.5 Integrated components

As reported in Ref. [40], one of the main advantages of the AM technologies is the free shape feature that can lead to an additional degree of freedom in the design of RF components, for instance, the integration of different functionalities (electromagnetic, mechanical, and thermal) in a monolithic component [45].

The previous survey has shown the evolution of the employment of AM process in the realization of guided microwave components, improving the quality of the realized parts, but this potentiality of AM is still not fully explored. A survey of some interesting and successful examples of this initial work is reported.

In Ref. [46], a mono-block Ku-band front-end, a combination of two integrated sub-assembly, a diplexer, and a 2 × 2 horn antenna array, has been presented. The components have been manufactured using the SLA technology and metalized with copper and tin with an electrolytic process. Measurement results agree with the simulated performances with a frequency shift of 150 MHz that can be related to the thickness of the metal deposition, not considered during the design.

In Ref. [38], a Ka-band feed horn integrated with a septum polarizer, manufactured with the SLM process using aluminum alloy. The septum polarizer, presented in the previous section, has been integrated with a smooth-wall horn. The measured values of return loss, cross-polarization, and isolation are comparable with the results of the septum polarizer alone, namely a return loss better than 25 dB, an isolation between the rectangular waveguide higher than 27 dB, and a cross-polarization better than 28 dB. The measurement results are comparable with those obtained with a standard manufacturing process in a split block layout.

In Ref. [47], an interesting study of an integrated feed system working in the frequency band in X-band (from 7.2 to 8.2 GHz) is presented. The device is composed of a circular horn antenna and an OMT. The authors compare the 3D printing of the feed system as a mono-block with the manufacturing in two symmetric pieces. Both the components have been printed using SLA and copper plated with a three-step metallization process. Although the two-pieces prototype has the advantages of an easier manufacturing process and the thickest metal coating, the monolithic component exhibits better RF performances.

A similar component has been developed and presented in Ref. [48]. The integrated Ku-band feed chain consists of a spline horn and an OMT and has been manufactured using the SLM process. The RF performances are compared with those obtained with standard machining processes. An X-band feed chain has been presented in the same paper. The developed component consists of a spline horn and a compact E-plane waveguide coupler. The feed chain is three times lighter than a comparable, conventional manufactured component. Perfect agreement with the simulated performance has also been obtained with these components.

In Ref. [49], an entire Ku-band feed cluster manufactured by SLM technology, suitable for SFB scenarios, is presented. The feed cluster operates in Tx- and Rx-frequency bands. The component consists of 18 feed chains composed of a horn, a transition, a single sidearm OMT and waveguide routing to provide interfaces for the measurement. The scattering parameters have been measured for both the frequency bands showing a return loss better than 19.5 dB and the isolation below −50 dB. The feed cluster exhibits excellent agreement between simulated and measured values of the pattern at 11.30 and 14–25 GHz.

In Ref. [50], a high gain K-band feed chain has been manufactured by SLM using aluminum alloy. The measurement results show return loss and port-to-port isolation better than 23 dB. The axial ratio is between 0.1 and 0.5 dB, while the cross-polar discrimination (XPD) is better than 30 dB. The measurements show good manufacturing accuracy with good rotational symmetry.

In Ref. [51], a passive front end for satellite communication, manufactured by SLM technology using aluminum alloy, is discussed. The component comprises four conical horns, four rectangular-to-circular waveguide tapers, two-stage 1×4 power dividers, and a WR-42 waveguide interface. The scattering parameters have been measured in the frequency band 19–21 GHz. Discrepancies between simulated and measured RF performances have been observed. They are mainly due to the dimensional tolerance and surface roughness of the fabrication process.

In Ref. [52], a leaky-wave antenna integrated with an OMT operating in K-band is presented. The antenna consists of a triple-ridge square waveguide perforated on its top wall with crossed slots. The component has been printed by SLA with the propagation axis aligned with the building direction to enhance the good symmetry of the antenna. Then, copper plating was applied. The measurements prove the high precision of the process. In particular, the 3D-printed prototype exhibits a return loss better than 16 dB, port isolation better than 40 dB, and a cross-polarization level below −35 dB.

A K/Ka-band dual-circular polarization antenna feed chain has been reported in Ref. [53]. The device has been designed with an AM-oriented approach. A prototype has been manufactured by SLM with aluminum alloy. The chain is composed of a dual-band orthomode junction (OMJ) which symmetrically extracts the K-band signal while the Ka-band one propagates inline to a series of circular steps connected to a Ka-band septum polarizer. Each of the four arms, where the K-band signal propagates, is integrated with a low-pass filter to isolate the channels from the Ka-band signals. Subsequently, these four arms are recombined in a turnstile junction connected to a K-band septum polarizer. The measured performances of the prototype show very satisfactory performance with values typically required for SATCOM application (i.e., isolation between the bands better than 50 dB, return loss, and isolation between the two polarizations of the same bands than 20 dB).

Finally, a complex RF component is presented in Ref. [54], where a Ku/K-band filter is integrated with an H-plane bend and 90° twist. The conceived geometry makes AM process the only feasible manufacturing technique. The design shows a 70% weight/volume reduction compared to a solution obtained assembling three different parts realized by standard machining. The idea has been stressed by considering the realization of three prototypes with different bend radii to make the device as smaller as possible. Good agreement between measured and simulated RF performance has been achieved even for the prototypes with a bend radius of 30 mm. The three components exhibit a rejection better than 60 dB and an insertion loss better than 0.2 dB. The return loss is better than 25 dB for the prototype with a bend radius of 40 mm, while it is better than 20 dB for the other prototypes.

6. Conclusions

After a brief introduction on the main AM processes, this chapter has summarized the huge work realized by the microwave community on the applicability of AM process on the manufacturing of RF components. For reader convenience, **Table 3**

Technology	Accuracy (μm)	Main frequency range (GHz)
SLM	<100	10–50
SLA	<50	10–90
FDM	100–200	2–20

Table 3.
Accuracy and exploited frequency range of the three AM processes reported in this chapter: SLM, SLA, FDM.

reports the obtained results in terms of accuracy in the relevant frequency range. The most used technology is SLM since it allows the manufacturing of all metal parts. SLA exhibits the best manufacturing accuracy and the largest bandwidth. FDM is used for cheaper components in a low-frequency range, due to the worst manufacturing accuracy. The study/research on AM for microwave components manufacturing is still ongoing, many thematic special sessions are organized within the main conferences in the microwave area. Large margins of improvement are expected in the near future from the manufacturing and design points of view in particular from the integration of different functions in the same realized part.

Conflict of interest

The authors declare no conflict of interest.

Author details

Mauro Lumia[1], Giuseppe Addamo[1*], Oscar Antonio Peverini[1], Flaviana Calignano[2], Giuseppe Virone[1] and Diego Manfredi[3]

1 Istituto di Elettronica e di Ingegneria dell'Informazione e delle Telecomunicazioni, National Research Council of Italy, Turin, Italy

2 Dipartimento di Ingegneria Gestionale e della Produzione, Politecnico di Torino, Turin, Italy

3 Dipartimento di Scienza Applicata e Tecnologia, Politecnico di Torino, Turin, Italy

*Address all correspondence to: giuseppe.addamo@ieiit.cnr.it

References

[1] Calignano F, Manfredi D, Ambrosio E, Biamino S, Lombardi M, Atzeni E, et al. Overview on additive manufacturing technologies. Proceedings of the IEEE. 2017;**105**: 593-612. DOI: 10.1109/JPROC.2016.2625098

[2] Gibson I, Stucker B, Rosen DW. Additive Manufacturing Technologies-Rapid Prototyping. 1st ed. Boston: Springer; 2010. p. 452. DOI: 10.1007/978-1-4419-1120-9

[3] Peverini OA, Lumia M, Calignano F, Addamo G, Lorusso M, Ambrosio EP, et al. Selective, laser melting manufacturing of microwave waveguide devices. Proceedings of the IEEE. 2017; **105**:620-631. DOI: 10.1109/JPROC.2016.2620148

[4] Neikov OD, Naboychenko SS, Dowson G. Handbook of Non-Ferrous Metal Powders. 1st ed. Amsterdam: Elsevier; 2009. p. 621. DOI: 10.1016/B978-1-85617-422-0.X0001-8

[5] Spierings AB, Herres N, Levy G. Influence of the particle size distribution on surface quality and mechanical properties in AM steel parts. Rapid Prototyping Journal. 2011;**17**: 195-202. DOI: 10.1108/13552541111124770

[6] Qian M. Metal powder for additive manufacturing. JOM Journal of the Minerals, Metals and Materials Society. 2015;**67**:536-537. DOI: 10.1007/s11837-015-1321-z

[7] Calignano F. Design optimization of supports for overhanging structures in aluminum and titanium alloys by selective laser melting. Materials and Design. 2014;**64**:203-213. DOI: 10.1016/j.matdes.2014.07.043

[8] About Additive manufacturing. Vat Photopolymerization [Internet]. 2021. Available from: http://www.lboro.ac.uk/research/amrg/ [Accessed: January 31, 2022]

[9] Peverini OA, Addamo G, Lumia M, Virone G, Calignano F, Lorusso M, et al. Additive manufacturing of Ku/K-band waveguide filters: A comparative analysis among selective-laser melting and stereo-lithography. IET Microwaves, Antennas and Propagation. 2017;**11**: 1936-1942. DOI: 10.1049/IET-map.2017.0151

[10] Khan S, Vahabisani N, Daneshmand M. A fully 3-D printed waveguide and its application as Microfluidically controlled waveguide switch. IEEE Transactions on Components, Packaging and Manufacturing Technology. 2017;**7**: 70-80. DOI: 10.1109/TCPMT.2016.2631545

[11] Coffey M, Verploegh S, Edstaller S, Armstrong S, Grossman E, Popovic Z. Additive manufactured W-band waveguide components. In: 2017 IEEE MTT-S International Microwave Symposium (IMS); 4–9 June 2017; Honolulu, USA. New York: IEEE; 2017. pp. 52-55

[12] Shen J, Aiken MW, Abbasi M, Parekh DP, Zhao X, Dickey MD, et al. Rapid prototyping of low loss 3D printed waveguides for millimeter-wave applications. In: 2017 IEEE MTT-S International Microwave Symposium (IMS); 4–9 June 2017; Honolulu, USA. New York: IEEE; 2017. pp. 41-44

[13] Menargues E, Garcia-Vigueras M, Debogovic T, Capdevila S, Dimitriadis AI, de Rijk E, et al. 3D

printed feed-chain and antenna components. In: 2017 IEEE International Symposium on Antennas and Propagation & USNC/URSI National Radio Science Meeting; 9–14 July 2017; San Diego, USA. New York: IEEE; 2017. pp. 1-2

[14] Hollenbeck M, Wamick K, Cathey C, Opra J, Smith R. Selective laser melting aluminum waveguide attenuation at K-band. In: 2017 IEEE MTT-S International Microwave Symposium (IMS); 4–9 June 2017; Honolulu, USA. New York: IEEE; 2017. pp. 45-47

[15] Addamo G, Peverini OA, Lumia M, Virone G, Tascone R, Calignano F, et al. Experimental research activity on additive manufacturing of microwave passive waveguide components. In: 2017 47th European Microwave Conference (EuMC); 10–12 October 2017; Nuremberg, Germany. New York: IEEE; 2017. pp. 496-499

[16] Tomassoni C, Peverini OA, Venanzoni G, Addamo G, Paonessa F, Virone G. 3D printing of microwave and Millimeter-wave filters: Additive manufacturing technologies applied in the development of high-performance filters with novel topologies. IEEE Microwave Magazine. 2020;**21**:24-45. DOI: 10.1109/MMM.2020.2979153

[17] Miek D, Simmich S, Höft M. Additive manufacturing of symmetrical X-band waveguide filters for wide-band applications based on extracted pole filter design. In: 2019 IEEE MTT-S International Microwave Workshop Series on Advanced Materials and Processes for RF and THz Applications (IMWS-AMP); 16–18 July 2019; Bochum, Germany. New York: IEEE; 2019. pp. 13-15

[18] Dahle R, Laforge P, Kuhling J. 3-D printed customizable inserts for waveguide filter design at X-band. IEEE Microwave and Wireless Components Letters. 2017;**27**:1080-1082. DOI: 10.1109/LMWC.2017.2754345

[19] Tomassoni C, Venanzoni G, Dionigi M, Sorrentino R. Compact quasi-elliptic filters with mushroom-shaped resonators manufactured with 3-D printer. IEEE Transactions on Microwave Theory and Techniques. 2018;**66**:3579-3588. DOI: 10.1109/TMTT.2018.2849067

[20] Tomassoni C, Bozzi M, Dionigi M, Venanzoni G, Perregrini L, Sorrentino R. Additive manufacturing of microwave components: Different approaches and methodologies. In: 2017 International Conference on Electromagnetics in Advanced Applications (ICEAA); 11–15 September 2017; Verona, Italy. New York: IEEE; 2017. pp. 848-851

[21] Salek M, Shang X, Roberts RC, Lancaster MJ, Boettcher F, Weber D, et al. W-band waveguide Bandpass filters fabricated by Micro laser sintering. IEEE Transactions on Circuits and Systems II: Express Briefs. 2019;**66**: 61-65. DOI: 10.1109/TCSII.2018.2824898

[22] Zhang F, Gao S, Li J, Yu Y, Guo C, Li S, et al. 3-D printed slotted spherical resonator Bandpass filters with spurious suppression. IEEE Access. 2019;**7**: 128026-128034. DOI: 10.1109/ACCESS.2019.2938972

[23] Li J, Guo C, Mao L, Xiang J, Huang G, Yuan T. Monolithically 3-D printed hemispherical resonator waveguide filters with improved out-of-band rejections. IEEE Access. 2019;**6**: 57030-57048. DOI: 10.1109/ACCESS.2018.2872696

[24] Périgaud A, Tantot O, Delhote N, Verdeyme S, Bila S, Pacaud D, et al. Continuously tuned Ku-band cavity

filter based on dielectric Perturbers made by ceramic additive manufacturing for space applications. Proceedings of the IEEE. 2017;**105**: 677-687. DOI: 10.1109/JPROC.2017. 2663104

[25] Périgaud A, Tantot O, Delhote N, Verdeyme S, Bila S, Baillargeat D. Bandpass filter based on skeleton-like Monobloc dielectric pucks made by additive manufacturing. In: 2018 48th European Microwave Conference (EuMC); 23–27 September 2018; Madrid, Spain. New York: IEEE; 2018. pp. 296-299

[26] Addamo G, Peverini OA, Calignano F, Manfredi D, Paonessa F, Virone G, et al. 3-D printing of high-performance feed horns from Ku- to V-bands. IEEE Antennas and Wireless Propagation Letters. 2018;**17**:2036-2040. DOI: 10.1109/LAWP.2018.2859828

[27] Reinhardt A, Möbius-Labinski M, Asmus C, Bauereiss A, Höft M. Additive manufacturing of 300 GHz corrugated horn antennas. In: 2019 IEEE MTT-S International Microwave Workshop Series on Advanced Materials and Processes for RF and THz Applications (IMWS-AMP); 16–18 July 2019; Bochum, Germany. New York: IEEE; 2019. pp. 40-42

[28] Maas J, Liu B, Hajela S, Huang Y, Gong X, Chappell WJ. Laser-based layer-by-layer polymer Stereolithography for high-frequency applications. Proceedings of the IEEE. 2017;**105**: 645-654. DOI: 10.1109/ JPROC.2016.2629179

[29] Teniente J, Iriarte JC, Caballero R, Valcázar D, Goñi M, Martínez A. 3-D printed horn antennas and components performance for space and telecommunications. IEEE Antennas and Wireless Propagation Letters. 2018;**17**: 2070-2074. DOI: 10.1109/LAWP.2018. 2870098

[30] Biurrun-Quel C, Lacombe E, Gianesello F, Luxey C, Del-Río C. Characterization of 3D-printed choke horn antenna for 5G backhaul applications. In: 2019 13th European Conference on Antennas and Propagation (EuCAP); 31 March–5 April 2019; Krakow, Poland. New York: IEEE; 2019. pp. 1-4

[31] Foged LJ, Giacomini A, Morbidini R, Saccardi F, Schirosi V, Boumans M, et al. Investigation of additive manufacturing for broadband choked horns at X/Ku band. IEEE Antennas and Wireless Propagation Letters. 2018;**17**:2003-2007. DOI: 10.1109/LAWP.2018.2868611

[32] Cailloce Y, Hourlay P, Lebrun F, Palacin B. Additive manufacturing of Ku band horn antennas for telecommunications space applications. In: 12th European Conference on Antennas and Propagation (EuCAP); 9–13 April 2018; London, UK. Stevenage: IET; 2018. pp. 1-4

[33] Chio T, Huang G, Zhou S. Application of direct metal laser sintering to waveguide-based passive microwave components, antennas, and antenna arrays. Proceedings of the IEEE. 2017;**105**:632-644. DOI: 10.1109/JPROC. 2016.2617870

[34] Shamvedi D, Danilenkoff C, Karam S, O'Leary P, Raghavendra R. 3D printed periodic structures in a horn antenna for side-lobe reduction using direct metal laser sintering. In: Loughborough Antennas & Propagation Conference (LAPC); 13–14 Nov. 2017; Loughborough, UK. Stevenage: IET; 2017. pp. 1-4

[35] Zhang S, Cadman D, Vardaxoglou JYC. Additively

manufactured profiled conical horn antenna with dielectric loading. IEEE Antennas and Wireless Propagation Letters. 2018;**17**:2128-2132. DOI: 10.1109/LAWP.2018.2871029

[36] Genc A, Basyigit IB, Goksu T, Helhel S. Investigation of the performances of X-Ku band 3D printing pyramidal horn antennas coated with the different metals. In: 2017 10th International Conference on Electrical and Electronics Engineering (ELECO); 30 November–2 December 2017; Bursa, Turkey. New York: IEEE; 2017. pp. 1012-1016

[37] Addamo G, Peverini OA, Virone G, Paonessa F, Manfredi D, Calignano F. 3D printing of Ka band Orthomode transducers. In: 2018 IEEE MTT-S International Microwave Workshop Series on Advanced Materials and Processes for RF and THz Applications (IMWS-AMP); 16–18 July 2018; Ann Arbor, USA. New York: IEEE; 2018. pp. 1-3

[38] Addamo G et al. Additive manufacturing of Ka-band dual-polarization waveguide components. IEEE Transactions on Microwave Theory and Techniques. 2018;**66**:3589-3596. DOI: 10.1109/TMTT.2018.2854187

[39] Kilian M, Hartwanger C, Schneider M, Hatzenbichler M. Waveguide components for space applications manufactured by additive manufacturing technology. IET Microwaves, Antennas and Propagation. 2017;**11**:1949-1954. DOI: 10.1049/IET-MAP.2016.0984

[40] Booth P, Roberts R, Szymkiewicz M, Hartwanger C. Using additive manufacturing for feed chain and other passive microwave components. In: 2017 11th European Conference on Antennas and Propagation (EUCAP); 19–24 March 2017; Paris, France. New York: IEEE; 2017. pp. 558-562

[41] Kohl P, Kilian M, Schinagl-Weiß A, Hartwanger C. Additive manufacturing developments for satellite antenna applications from C- to Ka-band. In: 2019 12th German Microwave Conference (GeMiC); 25–27 March 2019; Stuttgart, Germany. New York: IEEE; 2019. pp. 222-2250

[42] Addamo G, Peverini OA, Manfredi D, Calignano F, Paonessa F, Lumia M, et al. Electromagnetic and mechanical analyses of a 3D-printed ka-band integrated twist and orthomode transducer. In: 2019 IEEE MTT-S International Microwave Workshop Series on Advanced Materials and Processes for RF and THz Applications (IMWS-AMP); 16–18 July 2019; Bochum, Germany. New York: IEEE; 2019. pp. 31-33

[43] García-Vigueras M, Menargues E, Debogovic T, Silva J, Dimitriadis A, Capdevila S, et al. Mm-wave antennas and components: Profiting from 3D-printing. In: 2017 International Conference on Electromagnetics in Advanced Applications (ICEAA); 11–15 September 2017; Verona, Italy. New York: IEEE; 2017. pp. 1016-1020

[44] Deutschmann B, Jacob AF. Broadband septum polarizer with triangular common port. IEEE Transactions on Microwave Theory and Techniques. 2020;**68**:693-700. DOI: 10.1109/TMTT.2019.2951138

[45] Peverini OA, Lumia M, Addamo G, Calignano F, Virone G, Ambrosio EP, et al. Integration of RF functionalities in microwave waveguide components through 3D metal printing. In: 2017 IEEE MTT-S International Microwave Symposium (IMS); 4–9 June 2017;

Honolulu, USA. New York: IEEE; 2017. pp. 48-51

[46] Borgne FL, Cochet G, Haumant J, Diedhiou D, Donnart K, Manchec A. An integrated Monobloc 3D printed front-end in Ku-band. In: 2019 49th European Microwave Conference (EuMC); 1–3 October 2019; Paris, France. New York: IEEE; 2019. pp. 786-789

[47] Saeidi-Manesh H, Saeedi S, Mirmozafari M, Zhang G, Sigmarsson HH. Design and fabrication of orthogonal-mode transducer using 3-D printing technology. IEEE Antennas and Wireless Propagation Letters. 2018; **17**:2013-2016. DOI: 10.1109/LAWP.2018.2847654

[48] Kilian M, Schinagl-Weiß A, Kohl P, Sommer A, Hartwanger C, Schneider M. Additive layer manufactured waveguide RF components. In: 2019 49th European Microwave Conference (EuMC); 1–3 October 2019; Paris, France. New York: IEEE; 2019. pp. 790-793

[49] Kilian M, Schinagl-Weiß A, Sommer A, Hartwanger C, Schneider M. Ku-band SFB-cluster manufactured by additive manufacturing techniques. In: 2019 13th European Conference on Antennas and Propagation (EuCAP); 31 March–5 April 2019; Krakow, Poland. New York: IEEE; 2019. pp. 1-4

[50] Kilian M, Kohl P, Hartwanger C, Schneider M. High gain Ka-band ALM feed chain. In: 2019 IEEE MTT-S International Microwave Workshop Series on Advanced Materials and Processes for RF and THz Applications (IMWS-AMP); 16–18 July 2019; Bochum, Germany. New York: IEEE; 2019. pp. 37-39

[51] Zhang B, Li R, Wu L, Sun H, Guo Y. A highly integrated 3-D printed metallic K-band passive front end as the unit cell in a large Array for satellite communication. IEEE Antennas and Wireless Propagation Letters. 2018;**17**: 2046-2050. DOI: 10.1109/LAWP.2018.2824298

[52] Dorlé A, Gillard R, Menargues E, Van Der Vorst M, De Rijk E, Martın-Iglesias P, et al. Additive manufacturing of modulated triple-ridge leaky-wave antenna. IEEE Antennas and Wireless Propagation Letters. 2018;**17**:2123-2127. DOI: 10.1109/LAWP.2018.2848723

[53] Addamo G et al. 3D printing of a monolithic K/Ka-band dual-circular polarization antenna-feeding network. IEEE Access. 2021;**9**:88243-88255. DOI: 10.1109/ACCESS.2021.3089826

[54] Peverini OA et al. Integration of an H -plane bend, a twist, and a filter in Ku/K-band through additive manufacturing. IEEE Transactions on Microwave Theory and Techniques. 2018;**66**:2210-2219. DOI: 10.1109/TMTT.2018.2809505

Chapter 14

Microwave-Assisted Pyrolysis Process: From a Laboratory Scale to an Industrial Plant

Marco Frediani, Piero Frediani, Gianni Innocenti, Irene Mellone, Roberto Simoni and Gianpaolo Oteri

Abstract

One of the great challenges for the European Union (EU) is the "Circular Economy Package," and to achieve this goal, materials at the end of their life cycle must be recycled using a sustainable process. In this way, as a thermochemical treatment, pyrolysis represents a significant opportunity so long it leads to the recovery of both energy and chemical content of mixed, contaminated, or deteriorated plastics. An excellent history of an academic-industrial adventure started in 2008 at the Department of Chemistry of the University of Florence demonstrates the possibility of employing microwaves to recycle plastics to preserve their energy and chemical content. After that, Techwave started industrialization of the process in 2019, realizing a small-scale prototype followed by a full-scale pilot plant using different plastic materials (e.g., polystyrene, acrylonitrile-butadiene-styrene (ABS), and polypropylene). Nowadays, the plant may process 90 kg/h of plastics with a low formation of char and gas and an interesting amount of liquid useful as a source of chemicals or fuel because it has an LHV of 35–43 kJ/kg. The Microwave-Assisted Pyrolysis (MAP) is an industrial novelty in plastic recycling, and it looks very promising for a much more modern and innovative plastic waste recovery system.

Keywords: plastic, recycle, pyrolysis, microwave, polystyrene, ABS, polypropylene

1. Introduction

A real circular economy may be realized if products at the end of their life cycle are reutilized or transformed into raw materials giving them the possibility for another life. The energy balance among the content of the new products and that one required by the transformation must be positive; that is, the process must require less energy of those present in the products; otherwise, the process will have a deleterious effect on the environment. More positive is the balance less will be the contamination. This goal is one of the main challenges in which scientists are involved to solve the problems concerning both global warming and the end of mineral resources. These recycled sources also represent a strong driving force for developing sustainable industrial processes. Many industrial firms are now involved in environmental issues, and they are, even in a different way, engaged in solving this problem [1].

The world plastic production in 2019 was 368.0 x 10^6 tons with an increase of 2.5% concerning for 2018 [2] while in the same year, the European production was 57.9 x 10^6 tons between thermoplastic and thermosetting polymers. Among them, polyolefins are the most produced and employed material in everyday life for industrial, domestic, and technological applications [3]. They are thermoplastic polymers and are mainly used for packaging; their life cycle is very short, which means they must be disposed of in a short time after their production. A less important amount of them is employed to realize furniture, insulating materials, automotive parts, and so on, and their life cycle is considerably longer (10 years or more).

A very recent study by the Organization for Economic Co-operation and Development (OCSE) was published in the "Global Plastic Outlook" [4] where it is reported that less than 10% of plastics are recycled around the world. Considering that the world production is estimated to be 460.0 x 10^6 tons, higher than the total waste (353.0 x 10^6 tons) produced in the same year, many actions are required to dispose of these plastics.

Mechanical recycle of polymeric materials is easily through environmental-friendly and economic processes, as reported in **Figure 1**. In this way, renewed objects are realized by recycled plastic materials avoiding disposal of these plastics through landfilling or combustion and reducing the use of mineral oil for their production. This process, however, may be applied only when a single plastic material is available, and it is not contaminated or strongly deteriorated. In the other cases, different environmentally friendly routes may be followed, such as thermochemical processes, thus contributing to the realization of a circular economy and reducing the emission of greenhouse gases

Figure 1.
Pathway for mechanical recycling of thermoplastic materials.

(GHG). Polymer thermochemical processes may supply fuels and chemicals using end life plastic materials as a feedstock, and it may be an alternative to oil-based raw materials [5]. These research studies led to the development of new technologies able to convert waste into resources minimizing the environmental impact of their treatment, avoiding the production of by-products, or limiting the amount of secondary waste.

Chemical recycling technologies [1, 6] are likely to play a crucial role in the transition toward a circular economy and close the recycling of materials giving compounds such as hydrocarbons, available for the production of new compounds. These technologies are able to remove hazardous substances, eventually present in the waste, thus giving new recycled feedstocks. Agreements with materials and chemical producers the use of raw materials from secondary sources are necessary for developing sustainable, feasible, and cost-efficient chemical recycling.

Among chemical recycling technologies, pyrolysis is one of the most important, also referred to as thermolysis or chemical recycling. It represents the transformation of organic materials, under the effect of heat, in the absence of oxygen. Depending on the process conditions, pyrolysis typically yields a mixture of molecules in the form of liquid or wax as the main products. This liquid or wax can be refined to obtain chemicals or fuels, as well as solid and gas may be used for the production of energy.

In this way, waste or contaminated plastic materials may be transformed into oil for industrial purposes, reducing the request and the environmental impact of mineral oil. The potential for recycling is enormous. In 2018, in Europe, plastic production reached almost 62.0×10^6 tons, and all this plastic should be utilized at the end of its life cycle, in one way or the others. For the oil industry, the use of waste plastic as an alternative feedstock represents a new business.

From a chemical point of view, pyrolyzed plastics are a good raw base material for the oil industry so long the remaining impurities are removed from the plastics, and the oil can be used as feedstock for oil refineries. Pyrolysis involves the use of heat and anoxic conditions to break down plastic waste into compounds containing smaller molecules, yielding valuable hydrocarbons in the form of liquid, waxes, and gases. The end products of pyrolysis can be monomers, heating oil, refinery feedstock, transportation fuels, and chemicals.

Chemical recycling can be mainly used to recycle mixed post-consumer plastic waste when sorting single components is not economical. That is, pyrolysis is a very flexible method that allows the use of various feedstocks. Another method of chemical recycling is gasification. This process converts carbonaceous materials into gases. The main product of gasification is synthesis gas (syngas: CO, and H_2), which can be further processed into various final products such as gasoline, diesel, methanol, and synthetic methane. Among these processes, the microwave technologies have taken large attention due to their high efficiency in supplying the energy required for a plethora of industrial processes. The main performances of the use of microwaves as correlated to a classical heating system are resumed in **Table 1**. Some examples of products obtained from pyrolysis of different polymers using Microwave-Assisted Pyrolysis (MAP) are reported in **Table 2**.

This chapter aims to offer a comprehensive description of the path from laboratory research to the realization of an industrial plant able to transform up to 2,000 Kg/d of waste plastic into valuable products. These products are available to synthesize new plastic materials, and they may be certified as renewed plastics. However, for plastic waste-based pyrolysis products to become a reality on an industrial scale, ardent development in technologies, value chains, and supporting legislation is needed. Despite these hurdles, the missing links in the plastic recycling loop can be addressed and eventually fixed to establish an actual circular economy for plastics.

Microwave heating		Conventional heating	
Advantages	Volumetric heating	Detriments	Surface heating
	Short reaction time (minutes)		Long reaction time (hours)
	High heating efficiency		Heating efficiency is usually low
	The low thermal conductivity of polymers may be overwhelmed: easy heating of polymers		Hard heating of polymers: their thermal conductivity is low
	The gas formed does not contain combustion gas		The gas formed is contaminated by combustion gas (Direct heating)
Detriments	Electrical power is required	Advantages	Every fuel source may be employed
	Microwave absorber is required		Additives are not required; some heating carriers are sometimes used

Table 1.
Correlation between microwave and conventional heating.

Polymer	Products yield (Min-max) %			Ref.
	Solid	Liquid	Gas	
PE	0.4–0.6	80.2–83.9	15.7–19.2	[3, 7]
PP	9.1–12.0	56.5–74.4	13.3–34.4	[3, 5]
PS	0.9–16.7	73.5–96.1	3.0–9.8	[8–10]
PVC	13.9–53.2	10.2–70.8	13.3–75.8	[11]
PET*	15.6	41.3	38.7	[12]
PLA	2.6–7.6	28.4–57.7	38.5–69.0	[13]
Tyre	40.6–65.0	20.7–44.0	9.0–27.4	[14–17]
WEEE	14.2	76.6	9.2	[18]
MPB	24.9–59.9	31.1–43.7	9.0–42.5	[19]
CDP	11.1–30.1	22.3–47.6	28.7–46.2	[20]

*PE: Polyethylene; PP: Polypropylene; PS: Polystyrene, PVC: Poly(vynil chloride); PET: Poly(ethylene therephtalate); PLA: Poly(lactic acid); WEEE: Waste electric and electronic equipment; MPB: Multilayer packaging beverage; CDP: Corn-derived plastic bag. *Pyrolysis using classic thermal heating.*

Table 2.
Products obtained from microwave-assisted pyrolysys of plastics.

A circular economy is one of the drivers boosting chemical recycling because it may realize, more and more, so reusing all waste materials. Chemical recycling, such as pyrolysis, is therefore needed, but legislative issues are required for this big challenge of improving the development of this technology.

2. Scale-up of the MAP process

2.1 Preliminary studies of MAP process

Techwave s.r.l. is an Italian start-up established to realize an industrial plant from a patent able to transform, by means of microwave, plastic complex polymers into their original components, ready for a new use. The process is based on an Italian proprietary patent [21] of Cooperativa Autotrasportatori Fiorentini (CAF Scrl-Italy) developed and filed by a research team of Department of Organic Chemistry – University of Florence – Italy, in 2011, and converted into a European patent [3] acquired by Techwave in 2018.

Experiments were started in 2008 at the Department of Organic Chemistry "Ugo Schiff," University of Florence, where a set of trials were performed on tires and various types of plastic materials, ranging from polystyrene (PS) to polyethylene (PE), polypropylene (PP), poly(ethylene terephthalate) (PET), poly(vinyl chloride) (PVC), and their mixtures [22]. Similar experiments were also performed on Waste Electric and Electronic Equipment (WEEE) [18].

Pyrolysis studies were initially performed on a laboratory scale (**Figure 2**). The experiments were carried out on a batch oven, and 100–300 g/h of materials were processed. The oven was equipped with four external MW generators, each having an absorption of electric power of 2 KW capable of delivering up to 6 KW of microwave power inside the oven, operating at the frequency of 2.45 GHz.

During the experimental trial, many arrangements were tested to determine the influence of several parameters on the quantity and quality of the three products formed: a liquid, an uncondensable gas, and a char. In particular, it was verified how some critical factors such as the residence time inside the pyrolytic oven, the temperature reached, and the type of downstream fractionating system would affect the amount and the composition of the three products of the reaction. The liquid may be employed for the synthesis of new polymeric products (**Figure 3**), while the gas and char may be used as fuel or employed for other uses [1].

This first study showed MAP as a suitable technique for cracking various types of plastic even if they may be highly contaminated and absolutely heterogeneous. The results obtained led to the publication of various scientific papers and process patents

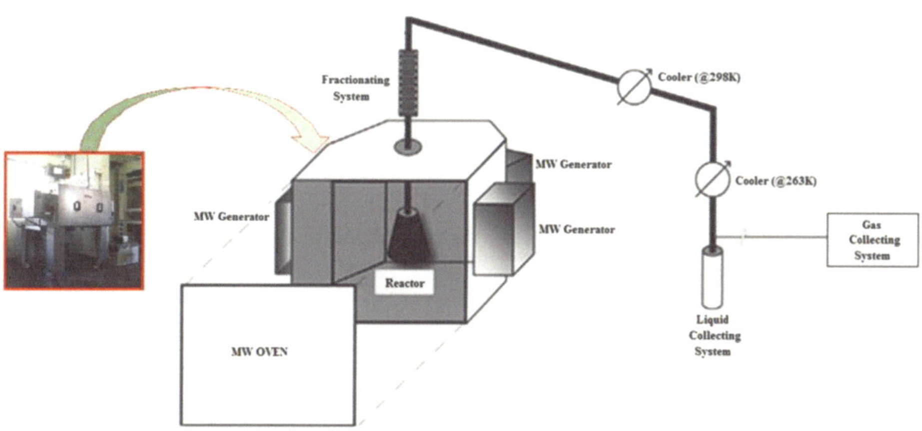

Figure 2.
Laboratory MW oven.

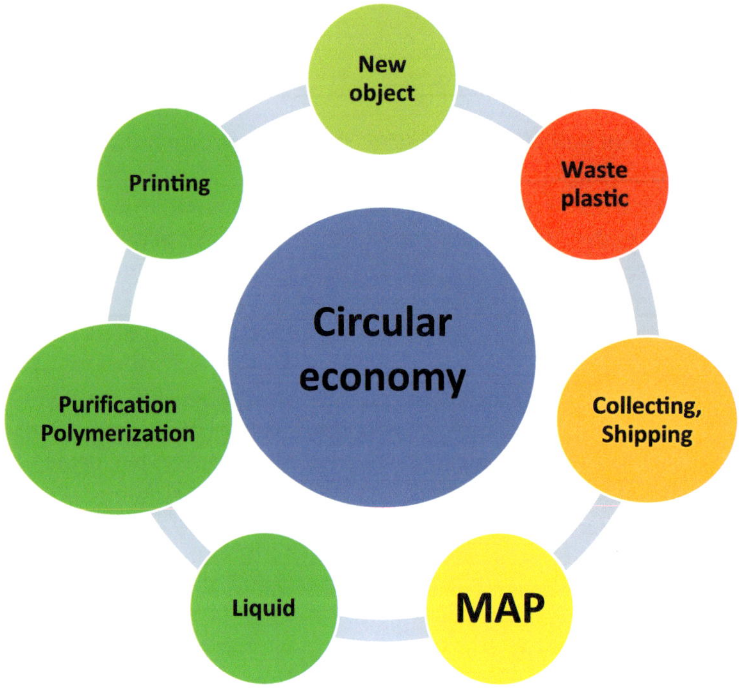

Figure 3.
The circular economy using a MAP process for polymeric waste materials.

(**Table 2**) and the identification of the guidelines for the scale-up of MAP technology to an industrial level [1].

2.2 The pre-industrial prototype

Therefore, a first development phase was run for this purpose, to design and then build a prototype that allows the scale-up of the MAP process, enabling the transition from the experimental laboratory studies carried out until today to a pre-industrial phase. The idea, born at the University of Florence and initially funded by one of Techwave's future partners, was materialized in 2016 with the construction of the first pre-industrial prototype (**Figure 4**).

This prototype was initially designed for the treatment of End Life Tires (ELTs) and was able to process [14–17] whole tire (approx. 8 kg) in 30 minutes. The prototype was equipped with electric heating to bring the material to the optimum temperature for the absorption of microwaves and four external MW generators, each supplied with electric power of 1 KW and able to deliver up to 3.2 KW of microwave power inside the oven, operating at the frequency of 2.45 GHz. Later with the birth of Techwave in 2018, it was decided to continue the scale-up of the MAP process by extending the technology to other waste plastic materials, even if mixed or contaminated. The results obtained during the first testing phase are described in the following paragraph.

2.3 Experiment with pre-industrial prototype for scale-up of MAP

The purpose of this initial phase was the validation of the data reported on the laboratory scale, both in qualitative and quantitative terms, from the pyrolysis of

Figure 4.
Pre-industrial prototype.

plastic wastes. It also aimed to determine the operating parameters necessary to treat flows of various types of plastic materials at the end of their life cycle, and PS was chosen as the first example. Polystyrene, like all plastics, is unable to absorb microwave energy and convert it into heat. It must therefore be blended with a microwave absorber to realize the pyrolysis. The materials used as a microwave absorber are commonly carbon or iron powder.

The experiments were carried out using expanded polystyrene (EPS) as starting material and focused on analyzing the ratio between the amount of polystyrene processed and the microwave power used. Therefore, an energy density close to that foreseen in the prototype was used to acquire the information necessary for the design. The process was also studied to correlate the overall yield of the process and the composition of the products to the operating parameters. The data obtained are reported in **Table 3**.

In all experiments, polystyrene was mostly converted into a liquid using mainly iron powder as a microwave absorber. Working at low power, the energy is not enough to start the pyrolysis process (EXP 1–3). Even by preheating the pyrolysis chamber through electrical resistances, the liquid yield is low (EXP 4). Instead, it can be noted that under the same reaction conditions, but using carbon as an absorber, the liquid increases considerably (EXP 5) confirming carbon as a better MW absorber. To obtain good yields of the liquid, it was necessary to increase the power (EXP 6–12), working with a ratio microwave power/PS of approx. 4.8–9.6 kW/kg. Analyses carried out on the liquid samples of the most significant tests showed aromatic hydrocarbons as the main compounds, among which single-ring aromatic compounds such as styrene, toluene, ethylbenzene, and α-methylstyrene were present in very high amounts. Styrene was the compound present in the highest percentage (approx. 55–70%). The results confirmed the previous studies carried out at the laboratory level [8–10], albeit working with a considerably lower MW power than that one previuosly employed in laboratory experiments (4.8–9.6 vs. 30 kW/kg), and made it possible to develop the application of the MAP process of PS on an industrial scale.

2.4 The industrial prototype plant

The experiments with the pre-industrial prototype were run to collect process information for the realization of the industrial plant. Thanks to the collaboration between Cognito Engineering srl and the Department of Chemistry of the University of Florence, in 2019, Techwave built its first experimental industrial prototype following the scheme reported in **Figure 5**. The plant was installed in its factory in Massa (Italy) (**Figures 6–8**).

Exp. N.	PS (kg)	MW Abs[*]	PS/ Abs	MW power (kW)	Total time (min)	Electrical time (min)	MW time (min)	Liquid (%)	Gas (%)	Solid (%)	Styrene in liquid (%)
1	1.26	Fe	2.0	1.2	30	0	30	0.0	0.0	0.0	0.0
2	1.26	Fe	2.0	1.2	62	20	42	0.0	0.0	0.0	0.0
3	1.26	Fe	2.0	2.4	90	28	62	0.0	0.0	0.0	0.0
4	1.13	Fe	1.8	2.4	80[a]	80	60	25.2	56.6	18.2	56.5
5	0.75	C	2.0	2.4	60[a]	60	45	82.4	15.3	2.3	57.5
6	1.13	Fe	2.0	4.8	90	0	90	64.6	24.8	10.6	69.5
7	1.01	C	2.3	4.8	60	0	60	56.7	24.4	18.9	55.6
8	0.98	Fe	2.0	4.8	80	25	55	71.9	17.9	10.2	55.5
9	0.98	Fe	2.0	4.8	90	90	60	71.9	18.9	9.2	54.7
10	0.49	Fe	1.9	4.8	75	75	30	44.1	24.3	31.6	62.1
11	0.49	Fe	1.9	4.8	90	90	45	62.9	21.8	15.3	59.6
12	0.98	Fe	2.0	4.8	80	80	45	55.6	22.4	21.9	68.2

[*] *MW Abs: Microwave absorber.*
[a] *Electrical and MW heating work at the same time.*

Table 3.
MAP of PS: operating parameters.

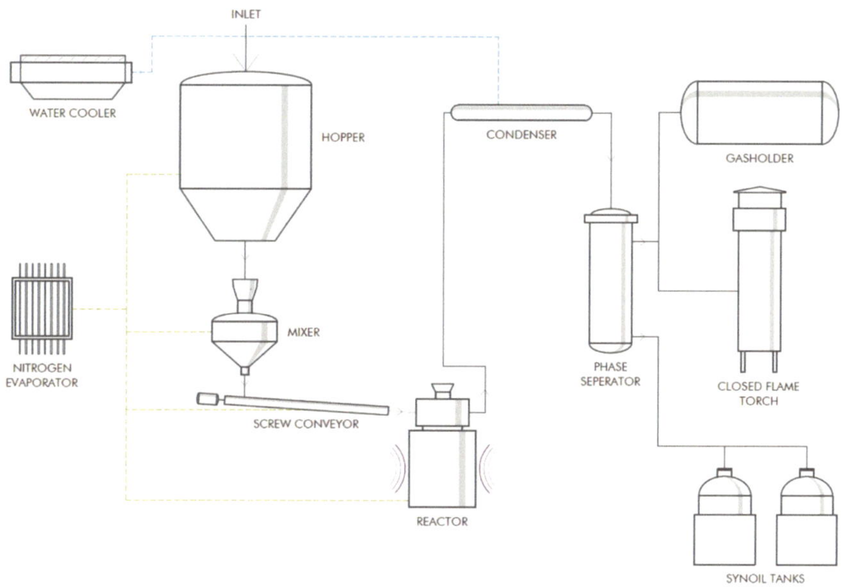

Figure 5.
Flow diagram of MAP process.

Figure 6.
Industrial prototype plant.

Figure 7.
Plastic container located in the upper part of the prototype.

Figure 8.
MW oven located in the bottom part of the prototype with, in front, the microwave guide.

The prototype was realized considering its possible introduction in two standard containers for easy shipping, even if its dimension may be strongly scaled up if required. Taking into account the dimension, it may be installed on one small ship. Waste plastic materials are largely present in the sea [23], and they may be collected and immediately disposed of through this plant or another plant close to this. The products formed may be employed to produce the energy required, while the excess may be sold on the market. Furthermore, a plant of this dimension may be installed in a municipal collecting and selection center of waste plastics, where it may be employed to pyrolyze the mixed plastics collected, avoiding their sending to a disposal plant. The prototype may be useful also for a large hospital to dispose of the contaminated waste plastics. The pyrolysis products do not contain dangerous contaminants because the biological products are destroyed during the process as reported in the literature [24, 25] while the chemicals formed may be sold for their commercial uses. In 2020, the plant started testing operations using EPS, ABS, or PP as plastic materials. Tests were carried out for 1 year, improving the results step by step, both in terms of plant efficiency and the quality of the secondary raw materials produced. The description of the tests carried out and the results obtained are reported in the following paragraphs.

2.5 Experiment with industrial prototype plant

The description shows how the tests for the MAP process were run in the industrial prototype with the plant working in semi-batch mode. The amount of plastic and carbon black for a single test was taken from the storage area. The carbon black required for the absorption of the microwave was manually introduced into the pyrolysis reactor, while the plastic material was added through the plastic loading system.

Experiments were carried out in an inert atmosphere (nitrogen), realized through various vacuum/nitrogen cycles. The carrier gas was not used to avoid the dilution of the uncondensable gas with the carrier gas. At the end of the purge operations, the plastic material was loaded into the pyrolysis reactor by a screw conveyor and the electrical resistances were switched on to preheat the reactor. Then, they were switched off, and the microwave generators were switched on.

The plastic thermal degradation formed hydrocarbon vapors conveyed to a cooling system. The higher boiling fraction was condensed and collected in the bottom of the cooling system, while the low boiling fraction, together with uncondensable gas, was sent to a torch. During the pyrolysis, further amounts of plastic material were added to the reactor constantly. At the end of the process, the microwave generators were switched off. When the plant was returned to room temperature, vacuum/nitrogen cycles were repeated as described above. At the end of this operation, the solid fraction was collected from the bottom of the reactor. The process was completed, and a new pyrolysis cycle could be started.

In the experiment, the microwave launch system was studied and modified several times until it reached the optimal configuration. The experiments let to identify and refine some operational parameters of the MAP process for improving the yields and quality of pyrolysis products. **Table 4** reports the operating parameters of the most significant tests. Microwave power is critical in MAP as it must provide enough energy to break the polymer bonds and start the thermal degradation process. Comparing the data in **Table 4**, an increase in MW power allowed to treat a double amount of plastic with the same reaction time (PS1, PS3, and PS4).

Entry[a]	Plastic[b] (kg)	Power (KW)	Time (min)	Liquid (wt%)	Gas (wt%)	Solid (wt%)
PS1	8	6;12[c]	210;15[c]	94	4	2
PS2	32	12	310	95	3	2
PS3	20	12	280	89	6	5
PS4	38	18	272	98	1	1
ABS1	75	18	221	67	23	10
ABS2	69	18	225	68	28	4
PP1	61	18	237	81	12	7
PP2	70	18	230	86	7	6

[a]*PS: Polystyrene; ABS: Acrylonitrile/Butadiene/Styrene rubber; PP: Polypropylene;*
[b]*Absorber 1.8 kg (in all tests);*
[c]*6 KW for 210 min then 12 KW for 15 min.*

Table 4.
Operating parameters of the most significant tests.

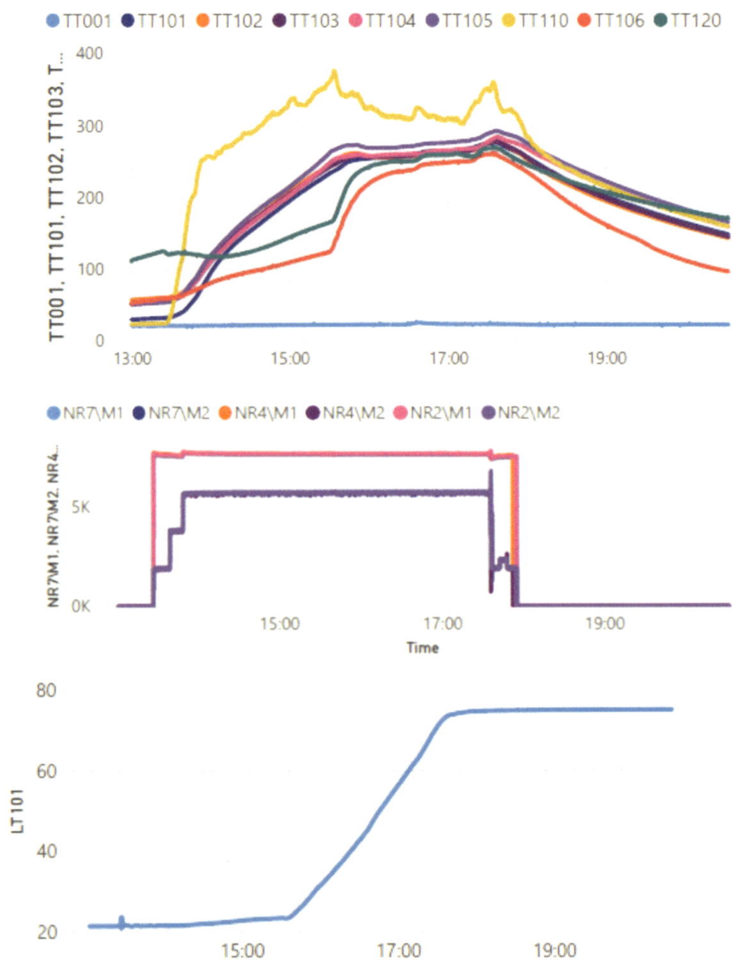

Figure 9.
From top to bottom: temperatures log, microwave power log, and level liquid log produced in MAP of ABS2.

Once the microwave launch system was fine-tuned, the design of experiments (DOE) was planned to optimize the plant productivity, using various types of plastic materials (PS, ABS or PP, **Table 4**). By way of example, the logs of temperatures, microwave power, and liquid level produced during the MAP of experiment ABS2 are reported in **Figure 9** to describe the evolution of the pyrolysis process. Although the temperature plays an important role, this parameter is detected with high uncertainty in microwave pyrolysis [26]. For this reason, the process was followed by monitoring the temperatures recorded by the probes located in several areas of the plant.

The probes from TT101 to TT105 were arranged on the reactor from top to bottom: on the bottom of the reactor was the TT110 probe. The temperature of the vapors generated during pyrolysis was recorded by the TT106 on the condensation pipeline, located after the exit from the reactor. The TT120 probe monitors the electrical resistances of the preheating reactor. All the probe temperatures were referenced to the TT001 corresponding to the ambient temperature.

When the reactor was heated at the prefixed temperature by the electrical resistances, these resistances were switched off while the microwave generators were turned on. Following the curves of the temperature probes on the reactor, it was possible to see how the temperature rose suddenly. During this phase, the carbon black absorbed the microwaves and transferred the heat to the plastic material, which starts to melt. As the process went on, the temperature curves of the reactor reached a value corresponding to the start of the thermal degradation of plastics and consequent generation of hydrocarbon vapors, as could be seen from the TT106 curve. At the same time, the LT101 level curve increased due to the condensation of the vapors and liquid was collected.

When the plant was in full operation, the TT101–106 curves remained constant, and the level LT101 of the liquid in the container continuously grew until the microwave generators were switched off, so the system was cooled, and all temperatures started to decrease when the industrial plant works in the best operating conditions, let to treat 33 kg/h of waste plastic.

3. Analysis of feedstock and products of the MAP process

As described above, the plant can carry out the MAP process of waste plastic materials at the end of its life cycle, from which three types of products were obtained (liquid, solid, and gas). During the experiment, samples were collected for each type of material for their characterizations. These analyses provided useful information to establish the quality and commercial value of the products formed. The following paragraphs report for each type of polymer tested, a description of analysis results, both for the starting material and the products obtained from the MAP.

3.1 Plastic materials used in MAP process

For each plastic material evaluated during the experiments, thermogravimetric analysis (TGA) and differential scanning calorimetric analysis (DSC) were carried out to obtain information on the polymer composition, stability and consequently predict the yield of the pyrolysis products. Specifically, TGA and DSC are two thermal analyses used to characterize polymers. TGA variations measure the weight loss of a sample subject to a constant increase in temperature to quantify the products formed, including gaseous emissions, while DSC measures phase transitions such as melting point and glass transition temperature.

3.1.1 Expanded polystyrene (EPS)

The polystyrene sample was shredded before being analyzed (**Figure 10**). Two thermal ramps were employed for the TGA/FTIR analysis, the first at a rising temperature of 20°C/min in an inert atmosphere followed by an isotherm in an oxygen atmosphere. The thermogram shows a single weight loss of 99.8 wt % related to the organic part of the polymer and a final residue equal to 0.2% attributed to inorganic impurities present (**Figure 11**).

The FT-IR spectrum of the gas released during the degradation of the polymeric component is reported in **Figure 12**. The spectrum of the gas released at a

Figure 10.
EPS sample before (left) and after (right) shredding for TGA analysis.

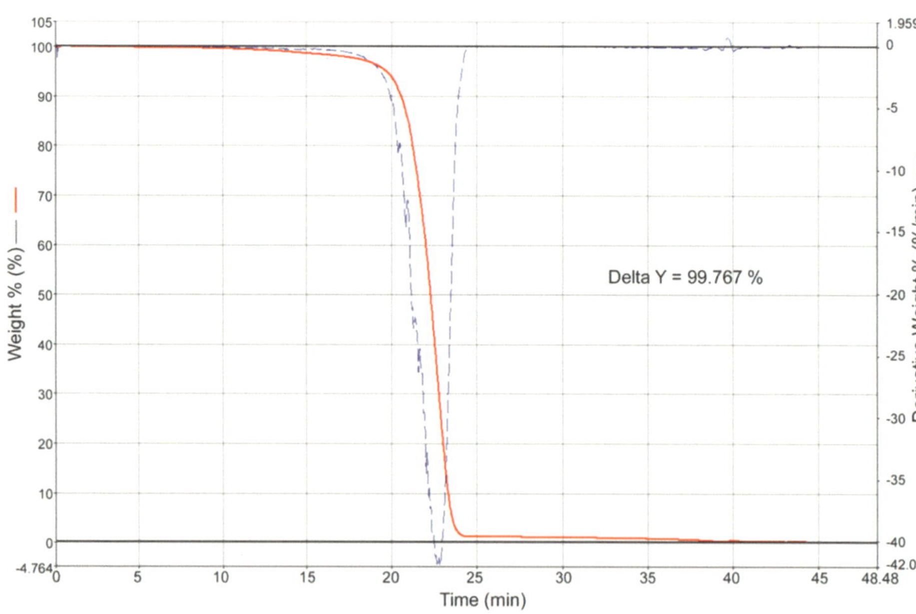

Figure 11.
TGA analysis of EPS.

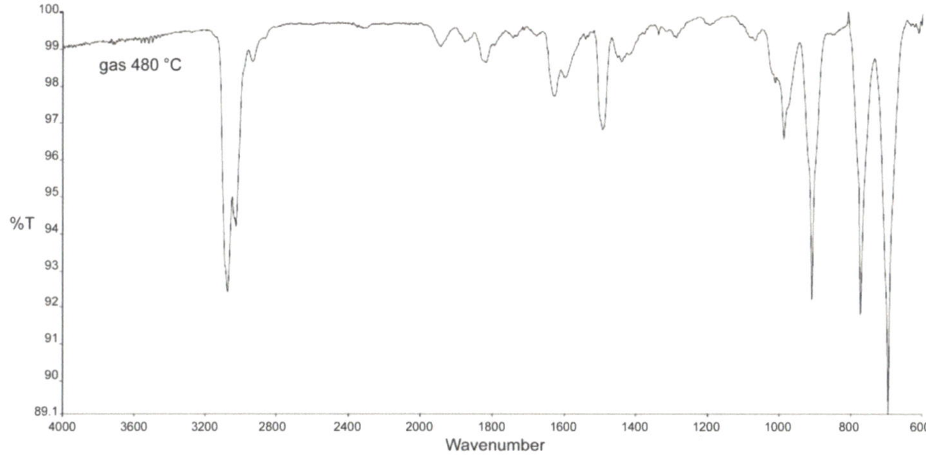

Figure 12.
FT-IR spectrum of the gases released during the TGA analysis of EPS.

temperature of 480°C shows the typical signals of hydrocarbons due to the degradation products of polystyrene. The degradation occurred by both end chains and random cleavage and gave only aromatic compounds such as styrene monomer and oligomers, benzene, and toluene. The spectrum shows the typical frequencies of aromatic compounds, such as the stretching of the C-H bond at 3000 cm^{-1}, the stretching of the aromatic double bond at 1600–1450 cm^{-1}, and the signals between 900 and 700 cm^{-1} of the aromatic C-C backbone.

The sample was also subjected to DSC analysis, using three thermal ramps in the temperature range from −10 to 200° C, alternating heating–cooling–heating, in an inert atmosphere. The DSC analysis (**Figure 13**) shows the presence of a peak at about

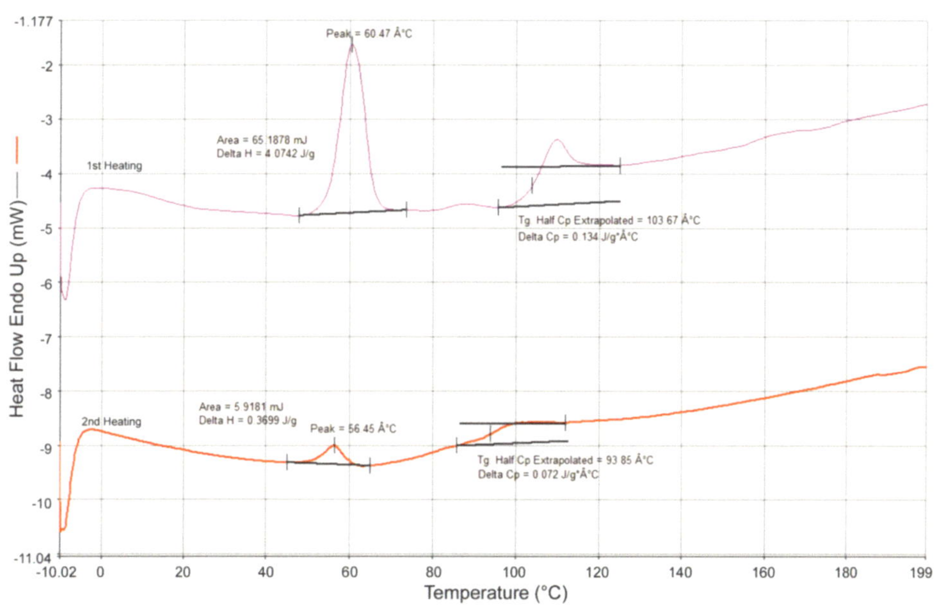

Figure 13.
DSC analysis of EPS.

60°C, which almost completely disappears in the second heating, probably due to the memory effect or to the presence of traces of volatile compounds in the sample, not visible in the TGA analysis. Furthermore, the sample shows a temperature glass transition at about 94°C, which agrees with the classical Tg of polystyrene. The TGA and DSC analyses show that the sample is essentially polystyrene with probable traces of volatile substances and a small percentage of inorganic impurities.

3.1.2 Acrylonitrile-butadiene-styrene rubber (ABS)

The sample was analyzed without any pretreatment. Two thermal ramps were adopted for the TGA analysis, the first at a rising temperature of 20°C/min in an inert atmosphere followed by an isotherm in an oxygen atmosphere. As can be seen from the thermogram (**Figure 14**), the following weight losses were recorded:

- at about 330°C, weight loss of 8.2 wt%;
- between 400 and 530°C, weight loss of 78.3 wt%, associated with the degradation of the polymeric component;
- at 700°C, weight loss of 2.5 wt% associated with the degradation of inorganic compounds;
- at 800°C after the switch in the atmosphere of O_2 weight loss of 3.3 wt%.

At the end of the analysis, a final residue of 7.7 wt% of the sample was present and associated with inorganic compounds that do not degrade up to this temperature.

The FTIR spectrum of the gas released by the sample during TGA analysis is reported in **Figure 15**. In the FTIR spectrum of the gas released at 330°C (black line) is present only the emission of CO_2, probably due to the degradation with the temperature of inorganic compounds present in ABS. At 440°C, the FTIR spectrum shows signals associated with degradation products of a carbonate containing ABS.

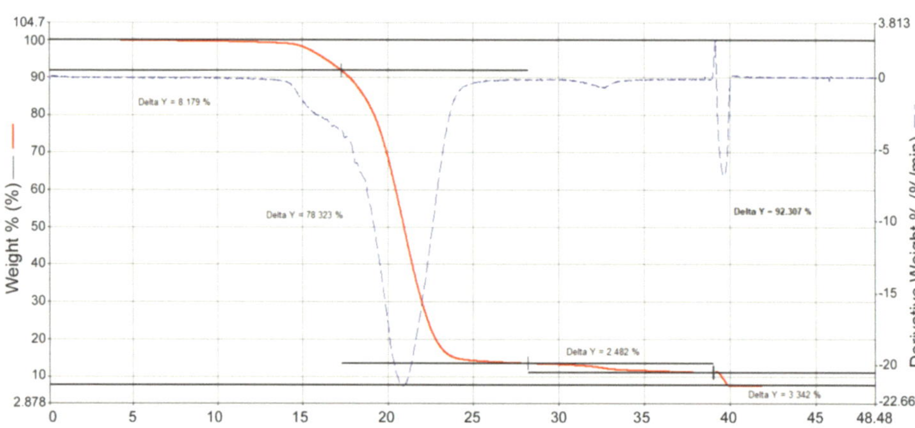

Figure 14.
TGA analysis of ABS.

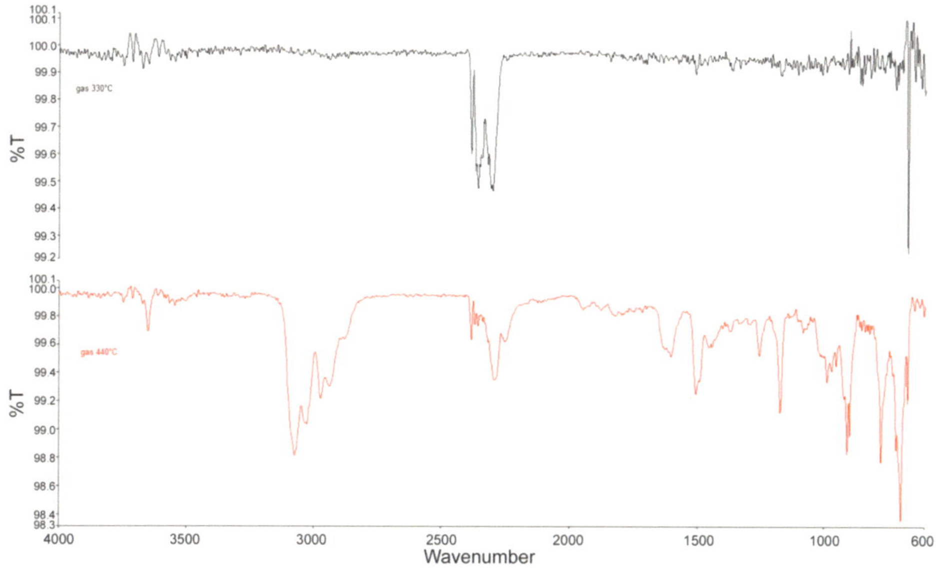

Figure 15.
FT-IR spectra of gases released at 330°C (black) and 440°C (red) during the TGA analysis of ABS.

As reported in the literature, the thermal degradation of ABS polymer begins at 340°C with the formation of the butadiene monomer. The aromatic compounds begin to be noticed at 350°C, a temperature at which the degradation of polybutadiene unities is still evident. As the temperature increases, the formation of styrene becomes more important, and at 420°C, the intensities of the C-H bands of butadiene and styrene are approximately equal. At higher temperatures, the presence of aromatics decreases in intensity while that one of butadiene is very strong. The evolution of acrylonitrile begins at about 400°C and ends at 450°C [27].

According to the above description, the spectrum of gas released at 440°C (**Figure 15**, red line) shows the stretching of aromatic C-H at 3100 cm^{-1}, and aliphatic C-H stretches at 2900 cm^{-1} and the stretching of the nitrile group at approx. 2200 cm^{-1}. Another sample of ABS was crushed and subsequently subjected to DSC analysis, running three thermal ramps in the temperature range from 25 to 300°C, alternating heating–cooling–heating, in an inert atmosphere. DSC analysis shows a glass transition at T = 108.25°C (**Figure 16**, red curve) associated with the styrenic portion typical of acrylic copolymers such as acrylonitrile-butadiene-styrene (ABS), styrene–acrylonitrile (SAN), or acrylonitrile–styrene-acrylate (ASA), a glass transition at T = 124.08°C associated with polycarbonate present and a glass transition at T = 162.42°C associated with ABS.

From the DSC and TGA/FTIR analyses, the sample is a polycarbonate containing ABS (78.3%) with small quantities of polypropylene.

3.1.3 Polypropylene (PP)

Propylene experiments were carried out using single-use masks as a sample of PP. The single-use masks were cut to separate the non-woven polypropylene from the elastic bands, and the two samples were analyzed separately by DSC. For each sample, three thermal ramps were performed in the temperature range from −10 to 300°C, alternating heating–cooling–heating, in an inert atmosphere. Measurement of the glass transition temperature, (Tg), of polypropylene is generally considered difficult

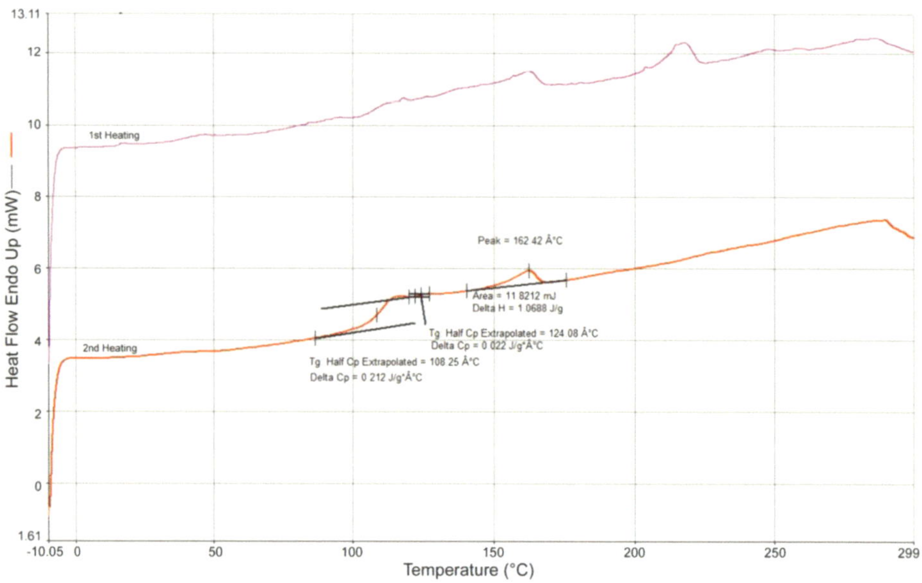

Figure 16.
DSC analysis of ABS.

to detect with DSC analysis because the transition is weak. The graph (**Figure 17**) instead shows a peak at 167°C corresponding to the melting temperature of polypropylene (green curve) and a peak at 221°C typical of the nylon used to realize the elastic bands of single-use masks (purple curve).

3.2 Pyrolysis products

The products formed in the MAP process of these plastic materials were liquid, gas, and char. In all cases, the liquid was the most important fraction both in quantity and commercial value. The characterization of this fraction for the most significant tests is reported below; the chemical–physical properties are shown in **Table 5**. Physical characteristics of liquids from MAP of PS were not affected by the variation of the MW power. The density recorded in all samples was almost the same and in a very narrow range among 0.92–0.95 g/cm^3. Furthermore, close values of ultimate analyses and the molar C/H ratios suggested that all the liquid samples had a similar composition, even if different process parameters have been employed. The C/H molar ratio in a range between 0.94 and 1.0 wt% confirmed the prevailing presence of the aryl compounds in liquids from PS pyrolysis.

The lower calorific value (LHV) of the liquids from MAP of PS also showed the same trend as the C/H molar ratio, with values close to styrene, benzene, and ethylbenzene (LHV of 39.6, 40.0, and 39.1 MJ/kg, respectively for PS1, PS2, and PS5). As foreseen from the literature and in agreement with the tests carried out both in the laboratory and on the pre-industrial prototype, the main products of PS pyrolysis are aromatic hydrocarbons such as benzene, toluene, ethylbenzene, styrene, and α-methylstyrene (**Table 6**).

Styrene is always the predominant compound in each liquid sample. From the analysis of data reported in **Table 7** it is possible to show how the amount of styrene (evaluated by chromatographic area of the GC/MS spectra) is correlated with the

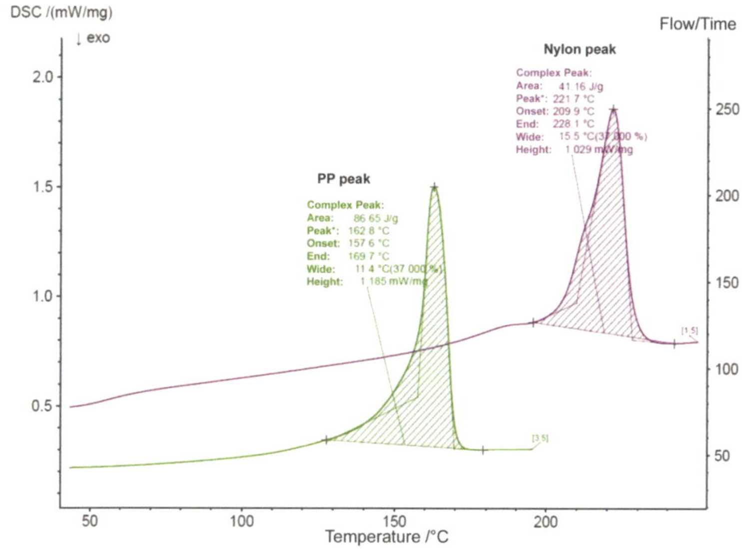

Figure 17.
TGA analysis of single-use mask.

Entry	Density	LHV	Elemental analysis (wt %)			C/H
	(g/cm^3)	(MJ/kg)	C	H	N	molar ratio
PS1	0.94	39.6	82.8	7.35	<0.50	0.94
PS2	0.95	40.0	89.2	7.60	<0.50	0.98
PS5	0.92	39.1	85.2	7.07	<0.50	1.00
ABS1	1.02	34.4	79.8	7.8	2.9	0.85
PP2	0.75	34.3	84.3	14.3	0.0	0.49

Table 5.
Physical characteristics of liquids obtained from MAP of plastics.

microwave power used. The highest amount of styrene was present in PS2 and PS4 (respectively 52.8 and 56.9%), corresponding to experiments where a higher microwave power was used. In contrast, using a lower microwave power (PS1, **Table 7**), the amount of styrene was drastically reduced. Also, as reported above for PS, the MAP of ABS gives a liquid fraction with a density between 0.97 and 1.02 g/cm^3 and C/H molar ratio of 0.85 wt% indicative of the prevailing presence of aromatic compounds. The liquid fraction was analyzed by GC–MS, a complete list of compounds identified through the NIST library is reported in **Table 8**. Aryl compounds, especially styrene, are present in large amounts in the liquid from MAP of ABS. In the presence of acrylonitrile, nitrile compounds such as 2-methylenpropionitrile were formed in a significant amount. MAP of PP gave a low-density liquid (0.746–0.760 g/cm^3) and a C/H molar ratio of 0.5 wt%, confirming the prevailing presence of aliphatic compounds in this liquid. The main compounds identified in the liquid from MAP of PP are reported in **Table 9**, they were branched saturated and unsaturated hydrocarbons, which wereformed through C-C bond cleavage of the PP backbone. In particular, there was

Compounds	Area (%)
Toluene	3.4
Ethylbenzene	7.9
Styrene	56.9
Cumene	1.2
α-Methylstyrene	8.0
1,3-Diphenylpropane	4.0
Bis(1,1′-(1-methyl-1,3-propanedyl))benzene	1.6
Stilbene	0.7
1,2-Diphenylcyclopropane	2.2
Bis(1,1'-(3-methyl-1-propenylidene))benzene	1.4
Others	12.7

Table 6.
Main compounds identified in liquid from MAP of PS4, by GC–MS.

Entry	Power (kW)	Styrene (wt%)
PS1	6;12	37.4
PS2	12	52.8
PS4	18	56.9

Table 7.
Correlation between styrene in liquid and MW power in the MAP of PS.

Compounds	Area (%)
2-Methyl-2-propenenitrile	12.5
Toluene	4.4
Ethylbenzene	8.8
Styrene	20.8
Cumene	3.8
α-Methylstyrene	12.5
6,7-dihydro-5-methyl-5H-cyclopentapyrazine	7.9
1,3-Diphenylpropane	7.2
Others	22.1

Table 8.
Main compounds identified in liquids from MAP of ABS, by GC–MS.

a prevalence of C_9-C_{12} hydrocarbons, and among these, the 2,4-dimethyl-1-hepthene was found as the main compound, in agreement with previous results [8]. Many other compounds were present in very low amounts and reported as others. The presence of aromatic hydrocarbons was attributed to cyclization and aromatization reactions of unsaturated compounds formed during the MAP process as reported.

Compounds	Area (%)
5-Methylhex-1-ene	0.8
2,4-Dimethyl heptane	12.5
1,4-Dimethylbenzene	1.2
1,3,5-Trimethylbenzene	0.5
1,2,4,5-Tetramethylbenzene	0.2
2,4,6,8-Tetramethylundec-1-ene	3.5
2,4,6,8-Tetramethylundecane	2.3
2,4,6,8,10-Pentamethyltridec-3-ene	0.6
Others	78.4

Table 9.
Compounds identified in liquids from MAP of PP, by GC–MS.

The facemasks were converted into the classical three products of a MAP (liquid, small amount of solid, and gas), having a composition close to those reported in the pyrolysis of PP. The possible contamination was absent in the products formed due to the pyrolysis conditions adopted.

4. Conclusions

The quality and quantity of end-of-life plastics are continuously improved while the amount of recycled material is always too low. Several firms such as AmSty and Agylix [28], ReVital Polymers, Pyrowave, and INEOS Styrolution [29], BASF, Quantafuel, and REMONDIS [30], Neste [31], and so on are involved in recycling plastic materials using thermochemical processes [1] using, mainly, classic heating method. This review shows the MAP process developed by Techwave as an industrial novelty proposed for the friendly and economical disposal of waste/contaminated plastics. In this way, valuable products, useful as a source of secondary raw material, close the cycle of a circular economy because plastic waste is converted in a feedstock for the production of new plastics.

MAP is a very interesting way to dispose of end life plastics because it does not produce waste from chemical recycling of plastics but three classes of products: a char, an oil, and a gas, available as fuel or the source of new materials for the synthesis of new products. The process is self-sustainable from an energetic point of view rendering the process economically sustainable. The Techwave proposal offers an industrial plant designed to obtain secondary raw materials from plastic wastes, thus closing the cycle for a circular economy and giving a strong advantage in terms of environmental protection and expansion of the recycled material market in a green and sustainable way.

Finally, the plant was realized as planned, so it is possible to introduce in two standard containers for its shipping. Furthermore, a plant of this dimension may be installed in a municipal collecting and selecting center of waste plastic where it may be employed to pyrolyze the mixed plastic present, avoiding the transport of this waste for disposal. The plant may be useful also for a large hospital to dispose of the waste/contaminated plastic present, taking into account that the pyrolysis products do not contain biological contamination because they are destroyed in the course of the process as reported in the literature [24, 25] while the products formed may be sold on the market. The dimension of the plant lets its installation on one small ship where the waste plastic materials present in the sea may be collected and immediately disposed. The products formed may be employed to produce the energy required for all operations, while the excess may be sold on the market.

Acknowledgements

Techwave srl is acknowledged for financial support. M.F. thanks the University of Florence for financial support.

Author details

Marco Frediani[1], Piero Frediani[1*], Gianni Innocenti[2,3], Irene Mellone[2,3], Roberto Simoni[2,3] and Gianpaolo Oteri[2,3]

1 Department of Chemistry, University of Florence, Florence, Italy

2 Techwave srl, Massa, Italy

3 COGNITO Engineering srl, Florence, Italy

*Address all correspondence to: piero.frediani@gmail.com

References

[1] Frediani P, Frediani M. Mixed or contaminated waste plastic recycling through microwave: Assisted pyrolysis. In: Ibrahim HAH, editor. Recent Perspectives in Pyrolysis Research. London: IntechOpen; 2021. p. 100179. 2021, DOI: 10.5772

[2] Plasticeurope. Plastic-the fact 2019 An analysis of European plastics production, demand and waste data. Available from: https://www.plasticseurope.org/en/resources/publications/4312-plastics-facts-2020 [Accessed June 25, 2021]

[3] Frediani P, Rosi L, Frediani M, Undri A, Occhialini S. Production of hydrocarbons from copyrolysis of plastic and tire material with microwave heating. WO2012110990,2012, PCT/IB2012/050747

[4] OCSE. Global Plastic Outlook: Economic Drivers, Environmental Impacts and Policy Options OECD, 2022. DOI: 10.1787/de747aef.en

[5] Do O, Pelucchi M, Van de Vijver R, Van Steenberge PHM, Dooge DR, Cuoci A, et al. The chemistry of chemical recycling of solid plastic waste via pyrolysis and gasification: State-of-the-art, challenges, and future directions. Progress in Energy Combustion Science. 2021;**84**:1009001. DOI: 10.1016/J.PECS.2020.100901

[6] Kaminsky W. Feedstock Recycling and Pyrolysis of Waste Plastics. New York: John Wiley & Sons; 2006, ISBN:9780470021521

[7] Undri A, Rosi L, Frediani M, Frediani P. Efficient disposal of waste polyolefins through microwave assisted pyrolysis. Fuel. 2014;**116**:662-671. DOI: 10.1016/j.fuel.2013.08.037

[8] Undri A, Frediani M, Rosi L, Frediani P. Reverse polymerization of waste polystyrene through microwave assisted pyrolysis. Journal of Analytical and Applied Pyrolysis. 2014;**105**:35-42. DOI: 10.1016/j.jaap.2013.10.00110

[9] Frediani P, Undri A, Rosi L, Frediani M. Waste/contaminated polystyrene recycling through reverse polymerization. In: Lynwood C, editor. Polystyrene Synthesis, Characteristics and Applications. New York: Novapublisher; 2004, ISBN 978-1-63321-356-2

[10] Bartoli M, Rosi L, Frediani M, Undri A, Frediani P. Depolymerization of polystyrene at reduced pressure through a microwave assisted pyrolysis, Journal of Analytical and Applied Pyrolysis. 2015;**113**:281-287. DOI: 10.1016/j.jaap.2015.01.026

[11] Undri A, Rosi L, Frediani M, Frediani P. Microwave pyrolysis of polymeric materials. In: Chandra U, editor. Microwave Heating. London: InTech; 2011. pp. 207-226

[12] Williams PT, Williams EA. Interaction of plastics in mixed-plastics pyrolysis. Energ Fuels. 1999;**13**:188-196. DOI: 10.1021/ef980163x

[13] Undri A, Rosi L, Frediani M, Frediani P. Conversion of poly(lactic acid) to lactide via microwave assisted pyrolysis. Journal of Analytical and Applied Pyrology. 2014;**110**:55-65. DOI: 10.1016/j.jaap.2014.08.003

[14] Frediani P, Rosi L, Frediani M, Undri A, Occhialini S, Meini S. Production of hydrocarbons from pyrolysis of tyres. WO2012110991, 2012, PCT/IB2012/050748

[15] Undri A, Meini S, Rosi L, Frediani M, Frediani P. Microwave

pyrolysis of polymeric materials: waste tires treatment and characterization of the value-added products. Journal of Analytical and Applied Pyrolysis. 2013;**103**:149-158. DOI: 10.1016/j.jaap.2012.11.011

[16] Undri A, Rosi L, Frediani M, Frediani P. Upgraded fuel from microwave assisted pyrolysis of waste tire. Fuel. 2014;**115**:600-608. DOI: 10.1016/j.fuel.2013.07.058

[17] Undri A, Sacchi B, Cantisani E, Toccafondi N, Rosi L, Frediani M, et al. Carbon from microwave assisted pyrolysis of waste tires. Journal of Analytical and Applied Pyrolysis. 2013;**104**:396-404. DOI: 10.1016/j.jaap.2013.06.006

[18] Rosi L, Bartoli M, Frediani M. Microwave assisted pyrolysis of halogenated plastics recovered from waste computers. Waste Management. 2018;**73**:511-522. DOI: 10.1016/j.wasman.2017.04.037

[19] Undri A, Rosi L, Frediani M, Frediani P. Fuel from microwave assisted pyrolysis of waste multilayer packaging beverage. Fuel. 2014;**133**:7-16. DOI: 10.1016/j.fuel.2014.04.092

[20] Undri A, Rosi L, Frediani M, Frediani P. Microwave assisted pyrolysis of corn derived plastic bags. Journal of Analytical and Applied Pyrolysis. 2014;**2014**(108):86-97. DOI: 10.1016/j.jaap.2014.05.013

[21] Frediani P, Rosi P, Frediani M, Undri M, Occhialini S. Produzione di idrocarburi da Co-Pirolisi di Materiali Plastici e Pneumatici con riscaldamento a Microonde" CAF scarl, Italian patent N. 1404131B1 (2011); Patent request N. FI2011A000029

[22] Undri A. Produzione di idrocarburi da pirolisi di gomme, Master thesis, University of Florence (Italy), 2010

[23] Available from: https://www.dw.com/en/plastic-pollution-in-oceans-growing-dramatically-wwf-warns/a-60692503#:~:text=WWF%20warned%20that%20plastic%20content,plastic%20waste%20in%20the%20ocean [Accessed February 26, 2022]

[24] Techwave private communication

[25] Available from: https://www.fraunhofer.de/en/press/research-news/2021/june-2021/closed-loop-recycling-pilot-project-for-single-use-face-masks.html [Accessed February 26, 2022]

[26] Bartoli M, Frediani M, Briens C, Berruti F, Rosi L. An overview of temperature issues in microwave-assisted pyrolysis. Processes. 2019;**7**:658. DOI: 10.3390/pr7100658

[27] Suzuki M, Wilkie CA. The thermal degradation of acrylonitrile-butadiene-styrene terpolymer as studied by TGA/FTIR. Polymer Degradation Stability. 1995;**47**:217-221. DOI: 10.1016/0141-3910(94)00122-O

[28] Available from: https://amsty.com/news/amsty-agilyx-announce-collaboration-to-build-advanced-recycling-facility [Accessed: February 25, 2022]

[29] Available from: https://www.ineos-styrolution.com/news/revital-polymers-pyrowave-and-ineos-styrolution-partner-to-launch-closed-loop-north-american-polystyrene-recycling-consortium [Accessed: February 25, 2022]

[30] Available from: https://www.basf.com/global/en/who-we-are/sustainability/whats-new/sustainability-news/2021/basf-quantafuel-and-remondis-want-

to-cooperate-on-chemical-recycling-of-plastic-waste.html [Accessed February 25, 2022]

[31] Available from: https://www.neste.com/releases-and-news/plastics/neste-successfully-completed-its-first-industrial-scale-processing-run-liquefied-waste-plastic [Accessed February 25, 2022]